Synthese Library

Studies in Epistemology, Logic, Methodology, and Philosophy of Science

Volume 406

The aim of *Synthese Library* is to provide a forum for the best current work in the methodology and philosophy of science and in epistemology. A wide variety of different approaches have traditionally been represented in the Library, and every effort is made to maintain this variety, not for its own sake, but because we believe that there are many fruitful and illuminating approaches to the philosophy of science and related disciplines.

Special attention is paid to methodological studies which illustrate the interplay of empirical and philosophical viewpoints and to contributions to the formal (logical, set-theoretical, mathematical, information-theoretical, decision-theoretical, etc.) methodology of empirical sciences. Likewise, the applications of logical methods to epistemology as well as philosophically and methodologically relevant studies in logic are strongly encouraged. The emphasis on logic will be tempered by interest in the psychological, historical, and sociological aspects of science.

Besides monographs *Synthese Library* publishes thematically unified anthologies and edited volumes with a well-defined topical focus inside the aim and scope of the book series. The contributions in the volumes are expected to be focused and structurally organized in accordance with the central theme(s), and should be tied together by an extensive editorial introduction or set of introductions if the volume is divided into parts. An extensive bibliography and index are mandatory.

More information about this series at http://www.springer.com/series/6607

Alberto Cordero
Editor

Philosophers Look at Quantum Mechanics

Springer

Editor
Alberto Cordero
CUNY Graduate Center & Queens College CUNY
The City University of New York
New York City, NY, USA

Synthese Library
ISBN 978-3-030-15658-9 ISBN 978-3-030-15659-6 (eBook)
https://doi.org/10.1007/978-3-030-15659-6

This Springer imprint is published by the registered company Springer Nature Switzerland AG.
The registered company address is: Gewerbestrasse 11, 6330 Cham, Switzerland

Editor's Foreword

Philosophers Look at Quantum Mechanics contains 16 essays based on outstanding keynote presentations made at venues of the International Ontology Congress (IOC) up to 2016. The selected essays are preceded by an introduction meant to provide an overview of the topics covered in the volume.

For 20 years now, IOC has held biennial meetings that promote interaction between scientists and philosophers interested in scientific ontology. While each edition has had a different focal point,[1] quantum mechanics has always been present. Operating from the departments of philosophy at Universidad del País Vasco (San Sebastián, UPV-SS) and Universitat Autónoma de Barcelona (UAB), during IOC's 25 years in existence, many of the grandees of contemporary philosophy have participated as principal speakers. One was Hilary Putnam, a dear friend and good supporter of IOC until his death in 2016, whose love of Spain had numerous roots.[2] Although his main presentations at IOC gatherings were about other topics, on several occasions, he offered informal discussions on quantum mechanics that touched on his research interests at the time.

In 1965, Putnam's paper "A Philosopher Looks at Quantum Mechanics" explained to a whole generation why the interpretation of quantum mechanics is a serious philosophical problem. The time was ripe for action. In 1964, John Bell's now-famous theorem had concluded that no physical theory of local hidden variables could reproduce all of the predictions of quantum mechanics. The world, Bell suggested, might be surrealistically different from what it seems to be at first sight. Abner Shimony and other philosophers joined forces with experimental physicists to study the impact of Bell's theorem. The resulting efforts built intellectual bridges between the two disciplines that astonish us to this day, fueling a

[1]These have included the idea of *physis* since antiquity to the present, scientific realism, evolutionary biology, the emergence of mind, the problem of infinity, and social ontology, among other topics.

[2]In particular, his father was Samuel Putnam, a prominent writer who did a very well-received English translation of *Don Quixote* in 1949.

renewed debate about the scope and limits of realism and understanding in scientific discourse.

The title of this volume pays tribute to the memory of Hilary Putnam. Philosophers have kept looking at quantum mechanics ever since, with growing technical skill and fruitfulness, helping the philosophical analysis of quantum physics to develop into one of the most sophisticated and productive areas in contemporary philosophy. As the essays included in this volume show, the foundations of quantum mechanics generate fruitful and exciting debates in contemporary philosophy that, luckily, have a forceful presence at IOC gatherings.

Acknowledgments The 12 editions of the International Ontology Congress reflected in this volume would not have been possible without the unfading support and enthusiasm of Víctor Gómez-Pin, Gotzon Arrizabalaga, and José Ignacio Galparsoro—IOC's miracle workers, efficiently assisted by Juan Ramón Makuso and Ima Obeso. Thanks go also to Bárbara Jiménez for her help with IOC archives and translation of the piece by Alain Aspect and Gómez-Pin included in this collection. At the institutional level, there is a huge debt of gratitude to the two host universities (UPV-SS and UAB) and to Pedro Etxenike Landiríbar (President of the Donostia International Physics Center, San Sebastian) for his help with many of the IOC activities on QM. Special thanks also to Robert Zuneska, M.A. (CUNY), for his generous technical assistance in the preparation of this volume. Finally, I wish to express my personal gratitude to the publisher's anonymous reader for helpful suggestions and to Springer's Project Coordinator Palani Murugesan for his valuable support during the final phase.

Credits All the essays included had presentations in venues of the International Ontology Congress. Three of the contributions have been previously published and appear here with permission of the authors and publisher (Springer); the details are as follows:

Simon Kochen: "A Reconstruction of Quantum Mechanics." *Foundations of Physics* Vol 45 (2015): 557–590.

Tim Maudlin: "The Universal and the Local in Quantum Theory." *Topoi*: Vol 34 (2015): 349–358.

Anton Zeilinger: "A Foundational Principle for Quantum Mechanics." *Foundations of Physics* Vol 29 (1999): 631–643.

New York, USA Alberto Cordero

Contents and Summaries

The contributions in this volume are a selection of outstanding papers presented as keynote addresses at some point between 1994 and 2016 in one of the biennial meetings of the International Ontology Congress (IOC) held in San Sebastian, Spain. The works included are grouped in six parts: Part I contains contributions about Bell's theorem and the debate on realism. Part II has papers on what the physical world is like according to quantum mechanics (QM). Part III concentrates on strategies of local scientific realism in the foundations of QM. Part IV considers arguments on individuals and individuation. Part V presents current proposals to revisit insights from the Copenhagen Interpretation. Part VI comprises proposals in favor of reconceptualizing QM.

Below is a list of the works included, along with their respective authors and summaries. The ordinal after "IOC" indicates congress number, followed by the meeting's year.

Philosophers Look at Quantum Mechanics

Chapter 1 Alberto Cordero: Introductory chapter: "Philosophers Look at Quantum Mechanics." This provides a rough map of the ideas and options discussed in the chapters that follow.

Part I: Bell's Theorem and the Debate on Realism

Chapter 2 Víctor Gómez-Pin: "Inseparable Twins" (IOC-III, 1998). A conversation with Alain Aspect about the philosophical aspects of current experimental work in the foundations of quantum mechanics, especially the experimental tests of John Bell's inequalities Aspect conducted in 1982, the last of which allowed for a choice between the settings on each side during the photons' flight.

Chapter 3 Peter Lewis: "Bell's Theorem, Realism, and Locality" (IOC-XI, 2014). Lewis argues that quantum mechanics is not a unified theory, and what Bell's theorem shows depends on which interpretation turns out to be tenable. He concludes that while the lesson of Bell's theorem could be that quantum mechanics is nonlocal, it could equally be that measurements have multiple outcomes, or that effects can come before their causes, or even, as the anti-realist contends, that no description of the quantum world can be given.

Chapter 4 Tim Maudlin: "The Universal and the Local in Quantum Theory" (IOC-XI, 2014). This contribution proposes that any empirical physical theory must have implications for observable events at the scale of everyday life, even when that scale plays no special role in the basic ontology of the theory itself. The fundamental physical scales are microscopic for the "local beables" of the theory and universal scale for the nonlocal beables (if any). This situation creates strong demands for any precise quantum theory. Maudlin examines those constraints and illustrates some ways in which they can be met.

Part II: Ontological Explorations of QM

Chapter 5 Harvey Brown: "The Reality of the Wavefunction: Old Arguments and New" (IOC-XII, 2016). Brown offers plausibility arguments for the reality of the quantum state and discusses what seem to be weaknesses in QBism as a philosophy of science. (QBism represents an attempt to solve the traditional puzzles in the foundations of quantum theory by denying the objective reality of the quantum state.)

Chapter 6 David Albert: "Preliminary Considerations on the Emergence of Space and Time" (IOC-XII, 2014). This chapter explores the idea that the wave function is the unique fundamental concrete physical stuff of the world *itself*. Albert focuses on two suggestions: (a) First-quantized nonrelativistic quantum mechanics is not a theory of the three-dimensional motions of *particles*, but of the $3\underline{N}$-dimensional *undulations* of a concrete physical *field*—the wave function itself—where N is a very large number that corresponds, on the *old* way of thinking, to the number of elementary particles in the universe. (b) This particularly radical coming-apart of the geometry (on the one hand) and the fundamental arena (on the other) is what's at the bottom of everything that's exceedingly and paradigmatically *strange* about quantum mechanics.

Chapter 7 Roland Omnès: "Decoherence and Ontology" (IOC-IX, 2008). Omnès discusses the consequences of quantum mechanics for our understanding of physical reality, particularly regarding how classical concepts are found to emerge from quantum laws; how commonsense logic stands out as a special case of quantum logic applied to macroscopic objects; how causality and locality are found to be "provincial" consequences of quanta; how tiny probabilities that would seem to turn reality into an appearance are so small that unreality does not matter; how quantum theory agrees with everything observed, except for a uniqueness that (alas) is the very essence of reality.

Chapter 8 James Cushing: "Bohmian Mechanics and Its Ontological Commitments" (IOC-III, 1998). Cushing comments on how the Bohmian option countenances a radically different ontology from the orthodox option that became standard in modem physics. In Bohmian mechanics the measurement process, which is inherently many-body in nature, is basically an act of discovery—there is *no* quantum-mechanical measurement problem. There is a well-defined criterion for a classical limit, so that there is no *conceptual* mismatch between the classical and quantum domains. Finally, insofar as all measurements are *ultimately position* measurements and quantum equilibrium ($P = |\Psi|^2$) obtains, Bohm's theory gives *complete* empirical equivalence with standard quantum mechanics. Ultimately, the choice between determinism and indeterminism in the fundamental laws of quantum mechanics is up to us.

Chapter 9 Albert Solé and Carl Hoefer: "The Nomological Interpretation of the Wave Function" (IOC-XII, 2016). Focusing on Bohm's theory, Solé and Hoefer assess the nomological interpretation, in which the wave function is interpreted as a parameter that defines the law of motion of the Bohmian particles. The authors motivate the nomological interpretation of the wave function on its own and by showing the drawbacks of the field interpretation. They then consider the main problems of the view recently discussed in the literature. Solé and Hoefer conclude with some suggestions regarding the relation of the nomological interpretation and the interpretation of the wave function that takes it to represent dispositional properties of Bohmian particles.

Part III: Local Scientific Realism

Chapter 10 Juha Saatsi: "Scientific Realism Meets the Metaphysics of Quantum Mechanics" (IOC XII, 2016). This chapter examines the epistemological debate on scientific realism in the context of quantum physics, focusing on the empirical underdetermination of different formulations (and interpretations) of quantum mechanics. Saatsi sketches a way of demarcating empirically idle metaphysics of QM from the empirically well-confirmed aspects of the theory in a way that withholds realist commitment to what $|\Psi>$ represents. He argues that such commitment is not required for fulfilling the ultimate realist motivation: accounting for the empirical success of quantum mechanics in a way that is in tune with a broader understanding of how theoretical science progresses and latches onto reality.

Chapter 11 Steven French: "Structural Realism and the Standard Model" (IOC-XI, 2012). This chapter argues for a local approach to scientific realism. According to French, taking the Standard Model seriously means taking the role of symmetries seriously and the way in which kinds and properties "drop out" of that framework. He claims that "ontic" structural realism, which holds that the world *is* structure, does just that. The option the chapter advances proceeds in the spirit of Cassirer and Eddington's efforts, who did not defend their structuralist conceptions on the basis of some commonality with earlier theories; rather they presented them as a way of making philosophical sense of quantum mechanics. French suggests to be a realist about the Standard Model one should be a realist about the symmetries and laws that it embodies and hence one should be a structural realist.

Part IV: Individuals, Individuation, and QM

Chapter 12 Peter Mittlestatedt: "The Problem of Individualism from Greek Thought to Quantum Physics" (IOC-IV, 2000). Individuals in the strict sense do not exist in quantum physics. Mittlestatedt argues, however, that unsharp observables, almost repeatable and weakly disturbing measurements, allow for the definition of unsharp individuals which is sufficient for all practical purposes. Many quantum physical experiments and the obvious existence of individuals in the classical world can be explained in this way. On the other hand, he stresses, if quantum mechanics is considered as universally valid, then there is no classical world in the strict sense. The chapter includes a Divertimento on an analogy between the motion of individual quantum systems and the motion of angels according to the treatment of Thomas Aquinas in his *Summa Theologica.*

Chapter 13 Otavio Bueno: "Weyl, Identity, Indiscernibility, Realism" (IOC-XI, 2012). This chapter reconstructs a technique originally formulated by Hermann Weyl to accommodate, in the foundations of quantum mechanics, aggregates of quantum particles despite these particles' apparent lack of identity. Bueno defends the importance of this technique and provides a variant of Weyl's original formulation by avoiding altogether the use of set theory. He then offers formulations of individuals and nonindividuals, inspired by considerations that Weyl made in the context of his theory of aggregates, and examine the status of nonindividuals with regard to debates about realism.

Part V: Copenhagen Insights Revisited

Chapter 14 Jeffrey Bub: "What Is Really There in the Quantum World?" (IOC-XII, 2016). This chapter argues for an information-theoretic interpretation that harks back to Bohr's original Copenhagen interpretation. The noncommutative theory formalized by Dirac and von Neumann is—Bub stresses—not just a new theory but a new *sort* of theory in which probability arises as a feature of the noncommutative algebraic structure and has a different significance to probability in other statistical theories. On the proposed approach, just as Minkowski geometry encodes generic kinematic constraints on spacetime configurations, the "intertwinement" of commuting and noncommuting observables in Hilbert space encodes generic kinematic constraints on probabilistic correlations between intrinsically random measurement outcomes. According to Bub, these nonclassical probabilistic constraints underlie new information-theoretic applications (e.g., to cryptography, computation, and communication). Quantum probabilities don't represent ignorance, he emphasizes, and they are not introduced because we don't or can't keep track of all the relevant variables. So what is really there in the quantum world? The proposed conception of the quantum world is in terms of probabilities of what you'll find if you measure an observable: (a) when a measurement is made, there is an agent-independent fact of the matter about what the outcome is; (b) the unitary dynamics applies universally, in principle, to systems of any complexity.

Chapter 15 Anton Zeilinger: "A Foundational Principle for Quantum Mechanics" (IOC-X, 2012). In contrast to the theories of relativity, quantum mechanics lacks a firm foundational principle to this day. This chapter proposes that the missing principle may be identified through the observation that all knowledge in physics has to be expressed in propositions and that therefore the most elementary system represents the truth value of one proposition, i.e., it carries just one bit of information. Zeilinger suggests that an elementary system can only give a definite result in one specific measurement, noting that the irreducible randomness in other measurements is then a necessary consequence. For composite systems, entanglement results if all possible information is exhausted in specifying joint properties of the constituents.

Part VI: Calls to Reconceptualize QM

Chapter 16 Simon Kochen: "A Reconstruction of Quantum Mechanics" (IOC-X, 2012). Kochen proposes a reconstruction of the formalism of quantum mechanics mathematically centered on a formulation of relational properties. To mathematically treat the extrinsic properties of quantum mechanics, he replaces the encompassing σ-algebra B(Ω) of properties by a σ-complex Q, consisting of the union of all the σ-algebras of the system elicited by different decoherent interactions, such as measurements. This change allows Kochen to define in a uniform manner the concepts of state, observable, symmetry, and dynamics, which reduce to the classical notions when Q is a Boolean σ-algebra, and to the standard quantum notions when Q is the σ-complex Q(H) of projections of Hilbert space H. Kochen then uses this approach to derive both the Schrödinger equation and the von Neumann-Lüders Projection Postulate. One feature of the reconstruction he offers is that the classical definitions of key physical concepts such as state, observable, symmetry, dynamics, and the combining of systems take on precisely the same form in the quantum case when they are applied to extrinsic properties. Kochen shows [contra Bohr] that once the relational character of properties is accepted, the definitions of the basic concepts of quantum mechanics are as real and intuitive as is the case for classical mechanics. In his view, quantum mechanics describes general interactions in the world, independently of a classical macroscopic apparatus and observer, arguing that the interactions we describe using a macroscopic apparatus could apply equally well to appropriate decoherent interactions between two systems in general. Kochen stresses that the aim of every theory is to predict the probabilities of the outcomes of interactions between systems, experiments being particular instances of such interactions. An experiment gives rise to a Boolean σ-algebra of events which reflects an isomorphic σ-algebra of properties of the system. Kochen derives elementary quantum mechanics by applying the natural classical definitions of the physical concepts to extrinsic properties, and then uses this derivation to resolve the standard paradoxes and problematic questions.

Chapter 17 David Wallace: "What Is Orthodox Quantum Mechanics?" (IOC-XII, 2016). Wallace proposes that the version of QM, as presented in standard foundational discussions (the so-called orthodox theory), relies on two substantive assumptions—the projection postulate and the eigenvalue-eigenvector link—that do not in fact play any part in practical applications of quantum mechanics. He argues for this conclusion on a number of grounds, but primarily on the grounds that the projection postulate fails correctly to account for repeated, continuous and unsharp measurements (all of which are standard in contemporary physics) and that the eigenvalue-eigenvector link implies that virtually all interesting properties are maximally indefinite pretty much always. Wallace presents an alternative way of conceptualizing quantum mechanics that does a better job of representing quantum mechanics as it is actually used, and in particular that eliminates use of either the projection postulate or the eigenvalue-eigenvector link. He reformulates the measurement problem within this new presentation of orthodoxy.

Contents

1 **Introduction: Philosophers Look at Quantum Mechanics** 1
Alberto Cordero

Part I Bell's Theorem and the Debate on Realism

2 **Inseparable Twins: A Conversation with Alain Aspect** 21
Víctor Gómez Pin

3 **Bell's Theorem, Realism, and Locality** 33
Peter J. Lewis

4 **The Universal and the Local in Quantum Theory** 45
Tim Maudlin

Part II Ontological Explorations of QM

5 **The Reality of the Wavefunction: Old Arguments and New** 63
Harvey R. Brown

6 **Preliminary Considerations on the Emergence of Space and Time** ... 87
David Albert

7 **Decoherence and Ontology** 97
Roland Omnès

8 **Bohmian Mechanics and Its Ontological Commitments** 107
James T. Cushing

9 **The Nomological Interpretation of the Wave Function** 119
Albert Solé and Carl Hoefer

Part III Local Scientific Realism

10 **Scientific Realism Meets Metaphysics of Quantum Mechanics** 141
Juha Saatsi

11 **Structural Realism and the Standard Model** 163
Steven French

Part IV Individuals, Individuation, and QM

12 **The Problem of Individualism from Greek Thought to Quantum Physics** 181
Peter Mittlestatedt

13 **Weyl, Identity, Indiscernibility, Realism** 199
Otávio Bueno

Part V Copenhagen Insights Revisited

14 **What Is Really There in the Quantum World?** 217
Jeffrey Bub

15 **A Foundational Principle for Quantum Mechanics** 235
Anton Zeilinger

Part VI Calls to Reconceptualize QM

16 **A Reconstruction of Quantum Mechanics** 249
Simon Kochen

17 **What is Orthodox Quantum Mechanics?** 285
David Wallace

Contributors

David Albert Department of Philosophy, Columbia University, New York City, NY, USA

Harvey B. Brown Faculty of Philosophy, University of Oxford, Oxford, UK

Jeffrey Bub University of Maryland, College Park, MD, USA

Otávio Bueno Department of Philosophy, University of Miami, Coral Gables, FL, USA

Alberto Cordero CUNY Graduate Center & Queens College CUNY, The City University of New York, New York City, NY, USA

James T. Cushing (deceased) Department of Physics and The John J. Reilly Center, Notre Dame University, Notre Dame, IN, USA

Steven French School of Philosophy, Religion and History of Science, University of Leeds, Leeds, UK

Víctor Gómez-Pin Autonomous University of Barcelona, Barcelona, Spain

Carl Hoefer Departament de Filosofia, Universitat de Barcelona, Barcelona, Spain
ICREA, Barcelona, Spain

Simon Kochen Mathematics Department, Princeton University, Princeton, NJ, USA

Peter J. Lewis Department of Philosophy, Dartmouth College, Hanover, NH, USA

Tim Maudlin Department of Philosophy, New York University, New York (NYU), NY, USA

Peter Mittlestatedt (deceased) Department of Physics, University of Cologne, Cologne, Germany

Roland Omnès University of Paris, Paris-Sud XI, Orsay, France

Juha Saatsi School of Philosophy, University of Leeds, Leeds, UK

Albert Solé Departament de Filosofia, Universitat de Barcelona, Barcelona, Spain

David Wallace Department of Philosophy, University of Southern California, Los Angeles, CA, USA

Anton Zeilinger Institute of Experimental Physics, University of Vienna, Wien, Austria

Chapter 1
Introduction: Philosophers Look at Quantum Mechanics

Alberto Cordero

Dedicated to Hilary Putnam.

This chapter provides background to the topics covered in the volume and gives a rough mapping of the papers included. Section 1.1 is on Bell's Theorem and the debate on realism. Section 1.2 considers non-realist responses to the puzzles of quantum mechanics (QM). Section 1.3 outlines the character of realist projects today. Section 1.4 looks at ongoing ontological explorations of the quantum state. Section 1.5 concentrates on fine-grain realist approaches to the nature of the quantum state. Section 1.6 is on individuals and individualization. Section 1.7 discusses a current revival of interest in Niels Bohr's insights on QM. Section 1.8 outlines some contemporary calls to reconceptualize QM. Section 1.9 ends the chapter with some personal suggestions regarding the scope and limits of realist interpretation.

1.1 From Solvay to Bell's Theorem

Debates about the intellectual content of quantum mechanics (QM) have been intense since the early days of the theory. Disagreements reached a peak in 1927 at a meeting in Brussels, the 1927 Fifth Solvay Conference, attended by most of the theory's founders, fifty per cent of whom had won a Nobel Prize or were on their way to get one. Niels Bohr and Albert Einstein were two of them. The issues at stake included ontological topics (e.g., what is the world like according to QM) and epistemological worries (e.g., challenges to the traditional idea that empirical success is a reliable marker of approximate truth for theoretical principles). Einstein and Bohr left the meeting more confident of their respective positions than ever.

A. Cordero (✉)
CUNY Graduate Center & Queens College CUNY, The City University of New York, New York City, NY, USA

A. Cordero (ed.), *Philosophers Look at Quantum Mechanics*,
Synthese Library 406, https://doi.org/10.1007/978-3-030-15659-6_1

However, the disputes initiated in Solvay could not be easily settled—they were "metaphysical." Realists are now more nuanced that they were then about the epistemic import of empirical success, but they still take the astounding success of QM as an indicator that at least *some significant part* of the quantum theoretical story is very probably true and responsible for its success.

Still, many of the central tenets of QM appear wildly bizarre when taken at face value. Those tenets include the notions that measurements do not generally reveal values had by target systems before measurement, physical objects generally lack "sharp-valued" properties, radical indeterminism seemingly creeps in measurement processes, and events can hold non-local correlations. Realist interpretations face consistency problems as well. In particular, there is tension with the world of ordinary experience (as illustrated by the Schrödinger's cat paradox[1]). There is also strain from the locality principle of relativity theory. These are just two of many perceived difficulties.

The "standard textbook" version of the theory (SQM) largely evades these difficulties by suggesting (typically in vague terms) that physical systems do not generally have intrinsic properties, and that, to the extent that they do, quantum mechanical objects cannot have precise conjugate dynamical properties simultaneously—i.e., they do not (and cannot) have complete classical sets of sharp-valued dynamical properties. Furthermore, on SQM, the act of measurement generally changes a system's state, prompting a concentration of the probability spectrum of the measured quantity around whatever value emerges from the process (Projection Postulate). Compounding the matter, SQM does not provide a clear physical interpretation of what it refers to as the "quantum state," which results in accounts that feel incomplete at best. From the textbooks in question, all one can infer about a system's dynamical properties are probabilities that performing measurements on it will yield one particular outcome or another. In these and other respects SQM is unlike other fundamental theories in science.

For example, SQM focuses on measurable quantities (observables) and quantum mechanical probabilities. The resulting probabilistic algorithms have extraordinary predictive success ostentatiously apparent in fields ranging from atomic and nuclear physics to the world of transistors, lasers, imaging, and much more; yet users of SQM learn surprisingly little about the physical ontology at play. Most practicing physicists seem satisfied with SQM's "practical" approach[2]. Since the early days of the theory, however, some distinguished thinkers have seen its circumspection as an empiricist deficiency, calling for richer explanatory alternatives. For decades, a seemingly reasonable response held an incompleteness thesis about QM, claiming that the theory provides an empirically successful but incomplete (limited, coarse-

[1]Schrödinger imagined a set-up in which a boxed cat is gassed to death if a particle decays, left alone if the particle does not decay. But the particle is neither decaying nor not-decaying, instead it is in a peculiar quantum state: a "superposition" of *both* decaying *and* not decaying. According to the Schrödinger equation, the cat evolves into a superposition of being *both* dead *and* alive.

[2]Classic presentations include textbooks like Lenard Schiff (1949, multiply reedited and still available) and Albert Messiah (1961, multiply reedited, also still available).

grained, statistical) description of physical reality. According to this response, SQM is weird, but deep at the fundamental level the world satisfies the traditional principles of locality, separability, dynamic completeness, and individuation (i.e., the world is profoundly classical). The most famous argument along these lines, published in 1935 by Albert Einstein, Boris Podolsky and Nathan Rosen (EPR), asserted that quantum systems interact in ways that *require* simultaneous sharp values for their conjugate properties (e.g., position and their momentum), a possibility disallowed by SQM. According to EPR, physical systems simultaneously have precise conjugate properties but their fullest quantum mechanical description misses aspects of the targeted reality, and so QM cannot be considered a "complete" theory of material systems.

This argument remained tenable until the 1960s, when now well-known theorems by Bell and by Simon Kochen & Ernst Specker uncovered some of the workings of contextuality in QM, exposing intriguing clashes between classical expectations and predictions derivable from QM. The disagreements were especially exciting because they appeared to open doors to empirically resolving previously metaphysical confrontations. Experiments conducted in the late 1970s contributed serious doubts about strictly classical interpretations of the laboratory results, which seemed to support QM against predictions drawn from the traditional ontology. Until roughly this time, raising foundational questions had been considered an otiose pursuit in physics: When Alain Aspect proposed his Bell experiment as a project for his *doctorat d'État*, his prospective supervisor asked with a worried face: "Have you a permanent position?" Happily, he already had one and was allowed to proceed (Aspect 2002).

With the Bell experiments, interest in the foundations of QM as a theory of matter strengthened and has kept growing ever since. Among the pressing philosophical questions are: *What is the nature of the world, given that it displays quantum phenomena? What makes something a "quantum-system"? What determines whether an interaction constitutes a measurement?* Ingenious work has been poured on questions such as these over the last half-a-century. As yet, however, no consensus answers are in sight.

The empirical superiority of QM over expectations from classical metaphysics gained strength in the 1980s through increasingly sophisticated measurements, conspicuously by Alain Aspect and other experimentalists. QM emerged the most clear winner. Not so SQM, however, as some highly respected analysts refused to take the results as supporting the theoretical physics of "observables" promoted by SQM. Bell, in particular, thought fundamental physics required theories that speak of what *is* rather than merely what is *observed*. He proclaimed that "beable" should replace the term "observable" in quantum physics (Bell 1973). Realist (beable-oriented) projects to reconstruct QM gained a new life and remain vibrant to this day.

Beable-oriented interpretations of QM have developed greater coherence in recent decades, but all the proposals available feel surreal. Some influential critics drop realist projects as superfluous. On the realist side, ontological weirdness is not a problem. For realists, the vital claim is that external reality—however bizarre—fundamentally contributes to the production of our knowledge, so that the information we gather is not exclusively based on fantasy and social construction: *there is a mind-independent world that constrains the production of knowledge*. So, the realist question is about the extent to which the ontological descriptions drawn from a theory can be taken as approximately true at face value.

In the biennial IOC meetings, works on Bell's Theorem have been a regular feature. This volume contains three contributions. In the first, Alain Aspect and Víctor Gómez-Pin discuss the Bell experiments as a turning point in physics and philosophy. Then, Peter Lewis considers how the Bell experiments have sometimes been taken to show that QM undermines scientific realism, moving on to argue that the matters involved are far from simple and we should avoid the temptation to simplify them. Tim Maudlin, in turn, considers the beables of QM and distinguishes between the local and non-local beables of the theory.

The Bell experiments have reinvigorated the debate on realism in the philosophy of physics. Reflecting on the vast empirical success of QM, realists try to justify taking at face value at least some parts of its theoretical content (i.e., content beyond the reach of ordinary human perception). By contrast, scientific non-realist positions deny either the need or the possibility of doing so[3]. The next sections discuss some leading options in each direction.

1.2 Non-realist Stances

From at least the days of Werner Heisenberg, the notion that quantum physics derails the ideal of scientific realism has been a recurrent theme. The Bell experiments are often portrayed as having antirealist import. As Lewis stresses in his contribution to this volume, matters are far more complicated, however. While post-Bell physics challenges traditional interpretations, what it accurately shows depends on which interpretations turn out to be tenable.

Nonrealist arguments in the literature rest on different considerations, from general, broad range (not specific to QM) to highly specific to QM.

1.2.1 General Empiricism

One general position builds on van Fraassen (1980). According to this view, anti-realism can be challenged, but otiosely, because searching for realist stances is superfluous. Empirical adequacy, followers urge, sufficiently accounts for the empirical progress and success displayed by scientific theories, including quantum physics.

[3]A more radical variety challenges the existence of any external reality—an option without takers among the contributors to this volume.

Realist reactions comprise several arguments, applicable depending how the distinction between observable and observable is drawn.

(a) One realist response is that the project of dichotomizing the empirical and the theoretical is a logical impossibility and a historical falsehood. Any epistemic primacy granted to the empirical level creeps up into the theoretical level, and so the alleged supremacy of empirical over theoretical adequacy cannot be maintained without begging the question.
(b) Another response is that challenging antirealism is not otiose: empirical adequacy explains a theory's success but not why it succeeds. For example, reacting to wave function antirealism, Harvey Brown complains in this volume (in the spirit of Christopher Timpson's 2008 critique) that quantum Bayesianist thinkers (QBists) explain why a physicist believes that matter is stable but cannot explain why it is.
(c) The third response is that realism about a theory, when consistently stated, makes the intended empirical domain intelligible by specifying structural and causal underpinnings that are free of specific doubts and beyond the reach of ordinary perception.

1.2.2 More Specific Non-realisms

A second non-realist option, more specific to the case of QM, also appears in van Fraassen's writings. It focusses on confirmational limitations of the particular theory. Some of his arguments (1980 on realism, and 1991 on metaphysics) present QM as a prospectively empirically adequate theoretical construct that is, unfortunately, marred by specific levels of underdetermination. The arguments offered assume that insufficient confirmation (not to mention error) kills all prospects of having theory-parts suitable for realist commitment, making fairly radical empiricist interpretation the most reasonable option. Realists reject this assumption as arbitrary, arguing that a theory can be approximately true by getting correct accounts, particularly at intermediate (as opposed to "fundamental" theoretical levels (see, e.g., Cordero 2017). James Cushing's paper in this volume, for example, takes a selective realist stance about some parts of quantum theory while also suspending judgment on whether the microscopic world is deterministic or indeterministic. The point here is that t is not necessary to be completely right to be scientifically truthful, selective realists emphasize.

1.2.3 Anti-Classicism

A more confined, non-realist line takes issue with specific claims associated with classical physics and traditional metaphysics—e.g., the ordinary view of the world as made of entirely local objects existing separately and independently of one another. This line is not properly "non-realist;" it simply focuses on claims that do not blend smoothly with experimental results like those highlighted, for instance, by Aspect and Gomez-Pin in this volume. Realist anti-classicism has a distinguished track record in quantum physics, from Bohr (1935) to John Bell (1964) to Kochen and Specker (1967) and beyond (see Simon Kochen's paper in this volume). As already noted the outcomes of Bell experiments seem at odds with any descriptive framework that respects the three traditional principles of single-measurement results, property determinateness, and locality (super-realism). If QM holds universally, then the realist view associated with classical physics is refuted.

Realists, in short, stress that derailing "classical" conceptions does not kill the project of realism, which is not committed to classicism or *any* given metaphysics—indeed, challenging and critically revising the received categories of understanding has been central to physics for centuries. Realist responses of this sort are already apparent in Bohr's (1935) relational reply to the Einstein, Podolsky and Rosen original EPR paper (1935). Bohr, whose writings blended instrumentalist and realist rhetoric unhelpfully, argues at points that the notion of a physical system having dynamical properties needs to be reconceptualized in accordance with QM. He does this by proposing (albeit obscurely) the idea we now call "quantum entanglement" and advancing a form of contextualism in which the dynamical properties of a system depend not only on the system by itself but also on its total physical environment (Lewis 2016). On Bohr's approach, the spin of an electron along direction z is defined only when the physical environment of the system is such that we can measure its z-spin. But the physical environment can never be such that we measure spin along different directions at the same time, and this leads Bohr to argue that QM is complete as it stands.

1.3 The Realist Outlook Today

Current realist projects revise one or more of the traditional principles of independence, locality, determinateness, and single-result measurement. Realists hope for truthful content at theoretical levels in QM, but theories contain falsehoods, and so a group of reformers, "selective realists," confine epistemic commitment to theory-parts rather than whole theories. As such, they fall under the umbrella of the *divide and conquer* variously developed in the late 1980s and 1990s by John Worrall, Philip Kitcher, Jarret Leplin, and Statis Psillos[4]. Taking a selective realist stance about QM involves claiming that the theory contains an abstract descriptive part that seems impossible to give up without compromising the predictive power of QM. This form of selectivism is compatible with adopting a non-realist or even skeptical stance about other contents. For example, realists about such substantive parts of QM as atomic particles and specific molecular structures in 3D-space and time may, simultaneously, be non-realists about, say, the "deeper" configuration space associated with the quantum state.

[4]See, in particular, Worrall (1989), Kitcher (1993), Leplin (1997), Psillos (1999).

The conditions for selecting theoretical posits (descriptive claims about theoretical entities, processes or natural structures) as prospectively truthful make a contentious matter, but there is some accord among realists. To be selected as prospectively truthful, a posit must: (1) be consistent (i.e., regarded as belonging to a possible world), and (2) be sufficiently warranted to compel belief (by current scientific standards). These conditions unsurprisingly invite two types of realist projects.

The first type is primarily ontological, made of efforts to detail consistent proposals about what the world is like according to the formulation of QM at hand. In broad terms, ontological projects ask what kind of world QM postulates. The emphasis is on intellectually exploring the world in which the theory is correct at face value—what the world is fundamentally like if a proposed version of QM (e.g., Bohm's mechanics) is true. The second type of selective realist exploration is primarily epistemological, made of projects that seek to determine in some principled way which parts (if any) of the theoretical stories licensed by QM qualify for realist commitment. These epistemological projects focus on justifying the descriptive claims derived from the various versions of QM, asking how justified they are. Ontological and epistemological explorations easily overlap.

One trend, advocated by Juha Saatsi (2015, 2016) and shared by some thinkers, argues that realists would be better off by providing local exemplars of the sense in which they want to commit to unobservable posits, without reducing that sense to any general definition of 'partial', 'approximate', or 'structural' truth. This approach shuns 'global recipes' that demand one should be realist about entities, descriptions or structures that feature across temporally related theories. The local approach favors accounts that leave one free to commit to such and such features, given theory T, where the features may be different when it comes to theory T. This collection contains two papers particularly sympathetic to this local trend, by Juha Saatsi and Steven French, respectively.

Saatsi's essay sketches a way of demarcating empirically idle metaphysics in QM from the empirically well-confirmed aspects of the theory in a way that withholds realist commitment to what the "wave function" Ψ represents, arguing that such commitment is not required for fulfilling the ultimate realist motivation. To Saatsi, the latter is to account for the empirical success of QM in tune with a broader understanding of how theoretical science progresses and latches onto reality. In a related vein, Steven French's paper considers the local selectivist strategy and suggests that to be a realist about the Standard Model of particle physics one should be a realist about the symmetries and laws that it embodies and hence one should be a structural realist.

1.4 Ontic Interpretations

In the 1920s, some distinguished theoreticians, notably Louis de Broglie and Erwin Schrödinger, argued that moving from classical to quantum theory could not go through intellectually without appreciating the objective wave aspect of both radiation and matter. Realism about Ψ had dedicated champions. As noted, however, until a few decades ago, the dominant interpretation of QM encouraged an instrumentalist interpretation to the wave function.

Realists argue for the physicality of Ψ from considerations of empirical success and present freedom from specific (as opposed to global, skeptical or metaphysical doubts regarding Ψ's ontic counterpart in nature. To realists, since QM both systematically accounts for physical systems in terms of the quantum state and those accounts are experimentally fruitful, Ψ represents something real. In this volume, Harvey Brown articulates fresh plausibility arguments for the physical reality of the quantum state and exposes what he sees as weaknesses in approaches that reject Ψ realism, particularly QBism. The word "wave function" fits especially well in the case of configurations in which the quantum state takes the form of something like a wave in 3D-space—albeit a peculiar one, as Albert explains also in this volume.

Realists about Ψ think the quantum state represents something ontologically fundamental and entirely independent of human beings. In their view, the wave aspects of physical systems express both the pervasiveness of quantum entanglement as an objective phenomenon and the need for a metaphysics that makes sense of such states. Realist talk about Ψ invites projects of revisionary metaphysics along several ontological options—each consistent proposal corresponding to a possible world in which what QM says is true. One major split among the positions concerns whether the dynamics of Ψ evolution amounts to just the linear law represented by Schrödinger's equation and its generalizations (an option taken by, e.g., decoherence theories) or the dynamics also includes a non-linear law that accounts for outcome selection (an option taken by, e.g., GRW-like theories). An "intermediate" position (represented by standard Bohmian approaches) proposes that there is more material stuff to the fundamental world than just Ψ; let us represent the extra component by "ς."

$$\text{World} = \Psi + \varsigma$$

At a deeper descriptive level, the proposals that take a realist stance about QM divide into those that bracket issues about the "deep nature" of Ψ and ς, and proposals that dare to speculate about them in finer detail (as outlined in the next section).

Among the programs of the first variety the following three stand out (see, e.g., Cordero 2001):

(A) The version of Bohmian QM discussed in this volume by Cushing is a direct offspring of the nonlocal hidden variables theory introduced by Bohm in the early 1950s, in which ς stands for localized particles. It radically challenges the projection postulate of SQM and reinterprets the latter's probabilistic algorithm. Not all versions of Bohm's approach take a realist stance toward Ψ, however; some deny that Ψ represents physical stuff. In this collection, Solé & Hoefer explore this selective non-realist line through a nomological interpretation of Ψ, in which the wave function does not represent a physical substance but has the character of a law. They motivate this option by showing the drawbacks of taking Ψ as a field, then consider the nomological interpretation on its own and outline problems recently discussed in the literature.

(B) The approach often called "Decoherence QM" elaborates on how the phenomenon of quantum decoherence helps bridge conceptual gaps that separate SQM and classical physics, as Roland Omnès explains in this volume. The Decoherence approach challenges the projection postulate while critically reinterpreting the probabilistic algorithm. It also seeks to explain—as Omnès indicates—how classical concepts can be found to emerge from quantum laws; how commonsense logic stands out as a special case of quantum logic applied to macroscopic objects; how causality and locality are found to be "provincial" consequences of quanta (as opposed to universal principles); how tiny probabilities that would seem to turn reality into an appearance are so small that such level of unreality does not matter; and how quantum theory agrees with everything observed, except for a uniqueness that (alas) is the very essence of reality. Omnès rejects Everett's many-worlds because, in his view, it would mean believing "quantum theory above the unique wonder of a reality we can contemplate every day," which looks to him as "the extreme of ideology." Wallace (2012) and many decoherentists disagree.

(C) Spontaneous collapse theories redescribe the standard Projection Postulate in physical realist terms. Detailed approaches of this sort were developed in the mid-1980s by G.C Ghirardi, A. Rimini, T. Weber, and P. Pearle, among others[5]. The resulting proposals articulate entirely physical versions of the collapse of the quantum wave function, leading to predictions that (at least in principle) disagree with those of linear QM.

There are, thus, at least three broad realist approaches to understanding the quantum world that interpret Ψ ontologically. Each yield an explanation of the domain covered by QM; all the approaches show some fertility. Bohmian mechanics encourages work of cosmological interest on superluminal signaling prior to the establishment of quantum equilibrium (as, for example, in Valentini 1991). Decoherence QM makes cosmological openings (as, for instance, in Gell-Mann and Hartle 1993). Novel predictions can be extracted from collapse theories, albeit not ones accessible in the laboratory as yet (see, e.g., Simonov and Hiesmayr 2016); additionally, spontaneous collapse theories promote new ways of looking at thermodynamics (Albert 2003).

Regarding internal coherence and unity, the early formulations of the noted proposals were all variously ad hoc and vague. However, descendant theories advanced in recent years show marked improvements. Although Bohm's initial theory was notorious for its artificiality, subsequent work has managed to provide physical motivation for most of the Bohmian rules. Much of what started as a patchwork of assumptions lacking internal coherence now drops out nicely from

[5]See, e.g., Ghirardi et al. (1986, 1990), and Tumulka (2006).

theoretical considerations in recent articulations. As underlined by Cushing in this volume, plausibility arguments offered, e.g., by Antony Valentini (1991) explain how random subquantum interactions drive systems to conditions of equilibrium, the probabilistic distributions spontaneously corresponding to those given by the standard rule. Valentini further establishes that, once the universal state satisfies the condition of quantum equilibrium, the wave function for any individual subsystem also satisfies this equilibrium condition for measured values. In his particular version of the Bohmian approach, the world possesses signal locality only in a contingent historical way, and then only after equilibrium is reached, being fundamentally nonlocal in its structure outside that regime. Turning to ontic versions of decoherence QM, recent offerings of the "Many Decoherence World" theory (an offspring of the "Many-Worlds" interpretation of the linear part of SQT, introduced by Everett in 1957 with serious conceptual difficulties) arguably display considerable improvement. In particular, revamped approaches motivates physically a preferred basis for the total state while otherwise lets classical features emerge naturally from the phenomenon of quantum decoherence (see, e.g., Wallace 2012). These proposals offer a literal reading of Schrödinger-cat situations, specifically the idea that our experience as observers does not correspond directly to the universal wave function but only to part of it—some "branch." From the Many-Worlds perspective, it is the various post-measurement branches of the wave function, not the total state, that correspond to the situations we experience, with different branches representing the different results observed in practice. So, what we perceive as an "instantaneous collapse" of the wave function is understood as part of the branch-rooted, branch-relative-reality character of the phenomenon we call "awareness." Turning to Ψ-collapse models, they too have made advances, particularly towards addressing their tensions with Lorentz invariance[6] and the symmetries of systems containing identical constituents. The collapse dynamics they offer now include resources for continuous spontaneous localization.

All these refinements make the three highlighted approaches experimentally discernible *in principle*—they explicitly describe different worlds. Unfortunately, however, the differences turn out to be exceedingly difficult to access. The debates on these ontic proposals remain strong.

1.5 Fine-Grain Explorations

A more ambitious variety of projects tries to reach into the nature of Ψ. They include approaches that present Ψ as a physical field in configuration space, as well as projects that take Ψ as a law of nature.

Ψ as a Physical Field: In this volume, David Albert considers the possibility that Ψ represents a field. Noticing that Ψ does not live in ordinary space but in

[6]See, e.g., Ghirardi, Grassi and Pearle (1990), and Tumulka (2006).

3N-configuration space, he explores the notion that the quantum mechanical wave-function is the unique, ultimate concrete physical stuff of the world itself. On this view, what is fundamental is Ψ, not particles, not space (see also Albert 1996). In realist terms, the fundamental physical object would be a real physical field that lives in configuration space (each point corresponding, in a way that needs further explication, to a configuration of particles). This "fine-grain" realism comprises monist and pluralist options.

(a) According to Ψ monism, all there is at the fundamental physical level is a field in high-dimensional configuration space. Monism need not be eliminativist about particles, which it can try to accommodate as derivative stuff, i.e. as "effective" (as opposed to fundamental) entities. For example, Everettians can make room for particles as entities that emerge out of the decoherent behavior of Ψ over time in coarse-grained spacetime. A strategy of choice here is to appeal to functionalism. In Alyssa Nay's (2013) formulation, there are functional particles in 3D-space just in case Ψ behaves over time so as to play the causal role of a system of N particles in a 3D-space. This claim cries for clarification, however: what is it about the field in configuration space that allows Ψ to ground the existence of a multi-particle system in 3D-space? To critics, Ψ monists seem overoptimistic about closing the apparent explanatory gap.
(b) The dualism advocated by most Bohmian theories provides examples of the pluralist option. The most common variety takes a realist stance about both Ψ and configuration space: Ψ is a physical field in configuration space, and particles are real physical objects. One question for this view concerns how the Bohmian particles in 3D receive behavioral guidance from a field in a radically different space. Configuration space realists try to articulate Bohmian mechanics so that at the fundamental level the theory posits no objects in ordinary space. A seminal development along these lines was Davis Albert's (1996) exploration of the possibility of reading Bohm's two equations as being about entities in 3N-dimensional configuration space. Not all Bohmian approaches are so ontologically daring, however.

Ψ as a Law: Many endorse nomological realism about Ψ and take configuration space as a convenient construct. On this view, Ψ is a real physical structure though not an actual physical field—rather like a law that governs the motion of Bohmian particles—and there is only one genuine physical space (the one in which the particles move). Ψ cannot be eliminated from the ontology because—as cases of entanglement of position illustrate—there is more to the quantum state than is carried by the states of the particles themselves. Accordingly, on this approach, Bohmian quantum theory is fundamentally about a configuration of particles in ordinary space, and Ψ is not what the theory is fundamentally about. As noted, in this volume Albert Solé and Carl Hoefer's contribution explore one nomological interpretation in which the wave function does not represent a physical substance and has the character of law; the authors interpret Ψ as a parameter that defines the law of motion of the Bohmian particles.

1.6 Individuals and Individuation

Characterizing quantum mechanical objects opens issues and problems that were absent in classical physics and traditional philosophy. The topics of individuals and individuation exemplify this aspect amply. The metaphysics of individuality in QM has long been the subject of lively arguments. As Steven French and Decio Krause (2006) point out, one can

> "take the claim that quantum physics is consistent with particles regarded either as individuals (with well-defined identity conditions) or as non-individuals which do not have such conditions. If so, there exists a kind of underdetermination of the metaphysics (of identity and individuality) by the physics."

In this volume, Peter Mittelstaedt and Otávio Bueno explore some of the issues involved. Mittelstaedt uses the complete set of phase space properties in an unsharp sense (corresponding to unsharp properties). The individual objects which can be thus determined are defined only unsharply. From Mittelstaedt's perspective, strict individuals do not exist, but unsharp observables in conjunction with almost repeatable and weakly disturbing measurements allow for the definition of unsharp individuals, which arguably suffices "for all practical purposes." According to Mittelstaedt, numerous quantum physical experiments and the existence of individuals in the classical world can be explained in this way.

Bueno, in turn, reconstructs a technique originally formulated by Hermann Weyl to accommodate aggregates of quantum particles despite the particles' apparent lack of identity. In defending this technique, Bueno provides a variant of Weyl's original formulation that avoids the use of set theory, offers formulations of individuals and non-individuals, and then examines the status of non-individuals concerning ongoing debates about realism.

1.7 Revisiting Insights from Copenhagen

Some recent projects in the philosophy of quantum mechanics take some of the ontological ideas of the Copenhagen interpretation as guiding principles. In this vein, for example, Jeffrey Bub and Itamar Pitowski (2010) argue that the common version of the measurement problem is a pseudo-problem brought on by the dogma that—as they put it—"the quantum state has an ontological significance." Their approach encourages interpretations of the quantum state as something inherently *informational*. In particular, we might understand the wave function as representing probabilistic information about the world (of what you will find if you measure a given observable), or as saying something about what rational degrees of belief we should have about it (e.g., where a proton is). The papers by Bub and Anton Zeilinger in this volume elaborate on options of this variety.

Rethinking Bohr's version of the Copenhagen interpretation, in his paper Bub proposes a conception of the quantum world in terms of probabilities of what one will find if an observable is measured. Quantum probabilities, he argues, don't represent ignorance, and they are not introduced because we don't or can't keep track of all the relevant variables. So, what is really there in the quantum world? Bub defends an information-theoretic interpretation that returns in certain ways to the original Copenhagen interpretation in Bohr's version.

The second contribution, by Anton Zeilinger, argues that the lack of an accepted Foundational Principle for QM can be remedied by taking seriously the thought that all knowledge in physics has to be expressed in propositions and that therefore the most elementary system represents the truth value of one proposition. An elementary system, Zeilinger stresses, can only give a definite result in one specific measurement, and so the irreducible randomness in other measurements follows as a necessary consequence. Making a measurement turns potentiality into actuality, Zeilinger states, but whether the system one measures has (say) a clear position or not before measurement, it exists. In his view, QM describes probabilities of possible measurement results.

More standard realists, by contrast, think that QM tells about more than probabilities, and achieves more than empirical adequacy.

1.8 Calls to Reconceptualize QM

One transformative lesson from the Bell and Kochen & Specker theorems is that QM doesn't fit into a classical framework. It seems that principles traditionally regarded as transparent to the intellect and long deemed essential to rational understanding—like the classical conditions of locality, uniqueness of measurement outcomes and the non-contextuality of dynamical properties—cannot be jointly upheld; at least one of them must go if the Bell experiments are accepted. Realists have several options.

(a) One alternative is to drop Bell's locality assumption, as do spontaneous collapse approaches and Bohmian hidden variables.
(b) Another possibility is to abandon the uniqueness of measurement outcomes. Many-worlds approaches take this option, explicitly stating that measurements do not, in general, have unique outcomes.
(c) A third choice is to drop non-contextuality. The two mentioned theorems assume that the dynamical properties of a physical system are independent of the rest of the world—i.e., "non-contextual." If, say, spin along a direction w is a contextual property, then there are no intrinsic w-spin properties. In Bohm's theory, in particular, the results obtained when one performs spin measurements are explained in terms of the position properties of all the constituents of the system. As Peter Lewis points out,

> "... a given configuration of underlying position properties could result in either a spin-up or a spin-down outcome because the dynamical laws obeyed by the underlying constituents are indeterministic. In that case, the measurement results are explicable, even though

> there isn't a distinct pre-existing property for each outcome. If so, the Kochen & Specker construction wouldn't succeed in ascribing contradictory properties to the system in question" (2016: 41).

On the other hand, if independence is given up, realists face a dilemma: they must either accept that choices of measurement (and more broadly future events) can causally influence existing properties, or postulate some unimagined common cause of both the system's features and the measurements to be performed on it.

The debate goes on. Some lingering frustrations find expression in much of the current philosophical work. One disappointment comes from enduring hopes of restoring intrinsic properties to the status they had in classical physics. Another frustration rests on the theory's inability to fully describe the outcome of quantum measurements, a snag that standard textbook versions tackle by introducing the Collapse Postulate and Born's Rule, which give QM the feel of a "black box" theory. The volume closes with papers by Simon Kochen and David Wallace that offer reformulations of the theory along each of the two lines just mentioned.

Kochen proposes a reconstruction of the mathematical formalism of QM that centers on a formulation of relational properties. One feature of the reconstruction he offers is that the classical definitions of key physical concepts such as state, observable, symmetry, dynamics, and the combining of systems take on precisely the same form in the quantum case when they are applied to extrinsic properties. Kochen argues (contra Bohr) that once the relational character of properties is accepted, the definitions of the basic concepts of QM are as real and intuitive as is the case for classical mechanics. He derives elementary quantum mechanics by applying the natural classical definitions of the physical concepts to extrinsic properties and then uses this derivation to resolve the standard paradoxes and problematic questions.

In the final chapter of the collection, Wallace contends that the version of QM presented in standard foundational discussions (the so-called "orthodox" theory) relies on substantive assumptions that do not, in fact, play any part in practical applications of QM. His primary targets are the projection postulate and the eigenvalue-eigenvector link. Wallace argues for closing the gap between the theory accepted as orthodox and the actual practice of QM. He proposes an alternative way of conceptualizing QM that, he reasons, both eliminates use of either the projection postulate or the eigenvalue-eigenvector link and does a better job of representing QM as it is actually used in physics.

1.9 Warranted Realism About What?

QM's success astonishes in the laboratory the realm of practical applications. But what is QM *credibly* about? Views vary widely, as we have seen. At one end, Bub and Zeilinger consider that Ψ represents probabilistic information about the world and says something about what rational degrees of belief one should have about it. Omnès seemingly agrees. The other contributors in the collection go for various degrees of selective realism and accept the limitations imposed by empirical underdetermination.

Importantly for selective realists, underdetermination does not spoil the entire story their ontic proposals tell about the world, just certain parts (including many at the most "fundamental" level). The proposals converge a great deal on restricted, unsharp, coarse-grained, functional ontologies specifiable at "intermediate" (but still theoretical) levels that tell of entities and processes with clear structural specifications, albeit without any pretension of fundamentality or completeness. Examples abound. They include convergences in the basic families of Many-Worlds, Bohmian Mechanics, and spontaneous collapse theories (Cordero 2001). At more ordinary levels, the ontological proposals share some coarse-grained theory-parts that seem suitable for realist commitment in virtue of their specific empirical success and freedom from reasonable doubt. Those parts contribute to the reliable scientific picture, extending it (although finitely) into theoretical levels of some depth. Here are some examples.

(a) Energy levels in molecules, atoms, nuclei, and fermions. When a bound system changes its energy state, photons are absorbed or emitted to make up the jump in energy. QM provides probabilities for these different transitions to occur, corroborated to high levels of accuracy in all sorts of microscopic systems. The resulting cartography of energy states has guided further exploration and understanding of the structure and functioning of elementary particles, nuclear physics, atomic physics, molecules, condensed matter and more in microscopic domains, including feats of scientific observation and manipulation of quantum systems.

(b) Spatial architecture of physical systems in the microscopic realm. Today, there is no specific doubt left about many geometric features of numerous aspects of the microscopic world. For example, presently, nobody has specific doubts that water molecules exist and have an electrically polarized angular structure ("V" shape), with an oxygen atom at the negatively-charged vertex and two hydrogen atoms forming the positively-charged arms at an angle of about 104° apart, and that in the ice state the molecules link up with high probability to form a "puckered hexagon." The realist point is that the theoretical entities and processes just mentioned (water molecules, their polarized parts, properties and interactions in "standard" environments) are well-supported and outstandingly fruitful in terms of corroborated novel predictions, external elucidation, and freedom of specific doubts.

(c) Dynamical unfoldings (temporal successions in quantum-mechanical processes). This aspect is exemplified by tunnel effects inferred from quantum theory, which have a temporal structure rich in novel predictions. Confirmational support comes from numerous directions in natural science. For instance, biochemists interested in the analysis of enzyme catalysis study the roles that non-statistical dynamical effects play, for instance, in proton tunneling that enhances reaction rates, typically by a factor of 1000, making it relevant to biological functions (see, e.g., Masgrau et al. 2006).

Although clearly beyond the reach of ordinary perception (and in that sense "theoretical"), these comparatively abstract, idealized, and not properly "fundamental" descriptions from QM seem nonetheless approximately correct (again, judging by their distinct novel predictions, external support, freedom from specific doubts, and high expectation of stable retention through theory-change). Being abstract, they are generally open to multiple realizations—e.g., by the various ontologies outlined in this paper. Similar assessments hold for myriads of other parts of the quantum mechanical map of the world. To repeat, the strongly warranted posits and stories do not generally lodge at the highest "fundamental" level, but they provide theoretical-level descriptions that seem as worthy of realist commitment as any in science (or so I'd like to suggest). There seems to be no reason for equating reality and fundamentality. But these are all contentious issues, of course.

References

Albert, D. Z. (1996). Elementary quantum metaphysics. In J. Cushing, A. Fine, & S. Goldstein (Eds.), *Bohmian mechanics and quantum theory: an appraisal (Boston Studies in the Philosophy of Science, 184)* (pp. 277–284). Dordrecht: Kluwer.

Albert, D. Z. (2003). *Time and chance*. Cambridge, MA: Harvard University Press.

Aspect, A. (2002). The naive view of an experimentalist. In R. A. Bertlmann & A. Zeilinger (Eds.), *Quantum [Un]speakables – From Bell to quantum information* (pp. 119–153). Berlin: Springer.

Bell, J. S. (1964). On the Einstein-Podolsky-Rosen paradox. *Physics, 1*, 195–200.

Bell, J. S. (1973). Subject and object. In J. Mehra (Ed.), *The physicist's conception of nature* (pp. 687–690). Dordrecht: Reidel. (Reprinted in *Speakable and unspeakable in quantum mechanics*. Cambridge University Press, 1987.).

Bohr, N. (1935). Can quantum-mechanical description of physical reality be considered complete? *Physical Review, 48*, 696–702.

Bub, J., & Pitowski, I. (2010). Two dogmas about quantum mechanics. In S. Saunders, J. Barrett, A. Kent, & D. Wallace (Eds.), *Many worlds? Everett, quantum theory, and reality* (pp. 431–456). Oxford: Oxford University Press.

Cordero, A. (2001). Realism and Underdetermination: Some clues from the practices-up. *Philosophy of Science, 68S*, S301–S312.

Cordero, A. (2017). Retention, truth-content and selective realism. In E. Agazzi (Ed.), *Scientific realism: The problem of objectivity and truth in science* (pp. 245–256). Cham: Springer Nature.

Einstein, A., Podolsky, B., & Rosen, N. (1935). Can quantum-mechanical description of physical reality be considered complete? *Physical Review, 47*, 777–780.

Everett, H., III. (1957). 'Relative state' formulation of quantum mechanics. *Reviews of Modern Physics, 29*, 454–462.

Van Fraassen, B. C. (1980). *The scientific image*. Oxford: Oxford University Press.

French, S., & Krause, D. (2006). *Identity in physics: A historical, philosophical, and formal analysis*. Oxford: Clarendon Press.

Gell-Mann, M., & Hartle, J. B. (1993). Classical equations for quantum systems. *Physical Reviews D, 47*, 3345–3382.

Ghirardi, G. C., Rimini, A., & Weber, T. (1986). Unified dynamics for microscopic and macroscopic systems. *Physical Review D, 34*, 440–491.

Ghirardi, G. C., Grassi, R., & Pearle, P. (1990). Relativistic dynamic reduction models—General framework and examples. *Foundations of Physics, 20*, 1271.

Kitcher, P. (1993). *The advancement of science*. Oxford: Oxford University Press.

Kochen, S., & Specker, E. P. (1967). The problem of hidden variables in quantum mechanics. *Journal of Mathematics and Mechanics, 17*, 59–87.
Leplin, J. (1997). *A novel defense of scientific realism*. New York: Oxford University Press.
Lewis, P. (2016). *Quantum ontology: A guide to the metaphysics of quantum mechanics*. New York: Oxford University Press.
Masgrau, L., Roujeinikova, A., Johannissen, L. O., Hothi, P., Basran, J., Ranaghan, K. E., Mulholland, A. J., Sutcliffe, M. J., et al. (2006). Atomic description of an enzyme reaction dominated by proton tunneling. *Science, 312*(5771), 237–241.
Messiah, A. (1961/2014). *Quantum mechanics. Mineola*. NY: Dover Publications.
Ney, A. (2013). Ontological reduction and the wave function ontology. In D. Z. Albert & A. Ney (Eds.), *The wave function: Essays on the metaphysics of quantum mechanics* (pp. 168–183). New York: Oxford University Press.
Psillos, S. (1999). *Scientific realism*. London: Routledge.
Saatsi, J. (2015). Replacing recipe realism. *Synthese, 194*(9), 3233–3244.
Saatsi, J. (2016). What is theoretical progress of science. *Synthese, 196*(2), 611–631. https://doi.org/10.1007/s11229-016-1118-9.
Schiff, L. I. (1949). *Quantum mechanics*. New York: McGraw-Hill Book Co.
Simonov, K., & Hiesmayr, B. C. (2016). Spontaneous collapse: A solution to the measurement problem and a source of the decay in mesonic systems. *Physical Review A, 94*(052128), 1–20.
Timpson, C. G. (2008). Quantum Bayesianism: A Study. *Studies in History and Philosophy of Modern Physics, 39*(3), 579–609. https://doi.org/10.1016/j.shpsb.2008.03.006.
Tumulka, R. (2006). A relativistic version of the Ghirardi–Rimini–Weber model. *Journal of Statistical Physics, 125*, 821–840.
Valentini, A. (1991) Signal locality, uncertainty, and the subquantum H-Theorem. Part I in *Physics Letters A 156*, 5–11. Part II in *Physics Letters A 158*, 1–8.
Van Fraassen, B. C. (1991). *Quantum mechanics: An empiricist view*. Oxford: Clarendon Press.
Wallace, D. (2012). *The emergent multiverse: Quantum theory according to the Everett interpretation*. Oxford: Oxford University Press.
Worrall, J. (1989). Structural realism: The best of both worlds? *Dialectica, 43*, 99–124.

Part I
Bell's Theorem and the Debate on Realism

Chapter 2
Inseparable Twins: A Conversation with Alain Aspect

Víctor Gómez Pin

Abstract During the third International Ontology Congress, held in October 1998, Alain Aspect, of the *Institut d'Optique in Orsay*, Paris, gave a lecture on the philosophical implications of his famous experiment. He is noted for his tests of Bell's inequalities with entangled photons, in which the settings are changed during the flight of the particles. From the early 1980s on, his experimental works on quantum entanglements ("Bell experiments"), single photons, laser cooling of atoms, and quantum simulations have earned him recognitions and awards, including the Gold Medal of the French National Centre for Scientific Research (CNRS), the Wolf Prize in Physics, the Balzan Prize on quantum information, the Niels Bohr Gold medal, and the Albert Einstein medal, among other awards. In 1998, in the context of the third International Ontology Congress, Víctor Gómez Pin, IOC's Chief Executive Coordinator, discussed with Aspect the philosophical aspects of experimental work in the foundations of quantum mechanics, especially the three experimental tests of John Bell's inequalities he conducted in 1982, the last of which notably allowed for a choice between the settings on each side to be made during the flight of the photons. The following is a transcription of their conversation, edited by Víctor Gómez-Pin and approved by Alain Aspect.

Editorial Note During the third International Ontology Congress, held in October 1998, Alain Aspect, of the *Institut d'Optique in Orsay*, Paris, gave a lecture on the philosophical implications of his famous experiment. He is noted for his tests of Bell's inequalities with entangled photons, in which the settings are changed during the flight of the particles. From the early 1980s on, his experimental works on quantum entanglements ("Bell experiments"), single photons, laser cooling of atoms, and quantum simulations have earned him recognitions and awards, including the Gold Medal of the French National Centre for Scientific Research (CNRS), the Wolf Prize in Physics, the Balzan Prize on quantum information, the

V. Gómez Pin (✉)
Autonomous University of Barcelona, Barcelona, Spain
e-mail: Victor.Gomez@uab.cat

A. Cordero (ed.), *Philosophers Look at Quantum Mechanics*,
Synthese Library 406, https://doi.org/10.1007/978-3-030-15659-6_2

Niels Bohr Gold medal, and the Albert Einstein medal, among other awards. In 1998, in the context of the third International Ontology Congress, Víctor Gómez Pin, IOC's Chief Executive Coordinator, discussed with Aspect the philosophical aspects of experimental work in the foundations of quantum mechanics, especially the three experimental tests of John Bell's inequalities he conducted in 1982, the last of which notably allowed for a choice between the settings on each side to be made during the flight of the photons. The following is a transcription of their conversation, edited by Víctor Gómez-Pin and approved by Alain Aspect.[1]

Víctor Gómez Pin (VGP) Perhaps we can start with an historical point. We know that Bohr became quickly aware of the need for a new set of categories of understanding suitable for quantum theory and that this requirement is at the origin of his more ontological version of the so-called "Copenhagen Interpretation." A second major step in the meta-theory was, perhaps, jointly provided by John von Neumann's development in 1932 of a rigorous mathematical formalism for the theory and the contrasting, more pragmatic, formalism published 2 years earlier by P. A. M. Dirac. In the resulting proposal, the change of state that takes place abruptly when a measurement is performed, the so-called "collapse of the wave function," has a formal explanation in terms of a system of postulates. In a certain way, the interpretations of von Neumann and Copenhagen are pedagogically complementary, since one could enrich the formalism of the first with the more "pictorial" explanations of the second. On the other hand, von Neumann sought to produce the first theorem of impossibility of hidden variables, as well as a formal consistency theorem aimed at showing the strength of his interpretation of quantum mechanics.

Alain Aspect (AA) Von Neumann thought he had demonstrated the impossibility of hidden variables and, in fact, we now know that von Neumann's theorem failed. And, interestingly, it was John Bell himself who showed that von Neumann's theorem was wrong. And it is by reflecting on this theorem that John Bell was able to articulate the train of thought that led to his theorem. It is pretty interesting historically.

VGP The predictions of quantum mechanics imply a violation of Bell's inequality, at least for some of the possible orientations for the spin detectors or, in any case, of the polarization. These predictions exclude a theory of hidden classical or "Einsteinian" variables, and the contextual theories that might be developed in response would save positing hidden variables at the high price of violating special relativity.

AA John Bell turned this difficulty into a scientific problem, in my opinion; a scientific problem is one that we can answer. The essential merit of Bell's work is that he showed that there was an experimental way to answer the question of

[1]Translation from the French assisted by Dr. Bárbara Jiménez, Universidad del País Vasco, San Sebastián.

whether classical hidden variables made a tenable hypothesis. At least in principle, the question can now be answered through an experiment.

VGP So, to use a trendy expression, he would have opened the door to experimental philosophy.

AA Yes, I willingly accept this expression.

VGP At the heart of the controversial concepts concerning your work are the concepts of realism and separability. These concepts have often been used in vague ways. It may be necessary, therefore, to ask what is meant by Realism. For example, Kant understood Realism differently than did the Aristotelians. As far as we are concerned, we may say that every system has a set of dynamical properties that belong to it even if no knowing subject is grasping them.

AA I answer naively and spontaneously "yes." It is my conception of Realism. I believe that Realism is the idea that systems have properties regardless of whether they are observed or not observed. So the whole question is to know what "property" means.

VGP Let us consider separability. Real things can be considered as localized and distinct entities (undivided in themselves and divided in relation to others—which does not exclude contiguity). They can, of course, come into interaction, but this interaction is not intrinsic to them and, when it occurs, it is limited by the impossibility of superluminal communication (in the context of the special relativity).

AA At first glance, we do not see how to do physics if we lack separability, because, how can I talk about the study of an object if I already do not admit that this object is separated from the rest of the world? At least I have to tell myself that this object interacts with the rest of the world, and does so through an interface that I can imagine involving action coming from the rest of the world. If I cannot imagine the object separated, at least in an abstract way, from the rest of the world, I do not even see how I can do physics on this object.

VGP Aristotle, distinguishing between the continuous, the contiguous and the consecutive, already formulated implicitly what might be called the principle of contiguity (or continuity-contiguity). Why before your experiments, one could be a quantum mechanics physicist and, at the same time, a classical realist?

AA What can I say? It seems obvious to me that before John Bell, we had not established at all the incompatibility between what we will call separability, a realistic and separable vision of the world (to caricature, I can say that was Einstein's vision) and then the Copenhagen's vision. Of course, these were opposing visions (shortly I'll talk about Bohr and Einstein to make the contrast vivid). So there was the formalism of quantum mechanics, which Einstein did not question. Then, to have an epistemological interpretation of this formalism, Bohr had developed his brand of the so-called "Copenhagen interpretation." But Einstein argued that there was another interpretation of the quantum formalism that was both realist and local. And

I think that, until Bell's theorem, we can honestly say that we could logically support Einstein's point of view—that is, support the view that the predictions of quantum mechanics were compatible with a realist and separable conception of physical systems. I also think Niels Bohr had the deep intuition that the two views were incompatible. But this incompatibility neither he, nor von Neumann, nor anyone else to my knowledge, managed to formalize before John Bell.

So, I'd like to make myself clear: there is the formalism on which everyone agrees and there is an interpretation that is the Copenhagen Interpretation, which was a majority view in the years 1930–1935. On the other side, there is Einstein, who could legitimately and logically argue for another interpretation of the formalism. And his interpretation goes a step further. Starting from predictions made within the framework of the agreed formalism, it reaches a conclusion that is precisely the opposite of Bohr's: Einstein showed that, because the formalism leads to the prediction of strong correlations, the only reasonable way to understand these correlations is by completing quantum mechanics.

VGP One recurrent criticism of your conclusion about the impossibility of Einsteinian separability insists on the possibility of a local and separable deterministic mechanism that would explain the experimental correlations. What would be your response to this criticism?

AA I think proponents of this objection appeal to the current limitation of the detectors employed in the experiments. Those detectors do not have 100% efficiency. So some photon pairs are not detected; and this leaves the door open, from a logical point of view, to models in which there would be a "conspiracy" of nature that would allow simulating quantum mechanics with a local hidden variable theory. So, to react to these criticisms, I believe first of all that, in pure logic, these people have every right to defend their position. However, I think their position is extremely fragile for a clear reason John Bell gave: no physical law prevents detectors from having better yields. And I must say that every year the detectors improve.So, it is safe to say that in a few years we will have better experimental access to the matter at hand; then we will be able to decide. At any rate, as Bell said, we do not see how the results of an experiment could, in one fell swoop, change qualitatively, simply because the performance of a detector has gone from 30% to 40%. And it would be extraordinary that simply by improving the efficiency of a detector, starting from something that was in agreement with quantum mechanics, all of a sudden it would no longer be consistent with quantum mechanics; it would be a tremendous change in the result. Bell said that if you go from an experiment in a certain scheme to a much more sophisticated experiment, based on another scheme, you can understand that the first gives a result in agreement with quantum mechanics and the second, much more sophisticated one, gives a result not in agreement with quantum mechanics. But when the experimental scheme is simply to improve the device a little, to think that we are going to change the results, that is something else—John Bell thought it unreasonable to imagine such a process. Nevertheless, in purely logical terms, the critics' hypothesis is tenable.

VGP From a layman's point of view, one sometimes gets the impression that scientists have not engaged in a sufficiently decisive way in the search for theories of hidden variables. Do you believe that, in the present state of affairs, it would be worthwhile to engage more in this way—for example, hidden variables but with supra-luminal interaction?

AA Two considerations come to mind. On the one hand, supra-luminal connection is present in Bohm's theory. I think that the attempts of Bohm and some people were not negligible at all; they were people who knew quantum theory very well and who knew theoretical physics very well; they were people who made considerable efforts. John Bell himself worked on such projects a great deal. On the other hand, I think there is no need to criticize the community of physicists who preferred to do something else. Indeed, physicists have had an extraordinary tool at their disposal: quantum theory. For almost 80 years, this tool has allowed us to accumulate some phenomenal successes—I mean, we explain the structure of matter and we have discovered such extraordinary things as the laser, the transistor, and so forth. It is normal that the vast majority of physicists who had been given a new tool—the extraordinary toy that is quantum mechanics—wanted to use it profusely; and then a small number of them questioned this tool. Certainly, quantum mechanics can be questioned philosophically, and I am not going to say that it is not legitimate; but there is also, on the other side, the empirical evidence that quantum mechanics works. In other words, the proof of the pudding is in the eating.

VGP Beware, however, of the fetishism of the results of quantum mechanics, when people often do not understand the theory itself, while remaining open-mouthed about its effects. But let us move on to another question. In previous interviews you gave about the reasons that drove you to your experiments to test Bell's inequalities, you mentioned both theoretical and practical reasons. Regarding the first, if I understand correctly you want to get as close as possible to a "thought experiment" situation (Gedanken Experiment). Could you elaborate on the need for an experiment approaching purity, so to speak?

AA Usually, in physics we do not question the basic principles. Everyone agrees on the theoretical and explanatory concepts in which we place ourselves. From there, all that I ask of an experiment is to be effective, to reach its goal. But here the situation is different: an experiment to check the violation of the Bell inequalities is intended to test a whole class of theories which are defined by a small number of postulates (realism, locality . . .), but these theories are not specified explicitly. As they are not specified explicitly, I cannot use one of them in particular to dissect the details of the experimentation. When we say "I'm going to test all local hidden variable theories," these theories taken together do not give me a precise description of what a polarizer does, for example. Allow me go into more detail about this. An ideal polarizer can give positive results. Polarization is, we will say, ordinary (perpendicular to the optical axis of the medium) or extraordinary (in the direction of the optical axis). This is the case of an ideal polarizer. The ideal scheme of the experiment rests on it. Now imagine that I have a polarizer that is not very good. In the normal context, a polarizer is not very good if when the light that impinges on it is in the extraordinary polarization, it registers "extraordinary" in 90% of the cases, but in 10% of the cases it will register "ordinary" and be wrong. When I do the usual physics with a polarizer, even if it is not perfect, I have a theory at my disposal; it is the quantum theory that allows me to say: well, for your imperfect polarizer that is the forecast. And I can compare this prediction with the experimental results. But when it comes to hidden-variable theories, I do not test a particular theory—I test all local theories with local hidden variables, and these hidden-variable theories do refer to a perfect polarizer, they do not give me a description of an imperfect polarizer. So, I have to have a perfect polarizer, because only a perfect polarizer is 100% or 0%; therefore, I can match all the hidden-variable theories because, by definition, these theories should give me 0% or 100%. Hence the need to have the purest experimental arrangement possible, because having no theory to describe the particular apparatus (with a possible degree of imperfection) that I have in front of me, it is necessary that the device be ideal and give only "yes" or "no" results.

VGP On a related matter, you felt it necessary to specify that the equipment employed in your tests is usable for purposes other than those we normally talk in "pure physics." Exploring the Bell's inequalities and their violation seems above all a fascinating theoretical issue. Even if the practical utility was null, the intellectual interest is enormous. Hence the question: is it not necessary to return to a conception of science that essentially ties it with the search for intelligibility?

AA My position on this matter is simple. The aim that has motivated me from the start is knowledge. I think we are always in this state of mind when we do basic research. If you are lucky enough to work in a field that deals with fundamental problems of knowledge,then you are in a very good position.In the university framework of basic research you can pursue knowledge for the sake of knowledge. But, on the other hand, if in the course of that pursuit I discover that something may have everyday utility, I think I must show that my results can also be used for something practical. If among the people who developed quantum mechanics there had not been many with that attitude, we would have missed a great deal—for example, better understanding the theory of electrons in solids and making transistors, discovering that we can understand stimulated emission and lasers, and so forth. I do not say that basic research must be driven by applications. However, I think—in my case categorically—that if, while doing basic research, I get an idea of application, it is my duty, if only to the taxpayers, to say: "attention, here we have interesting applications that will, perhaps, change telecommunications."

VGP Of course. But, perhaps, in our culture, there is the question of the hierarchy between the two aspects.

AA Yes; but, personally, I do not want to treat this question in hierarchical terms. I prefer to treat it in terms of taste. That is to say, some people prefer to pursue knowledge for knowledge's sake, and some people pursue knowledge because they are stimulated by the problems of applications. It is a question of taste.

VGP Bernard d'Espagnat had suggested in his philosophical writings that experiments such as yours made it possible to pass from "philosophical" convictions on non-realism and non-separability to scientific certainty. But D'Espagnat adds something interesting, namely, that the information that one draws from experiments would be purely negative—it would tell us what reality is not, thus eliminating philosophies like Democritus', but we could not turn that into positive information. What do you think?

AA I only agree at the logical level, and will tell you why: Bell's inequality does not show the traits of a non-separable nature or non-realist physics. In terms of logic, d'Espagnat is right. But in terms of facts and empiricism, and the impact Bell's inequality has on the development of, for example, quantum optics, it has a positive impact. It has a positive, not a negative consequence. Let me explain. The results we found not only violate Bell's inequalities, but they are also extraordinarily precise and in agreement with the predictions of quantum mechanics. Now, these predictions have been made with a quite strange situation in mind—the famous entangled states that Einstein and Schrödinger had exposed, which are quite unbelievable. Now, these extraordinary properties of the entangled states predicted by quantum mechanics have proved to be realizable in the laboratory. This means that we are able to produce these entangled states that no one had ever realized before, and verify that they have these extraordinary properties. One application we develop from there marks a new technological line: quantum cryptography, which attests to a conceptual revolution in cryptography. We are developing the possibility of producing states in a single photon that can revolutionize optical telecommunications and things like that. Why has this happened? Because, in making such delicate experiments, we have, in a positive way, drawn attention to the fact that these incredible states, these entangled states—of which everyone said "yes"—are the idiosyncrasies of quantum mechanics. In practice, however, we never see them in the ordinary world. In fact, we drew attention to the fact that these entangled states could be really produced and show their extraordinary peculiarity. So, in this sense, all that is demonstrated is not negative. Empirically, we had an extremely positive impact, as exemplified by quantum cryptography based on the Bell inequalities.

Do you see the conceptual change? In this new cryptography, the guarantee of the secrecy of your communication is not based on adversary's technology being less advanced than yours: it is based on a fundamental law of nature, rooted in quantum mechanics. You see? It is pretty amazing! So, when we say that the impact of these experiments was negative, I would like to show that these extraordinary quantum properties that can be really demonstrated experimentally using sources such as photon pairs had a positive impact. People began to think that these entangled states were usable developments. Quantum entanglements were not a dream of the theorists; we could really produce them, and use them.

VGP Bell's Theorem operates at a deeply abstract level; there is a logical and disembodied rigor, almost like a formal game. For many who approach it for the first time, the question of the "link" inevitably arises: what does all this have to do with the correlation of photon polarizations from the same atom? There is something striking about it. Basic logical presuppositions and presuppositions concerning nature (locality, objectivity, determinism) which seem at first sight elementary (almost axiomatic in the Aristotelian sense) seem to fail here. And all this springs from an experimental test which is not elementary. When we tell the story of the problem, John Bell's role always appears ambiguous. On the one hand, he is indisputably at the center. On the other hand, he often appears as a theoretician, and you have pointed out somewhere that he reasoned mostly about "thought experiments" and that the conflict he highlighted (a contradiction between predictions of quantum mechanics and the theory of hidden variables) is, above all, "numerical". This use of the word "numerical" brings us to a problem, as old as Pythagoreanism, whose philosophical importance is enormous. John Bell is, in a way, a mathematician who made us think in terms of Schrödinger's equation—which, although only a postulate, proves not only effective at accounting for the facts but, in a particular case, allows us to find knowledge as old as the harmonies of the Pythagoreans.What does a physicist such as yourself think of this question: does mathematics match Nature, and if so why?

AA I have no competence to answer this question. However, I have the right to have an opinion as an individual, anyway. What you said about the fact that the distinction was numerical gives me the opportunity to insist on a feature of Bell's theorem: it is not just that the hidden-variables theory predicts one number and then quantum mechanics predicts another number. In my opinion, there is something more—the fact that it is likely to be tested experimentally. Bell says that a certain magnitude must be smaller than, say, 2. The mere fact that we have put an inequality—a barrier—allows us to do experiments like mine, because if I now do an experiment and get 2.3, I know that the precision of my experiment must be sufficient to distinguish between 2.3 and the barrier—the limit that John Bell put. I think that the inequality means that I have to find a result that is outside the proposed limit. So, the key question that makes my experiment credible or not is going to be the margin of error of my result, which will be the accuracy of my result. And if the accuracy is sufficient for me to have the result falling on the bad side of the barrier for hidden variable theories, then the result is credible. It means that results on this side of the border will allow me to eliminate all theories of local hidden variables.

There are two types of conflict. There are logical conflicts between numbers; that is, one can try to demonstrate logically, in an abstract way, that quantum mechanics is incompatible with hidden variable theories. And for that, it is enough to find two numbers which are different from each other, but these two numbers can be very close. For example, on the matter of pure logic and theoretical incompatibility, if the hidden variables explored yield, say, π and then quantum mechanics predicts 3.14, the fact that 3.14 is not equal to π would suffice to show that the hidden variable theories are not identical to quantum mechanics. But John Bell has demonstrated that there is a boundary which is far enough from the predictions of quantum mechanics, and this gives ample space to make an experiment. What Bell's reasoning allows us to say is that if your experimental results, although not ideal, are good enough, then you can rule in favor or against quantum mechanics. In pure mathematics, 3.14 is not equal to π; but, for the physicist, 3.14 may be practically equal to π. So, importantly, hidden variable theories predict that the result must be no bigger than 2, but quantum mechanics predicts 2.5; so, if quantum mechanics predicts 2.7 and your experiment is not perfect, 2.7 is still greater than 2. So, there is a barrier and, beyond this barrier, we have all the relevant margin granted by the experiments. The experiment must be precise enough for its result to fall to the right of the barrier; but it does not need to be ideal, the ideal experiment does not exist. Because there is an inequality there is a very peculiar status that opens the way to the sought experiments; that is to say, it is not simply a logical conflict between two concepts; it immediately opens the way to experimentation.

VGP And what about mathematics matching reality?

AA Yes, the mathematization of nature. I think that is the greatest mystery there is. I have this rather naive remark to make: perhaps physics limits itself to what can be represented by mathematics. That is, perhaps every time a field escapes mathematization—like, for example, painting—physicists say, "... obviously, it is not physics." So all of this poses a problem of demarcation; biology a century ago was probably outside the boundary, but biology is moving towards mathematization and rigorous theorizing more and more nowadays. So biology is joining the realm of what is mathematizable.

VGP But, unlike painting, music is under the imprint of mathematics.

AA Indeed, music is a special case, because indisputably the psychology of individuals is outside mathematics.

VGP And yet, the trend is to mathematize that too.

AA Yes, but it is a false mathematization. For example, it is obvious that if we want to make epidemiological or statistical types of surveys, we use mathematics but we know that it is not mathematization—we are not theorizing mathematically. It is as if we were saying that when someone investigates the typology of individual words he replaces psychology with a theory of sound waves. Psychology is not mathematized. So there is a whole field of what exists, a whole field of the world around us which, unquestionably, cannot be represented by mathematics.

VGP You know what the answer of the Pythagoreans would be: mathematics is hidden there. But we must leave that question open and come back to John Bell. In one of his interviews, he stated that he would like to be able to have a realist view of the world, to speak of the world as if it was really there even when nobody observed it. Bell believed that there would be a world after his death, adding that all physicists tend to accept this point of view when they are "pushed into a corner" by philosophers. There is something moving in re-reading this statement when Bell is no longer here. When philosophers press you, the realist answer might be preferable, but not because the philosopher necessarily expects a realist answer. The properly metaphysical question is naive and needs a short pre requisite. Any student who sees himself for the first time confronted with Kant's *Critique of Pure Reason*, formulates for himself the question that John Bell mentions as soon as he begins to grasp the argument: will there be a people any more when we are no longer here to bear witness of its presence? One cannot avoid this question that is at once naive and radical.

AA Yes indeed. I believe this is so, especially for a physicist. Because, in the end, it seems to me that the physicist chooses to do physics because he thinks that there is an intelligible world. In other words, I believe that the physicist, when he first imagines his life as a physicist, sees himself as someone who has an object outside of himself, say a clock, and will open it to understand how it works inside. So I think the physicist, perhaps more than others, has this naive spontaneous belief that there is a world external to him and that his role is to discover the way that world works, but he must do this by altering it as little as possible. That is, as an observer physicists must be as discreet as possible. And so the physicist has the vision that, in principle, the world works and is out there even if the observer is not there to watch.

We must realize that, to make progress, this vision had to be overtaken by Werner Heisenberg, Niels Bohr and the people of the Copenhagen school. Conceptually, these people managed to make progress by recognizing that the interaction between the observer and the observed world plays an essential role. I do not conclude that the world does not exist when the observer is not there. My claim is only that, in a certain state of development of the sciences, we had to accept to take into account the interaction between the observer and the system measured to make progress. We had to accept that interaction because the previous vision—according to which the observer role could be made infinitely discreet—blocked progress. The fact that it was necessary to cross this barrier and say: "attention, there is no infinitely innocuous measurement: any measurement involves a minimum of interaction with the object undergoing measurement." Crossing the barrier helped physics to make progress. I do not conclude that the world does not exist if there is no observer. For me, any trace left in the Universe is a form of observation. Feynman described this point very well in his *lectures on Physics* (III-2): If a tree falls in a forest and there is nobody there to hear it, does it make noise? A *real* tree falling in a *real* forest makes a sound, of course, even if nobody is there. Even if no one is present to hear it, there are other traces left. The sound will shake some leaves, and if we were careful enough we might find somewhere that some thorn had rubbed against a leaf and made a tiny scratch that could not be explained unless we assumed the leaf were vibrating. So, although there was no one to listen, this noise left traces; so, it existed. I do not attribute a particular role to the transcendental human observer, and I will tell you something very strong and provocative: in my experience, my photo-checker is as strong as the transcendental subject when it comes to performing an act of measurement.

VGP But some problems remain; for example, can non-locality be considered compatible with science? René Thom, the topologist and inventor of catastrophe theory, was radical on this matter. He counted himself among those who consider that a non-local theory cannot even be considered scientific *strictu senso*, because—

he thought—one cannot act and know except locally. Einstein thought something analogous. Do you subscribe?

AA Initially, I am completely seduced by this thought; but experience forces us to accept that the world is more complicated than that. On the other hand, I think we must give Einstein justice; he did not know Bell's theorem. At the time, one could legitimately think that the world was local and separable. Today we know none of that is true—we have proved experimentally that some situations cannot be described using the naïve concept of separability. I think our "ordinary" concept of separability—Einstein's concept—was too naive. We have to admit that the world as we observe it in the laboratory must be described by concepts more nuanced than what Einstein had in mind. On the other hand, today we still retain the idea that the world is local and separable in the sense of direct action. That is, even with twin photons, even with entangled states, even with quantum correlations, I know that I cannot turn a knob here and act instantly in any usable way at the other end of the Universe. By "usable way" I mean that I cannot transmit energy instantly (for example, information usable by my correspondent to make a decision). So, in this sense, I am preserving some form of locality.

Then again, we know that there are nevertheless non-separable objects in the world—conspicuously, the EPR pairs produced in the laboratory; and these objects behave as an inseparable whole, though obviously each EPR pair is spread out over distances that may be extremely large. In Nicolas Gisin's experiments in Geneva, the separation is 10 km; 20 years ago the maximum separation realized was just 15 m. Well, one must consider an EPR pair as a non-separable object, because, if we start to think of it as a separable object, we make predictions that are at odds with what we observe.

So, the concept of separability needs refinement. There is operational separability, which I think continues to obtain: that is to say, by turning a button here I cannot instantly turn on a light in New York. On the other hand, if I seek to represent the pair of inseparable photons, well, I have to accept that the two continue to make a whole and that I cannot cut this system into slices. Moreover, the quantum formalism does not allow me to cut the two-photon system into two spatial parts and say that on one side there is an object with one property and on the other side there is another object with another property. According to quantum non-separability. I have only one unique object, this two-photon object that has a global property that I cannot attribute to one or the other component.

VGP In an earlier interview you said that, if Einstein were alive, he would have reacted very intelligently to your experiments. Others have been less cautious. After the tests by the Gisin team in Geneva, one of the headlines read: "Einstein loses in photon test." John Bell himself, who for a long time remained cautious, did not hesitate to proclaim the need to return to a pre-Einsteinian relativity. From the configurational myths, this issue is essential, because Einstein is something more than just a scientist whose theories have philosophical implications. He is, in a way, the Picasso of science, more than a scientist—a basic referent that plays the symbolic role of a taboo. Whatever the outcome of the quantum mechanical tests you and others are conducting, the result seems inevitably fraught with consequences not just for Einstein but for thinkers as different as Descartes and Aristotle, whose principle of contiguity seems deeply shaken. In short, then, could Einstein say something specific about the results of your experiments?

AA Does not the history of thought show us situations analogous to that? For example, Bergson—I think he was smart—could not accept that the concept of time was not an absolute notion, he could not accept the relativity of time. Bergson thought he had shown that relativity was impossible because he did not accept the twin paradox. Anyone, however intelligent, may at some point find himself stranded. So, no one can know if Einstein would have gone beyond and managed to find something extraordinary in response to my experiment. Picasso passed through several periods in his life; but if he had not had a blue period, a pink period, we would not know that there were several periods in his life. Perhaps Einstein, in the light of Bell's theorem, would have imagined something else; but we do not know it.

VGP But if he had imagined anything else, he would have been forced to question some of his essential presuppositions anyway.

AA But I do not know if these are essential presuppositions. Let us say that separable realism, as traditionally formulated, had to be questioned; but perhaps Einstein would have found a form of separable realism which would have been less strict than the one he started with and which would have been acceptable. After all, there are formal ways out. For example, if you agree to place yourself in a space other than the real (three-dimensional) space the world becomes separable again: in a six-dimensional space you can have locality for these entangled states. I do not understand locality in a six-dimensional space because when I try to plunge into a three-dimensional space it becomes non-local again. But this it does not deny that, formally, one can have locality in a six-dimensional space. You have people like Feynman who have been able to write—albeit I do not know if they believed it—articles showing that it is logically possible to have escape routes to certain stalemates by accepting negative probabilities—if we accept the negative probabilities, we can save locality.

VGP But, does not negative probability shock you?

AA It totally shocks me. It is completely incompatible, because to say that a result occurs with a negative probability is to say that it is erased from the rest of the universe. So, for me, it does not make sense (an event that took place, it cannot be erased from the Universe). Neither does six-dimensional locality make sense to me. I don't think it is a solution, but it provides a formal resolution. There are two kinds of issues here: formal issues, such as those I am vaguely enunciating; and empirical issues, which say "pay attention, locality is not empirically violated." Although EPR states are non-separable, we can demonstrate that they cannot be used to transmit energy instantaneously, or even usable information. In other words, non-locality and non-separability are there, are a given, but I cannot use them. It is extraordinary! Well, you can use separability by saying "empirically, separability is not violated; I cannot send messages to the other end of the Universe instantly."

Chapter 3
Bell's Theorem, Realism, and Locality

Peter J. Lewis

Abstract Bell's theorem is sometimes taken to show that quantum mechanics undermines scientific realism. If so, this would be a striking empirical argument against realism. However, Maudlin has claimed that this is a mistake, since Bell's theorem has precisely one conclusion—namely that quantum mechanics is non-local. I argue here that matters are more complicated than Maudlin acknowledges: quantum mechanics is not a unified theory, and what Bell's theorem shows of it depends on which interpretation turns out to be tenable. I conclude that while the lesson of Bell's theorem could be that quantum mechanics is non-local, it could equally be that measurements have multiple outcomes, or that effects can come before their causes, or even, as the anti-realist contends, that no description of the quantum world can be given.

Various people have claimed that quantum mechanics undermines (some form of) scientific realism. For example, Bohr writes that "there can be no question of any unambiguous interpretation of the symbols of quantum mechanics other than that embodied in the well-known rules which allow [us] to predict the results to be obtained by a given experimental arrangement" (1935, 701).[1] If quantum mechanics *does* show that scientific realism is false, this is highly significant. If a particular set of phenomena undermines realism, this provides an argument against realism even Carnap could respect. It turns the realism/anti-realism issue from a metaphysical *scheinproblem* to a real empirical question—an internal question, in Carnap's idiom.

There are of course several other kinds of arguments against realism, most prominently those appealing to underdetermination (van Fraassen 1980) and to the pessimistic induction (Laudan 1981). The latter is even an empirical argument:

[1] Bohr's position is subtle, and he wouldn't qualify as anti-realist on every construal, but I think he would deny that quantum phenomena are explained via a description of the micro-world.

P. J. Lewis (✉)
Department of Philosophy, Dartmouth College, Hanover, NH, USA

A. Cordero (ed.), *Philosophers Look at Quantum Mechanics*,
Synthese Library 406, https://doi.org/10.1007/978-3-030-15659-6_3

experience shows that once-successful theories are eventually overturned.[2] But even so, the empirical argument is at the meta-level—it involves data about theories, not data about the world. It would be quite a different thing if studying certain phenomena in the *world* could provide a direct argument against realism. The goal of this paper is to evaluate the claim that quantum phenomena undermine realism, and in particular, the role of Bell's theorem in any such empirical argument.

It is often unclear what is meant by "realism" in these debates. So let's start with a statement of scientific realism: *scientific theories aim to describe the world, where those descriptions aim to explain the phenomena covered by the theory.*[3] The description here, of course, is just of a particular aspect of the world at a particular level of description; no theory is a theory of *everything*. This definition is deliberately broad, but it rules out e.g. constructive empiricist views, according to which scientific theories aim merely to predict the phenomena. Constructive empiricist accounts of science deny the descriptive role of theories of the micro-world, and curtail their explanatory role insofar as the explanations invoke the descriptions.[4]

So, for example, on a realist construal, evolutionary biology aims to describe real processes in the world that explain the variety of living things we observe. Chemistry aims to describe the elements in the world and their forms of interaction to explain, for example, why iron rusts. And quantum mechanics aims to describe physical entities at the atomic scale and the laws governing them, to explain, for example, how a laser works.

At least, that's what quantum mechanics looks like from a realist perspective. But the claim to be considered here is that quantum mechanics reveals the limits of the realist project: one cannot look to quantum mechanics for a description of the micro-world, or hope to explain quantum phenomena in terms of such descriptions.

In outline, the narrative goes something like this. Einstein notes that sometimes the outcome of a quantum measurement on a particle is not represented in the theory, even when that outcome can be predicted with certainty (Einstein et al. 1935). He concludes that quantum mechanics is incomplete as a description of the world. In response, Bohr (1935) argues that quantum mechanics is in fact a complete description, in the sense that no further ascription of properties to particles is

---

[2]Some versions of the underdetermination argument are also empirical, insofar as they appeal to actual underdetermination in the history of science rather than hypothetical underdetermination; indeed, quantum mechanics is arguably an excellent candidate for actual underdetermination (Barrett 2003, 1211). But again, this argument involves data about theories, not data about the world.

[3]This is a slight reworking of van Fraassen's minimal formulation of scientific realism (1980, 8), designed to highlight the roles of description and explanation.

[4]Van Fraassen (1980, 23). Note that van Fraassen alludes to quantum mechanics here in his plea for limits on explanation.

possible. Bohr's argument for this conclusion may not be particularly compelling,[5] but the conclusion itself is vindicated by Bell's theorem (Bell 1964). So Einstein is proved wrong, in the sense that our expectations of descriptive completeness in a theory cannot be met at the micro-level.

This is, I think, a fair (if brief) summary of something like the received view among physicists.[6] But in a recent paper, Tim Maudlin has argued forcefully that the received view is wrong: in particular, it misrepresents what Bell did:

> Early on, Bell's result was often reported as ruling out determinism, or hidden variables. Nowadays, it is sometimes reported as ruling out, or at least calling in question, realism. But these are all mistakes. What Bell's theorem, together with the experimental results, proves to be impossible (subject to a few caveats we will attend to) is not determinism or hidden variables or realism but locality, in a perfectly clear sense. (Maudlin 2014a, 1)

In other words, Bell's theorem has nothing to tell us about realism or the descriptive completeness of quantum mechanics; it does not show that Bohr was right and Einstein was wrong. What it shows is that quantum mechanics is non-local, no more and no less.

What I intend to do in this paper is to challenge Maudlin's assertion about the import of Bell's proof. There is much that I agree with in the paper; in particular, it does us the valuable service of demonstrating (hopefully once and for all) that Einstein's objections to quantum mechanics have nothing to do with its (supposed) indeterminism. But I do think that Maudlin's conclusion is overly cut-and-dried. Quantum mechanics (as Maudlin would be the first to admit) is far from a unified edifice, and what Bell's theorem shows depends on what version of quantum mechanics you look at. In particular, I'll try to make the case that there's an interesting, if ultimately uncompelling anti-realist construal of the import of Bell's theorem. And I also want to suggest that locality isn't quite as decisively defeated as Maudlin claims.

Bell's theorem is easy to set up; here I follow Mermin (1981). Consider a pair of particles produced in the entangled spin state

$$|S\rangle = \frac{1}{\sqrt{2}}\left(|\uparrow\rangle_1|\downarrow\rangle_2 - |\downarrow\rangle_1|\uparrow\rangle_2\right),$$

where the spins are relative to the z-axis.[7] And consider two axes v and w that make an angle of 120° with the z axis (and with each other). Then if the spins of the two particles are measured relative to the same axis (both v, or both w, or both z),

---

[5] What he says is that an assignment of properties to a quantum mechanical system can only be made relative to a particular choice of measurements on the system, and hence no unique property ascription is possible (1935, 700). But it is hard to motivate this claim absent a proof like Bell's.

[6] Maudlin (2014a) laments this, but both he and Werner (2014) suggest that some view like this is common.

[7] In fact, this state takes the same form when the spins are expressed relative to any other choice of axis too.

quantum mechanics predicts that the results will always disagree: one is spin-up and the other is spin-down. And if the spins are measured relative to two different axes, then quantum mechanics predicts that the results will disagree 1/4 of the time and agree 3/4 of the time.

Einstein's complaint about these predictions is that if the spins of the two particles are measured sequentially, relative to the same axis, then the result of the first measurement allows you to predict with certainty the outcome of the second, even though nothing in state $|S\rangle$ tells you the spin of either particle. Einstein concludes that quantum mechanics is incomplete as it stands, insofar as there are physical states of affairs, such as the one that produces the spin result for the second measurement, that are not represented in the theory (1935, 780).

Put this way, it doesn't seem too tall an order to complete quantum mechanics: one simply needs to add some kind of representation of the missing states of affairs. These states of affairs (for the set-up considered here) are the spins of the two particles along each of the three possible measurement directions. So, for example, we could represent the spins of particle 1 along the v-, w- and z-axes using the triple (up, down, down), and the spins of particle 2 along these axes using the triple (down, up, up). Note that in this example the spins of particle 2 are the opposite of the spins of particle 1. This ensures that when the spins of the two particles are measured along the same axis, the results always disagree, as quantum mechanics predicts. But what if the spins are measured along different axes? How can we ascribe spin values to the two particles to ensure that if the spins are measured along different axes, the results agree 3/4 of the time? What Bell's theorem shows is that this task is impossible: the best you can do is agreement 2/3 of the time.[8]

Taken at face value, Bell's theorem seems to show that quantum mechanics is impossible—that no physical model could in principle produce the distribution of measurement outcomes predicted by the mathematical algorithm at the heart of quantum mechanics. But quantum mechanics is well confirmed; this distribution of measurement outcomes is *actually observed*, and what is actual cannot be impossible! So the way to read Bell's theorem is as a reductio: since Bell's proof leads to an absurd conclusion, one of its assumptions must be false.

The question, of course, is *which* assumption is false. It isn't obvious what physical assumptions are required to derive Bell's conclusion, and different authors divide up the premises in different ways. Perhaps the most straightforward way to proceed is to explicitly construct a theory that generates the predictions of quantum mechanics, and then *see* how it evades Bell's theorem.

Maudlin, as noted above, thinks that the lesson of Bell's theorem is that the world is non-local in a precise sense. That is, he thinks that the premise of Bell's theorem we should deny is the following locality assumption: "procedures carried out in

[8]Note that if particle 1 has the spin properties (up, down, down) and particle 2 has the properties (down, up, up), then for measurements along different axes, the results agree 2/3 of the time. The same goes for all the other possible spin property assignments, except for the pair (up, up, up) and (down, down, down) for which the results never agree. So no assignment of spin properties to particles can produce agreement more than 2/3 of the time.

one region do not immediately disturb the physical state of systems in sufficiently distant regions in any significant way" (2014a, 8). And indeed, there are versions of quantum mechanics that violate this assumption—most notably Bohm's theory, which has been actively championed by both Maudlin (1995) and Bell (1982).

The way that Bohm's theory evades the conclusion of Bell's theorem is that it adds a non-local dynamical law via which a measurement performed on one of an entangled pair has an instantaneous effect of the state of the other. More precisely, Bohm's theory "completes" the quantum mechanical description provided by state |*S*> by ascribing a *position* to each particle, and postulating a new dynamical law via which those positions change over time. Notably, the law is such that the motion of one particle depends on the positions of *all* the particles in the system. When the spin of particle 1 is measured, it moves along the axis in which it is measured—up if the result is spin-up, and down if the result is spin-down.[9] Then when the spin of particle 2 is measured, its motion depends on the current position of particle 1, and hence on the outcome of the measurement on particle 1. This provides us with a physical state of affairs explaining the outcome of the second spin measurement, as Einstein demanded. And it provides us with a way of explaining the correlated spin results that are seemingly ruled out by Bell's theorem. But it does so at the cost of introducing instantaneous action at a distance into fundamental physical law.

Locality is an explicit assumption in the proof of Bell's theorem, and it is uncontroversial that one can evade Bell's conclusion by postulating non-locality in the world. Bohm's theory takes this route, as do spontaneous collapse theories like GRW, which also postulate a non-local dynamical law. What is controversial is Maudlin's claim that the *only* way to evade Bell's theorem is via non-locality: this is the content of his claim that Bell's theorem rules out locality. To establish this, we need to convince ourselves that there are no other ways around Bell's theorem.

In particular, Maudlin claims that denying realism is not an option here. Against this, Werner argues that one can construct a local quantum mechanical theory if one is willing to violate an assumption he calls 'classicality' or 'realism', where "'realism' is the mathematical assumption 'The state space is a simplex'" (2014, 7). He argues that operational quantum mechanics violates this assumption, and hence provides a way to construct a local quantum mechanical theory. If Werner is right, then Bell's theorem has an equal claim to challenging realism as it does to challenging locality.

However, Maudlin rebuts this charge on the grounds that Werner cannot identify anywhere in Bell's reasoning where realism in this sense is presupposed (2014b, 2). Furthermore, Maudlin argues that operational quantum mechanics is not a counterexample to his thesis: it is not a local, non-realist account of quantum mechanics, because in fact operational quantum mechanics, too, violates locality. In particular, Maudlin focusses on Werner's claim that the physical state of a system is "the quantity which allows us to determine the probabilities for all subsequent operations and measurements" (2014, 3). Since a measurement on

[9] I assume that the spin is measured by passing the particle through a Stern-Gerlach device.

particle 1 instantaneously changes the probabilities for subsequent measurement on particle 2, it must change the physical state of particle 2, and hence operational quantum mechanics is non-local after all.

However, I think Maudlin is being unfair here. Immediately after the passage quoted above, Werner clarifies that he takes the physical state to be "'epistemic' rather than 'ontic'" (2014, 3). That is, what Werner (perhaps misleadingly) calls the "physical state" should be taken as a representation of our knowledge (in some sense), rather than a description of the world. Further, Werner mischaracterizes the assumption being denied here—and Maudlin takes him at his word. It is not just that the state space is a simplex; indeed, denying an assumption about the structure of the state space doesn't amount to denying *realism*. What makes operational quantum mechanics *operational* (as opposed to realist) is that the quantum state is taken as a formalism connecting preparation events with probabilities over measurement outcomes, without any commitment to the state representing or describing the micro-world. Indeed, Werner later notes that the assumption which operational quantum mechanics rejects can also be characterized as "commitment to ontology at the level of quantum particles" (2014, 7).

If the assumption to be challenged is that an adequate quantum mechanical theory should describe the world (as opposed to merely describing our knowledge), then plainly it is an assumption in Bell's proof. Einstein complains that standard quantum mechanics cannot be taken as a complete description because it does not represent the physical states of affairs underlying certain predictable measurement outcomes. Bell shows that any attempt to represent such states of affairs cannot account for the observed distribution of measurement results. The point of operational quantum mechanics (as I understand it) is to *deny* the requirement that a prediction must be explained via a physical state of affairs that is described by our theory.

However, if this is the correct way to view operational quantum mechanics, then Maudlin has an argument against it. His argument is based on Einstein's "criterion of reality" from the EPR paper: "If, without in any way disturbing a system, we can predict with certainty (i.e. with probability equal to unity) the value of a physical quantity, then there exists an element of physical reality corresponding to this physical quantity" (1935, 777). It is this criterion that requires that there be a physical state of affairs explaining the result of the measurement on particle 2 when the two particles have their spins measured in the same direction, since after particle 1's spin has been measured the spin of particle 2 can be predicted with certainty.

Furthermore, Maudlin claims that Einstein's criterion of reality is *analytic*: it is "just not the sort of thing that can coherently be denied" (2014a, 7). The reason is that "the physical behavior of a system depends on its physical state: if a system is certain to do something physical, then *something* in its physical state entails that it will do it" (2014a, 7). If Maudlin is right, then I cannot construe operational quantum mechanics as denying that the outcome of the measurement on particle 2 has a physical explanation, because such a construal is incoherent.

I doubt that Einstein's criterion of reality is really analytic, though. It seems perfectly conceivable that an event could be predicted with certainty even when there is nothing physical that brings that event about. That is, it seems perfectly conceivable that an event just happens, without a physical cause, and yet happens with certainty. Indeed, Maudlin is perfectly sanguine about fundamentally probabilistic laws (e.g. in spontaneous collapse theories), according to which there is in general no physical reason why *this* result is obtained (as opposed to *that* result) when the probabilities differ from zero and one. Why should things be different when the probabilities *are* zero and one?

But let's grant for the sake of argument that Einstein's criterion can't be coherently denied. Still, even if we can't deny that there's some physical element of reality behind any probability-1 event, it doesn't follow that it is *knowable* or *describable* by us. There is nothing that guarantees that the physical world is epistemically accessible and amenable to capturing in a unified theoretical model. Perhaps every probability-1 event has its own *sui generis* physical explanation. That would be unfortunate for us as theorists, but it is hard to see how it could be ruled out a priori.

Indeed, the resistance to the demand for explanation is a classic anti-realist move. Why does a measurement on a system yield a particular result with certainty? When the theory of the system in question appeals to microscopic entities, anti-realists of a certain stripe (e.g. van Fraassen) may refuse the demand for an explanation that goes beyond the prediction itself. One may not find this form of anti-realism philosophically attractive, but it surely not analytically false.

This, it seems to me, is precisely the move made by operational quantum mechanics. When particle 1 is measured, then the tools of standard quantum mechanics allow us to predict the spin of particle 2 with certainty. But there is no need to posit a physical state of affairs—a property of a physical particle—to explain this measurement outcome. So Bell's proof is blocked before it starts: there is no call for a physical explanation of spin results, and so no need to ascribe spin properties to particles. Furthermore, without a physical explanation of measurement results, there is obviously no *non-local* explanation, so locality (in this sense) is saved.

I think, then, that one can take anti-realism to be a potential lesson of Bell's theorem in just the same sense that one can take non-locality to be a potential lesson. That is, there are accounts of quantum mechanics that evade Bell's theorem by appealing to non-local causal mechanisms, and there are accounts that evade Bell's theorem by denying that our physical theories *describe* the micro-world.

Of course, this is not to say that either of these ways around Bell's theorem is attractive; there are reasons to dislike them both. Non-locality, as is well known, stands in conflict with special relativity. Special relativity tells us that simultaneity is frame-dependent—that there is no objective matter of fact about whether two events are simultaneous or not. But instantaneous action at a distance *requires* such an objective matter of fact, since it requires a fact of the matter about which distant events are simultaneous with *this* one. One can of course deny that special relativity is an adequate theory, and add a preferred frame to it to define absolute simultaneity, but this is certainly a theoretical cost.

Similarly, anti realism is not an attractive option. Physics is in the explanation business, and routinely denying the call for explanation seems tantamount to giving up. Furthermore, one might be suspicious of my claim that operational quantum mechanics saves locality. In the absence of a descriptive theory of the micro-world, one cannot identify non-local mechanisms in the world, but neither can one assure oneself that only local mechanisms are involved. Indeed, if we could show that only a non-local mechanism could in principle account for the observed effects, then even if we don't regard quantum mechanics as descriptive, we might still conclude that the physical world (which fails to be described by the theory) embodies non-local causation.[10]

So even if Maudlin's claim that Bell's theorem tells us nothing about realism is wide of the mark, he still might be right that Bell's theorem shows that the world is causally non-local. But are there other alternatives—accounts of quantum mechanics that are neither non-local nor anti-realist? Arguably, there are. Consider first Everettian (many worlds) accounts of quantum mechanics. When particle 1 has its spin measured, this induces a process whereby the particle, the measuring device and everything that becomes correlated with it splits into two branches. In one branch the particle is spin-up, and in the other it is spin-down. Similarly, two branches are formed when particle 2 has its spin measured. So the measurement of particle 1 does not entail that the spin of particle 2 can be predicted with certainty, because there is no unique spin result for particle 2 to be predicted.

This is a little quick, though. Human observers split into branches too. If you find yourself in the spin-up branch for particle 1, you can predict with certainty that particle 2 will be spin-down (in your branch). But arguably, at least, this correlation between the branch-relative spins doesn't require any non-local mechanism to enforce it. The global quantum state of the system means that if you travel to the location of particle 2, you will find yourself in the spin-down branch for particle 2, but this causal mechanism (your travel) takes place at ordinary sub-luminal speeds. Admittedly, the explanation appeals to the global state of the system, but this arguably requires non-separability (holism) rather than non-locality (Wallace and Timpson 2010).

So the many worlds theory at least looks like a perfectly realist, fully local account of quantum mechanics. If it is tenable, then another potential lesson of Bell's theorem is that the assumption that each quantum measurement results in a *unique* outcome is false. But it is perhaps not fully clear that the many-worlds theory is local, since the relationship between non-separability and causal locality is a tricky one.[11] Furthermore, many worlds theories have notorious difficulties handling probability: how can we say that one outcome is more probable than another if both actually occur? A good deal of progress has been made recently in addressing

[10] I'm not sure whether such an argument would really go through. If it is conceivable that every measurement outcome has its own *sui generis* physical explanation, then there might be no underlying causation, at least on a regularity view of causation. In which case the question of locality becomes moot.

[11] Maudlin contends that "a tremendous amount of interpretive work" would be needed to decide whether the many worlds theory is really local (2014a, 23). But Wallace and Timpson (2010) claim to have done the requisite work and shown that many worlds quantum mechanics is causally local.

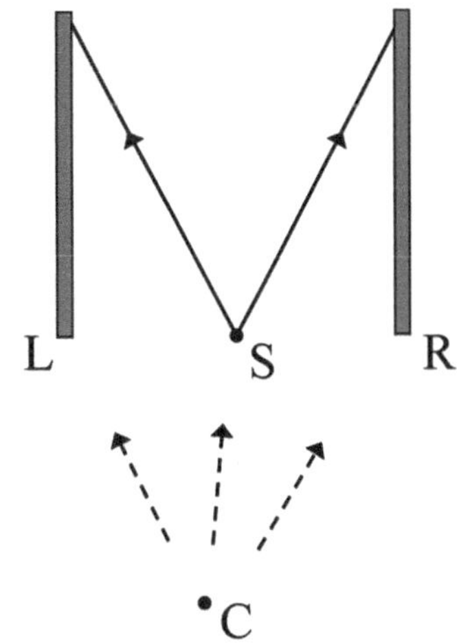

Fig. 3.1 Bell experiment with a common cause

this problem (e.g. Wallace 2010), but it is still less than clear (to me, at least), that the many worlds theory can really deliver the empirical probabilistic predictions of standard quantum mechanics (Lewis 2010).

The second alternative worth considering is what Bell calls *superdeterminism*.[12] Bell assumed in his proof that the properties of the particles are independent of the measurements that will be performed on them. This seems like a perfectly innocuous assumption: after all, the measurements can be chosen however we like *after* the particles have been created. But if it could somehow be called into question, then another route to bypassing Bell's conclusion would be opened up: if the properties the particles have are dependent on the measurements that will be performed on them, it is trivial to arrange the actual possessed spin values of the particles so as to reproduce the observed quantum mechanical predictions.

How could this independence assumption be violated? Consider the space-time diagram of the Bell experiment in Fig. 3.1. Here the particle trajectories are the diagonal lines, and the measuring devices are the vertical bars. The particles are emitted at S, and the measurements to be performed by the measuring devices are chosen at L and R. Note that there is no way that the choices at L and R can *directly* affect the particle properties at S without some non-local causal influence. So if we want to keep things local, it looks like we have to posit a common cause C that influences both the measurement choices and the particle properties.

But notice how *powerful* such a cause would have to be. It would have to be capable of correlating *anything* that could be used to set the measuring devices with the properties of the particles—coin-tosses, human choice, the air temperature in Llandudno, or whatever. As Maudlin notes, "such a purely abstract proposal cannot be *refuted*, but besides being insane, it . . . would undercut scientific method"

[12] In Davies and Brown (1986, 47).

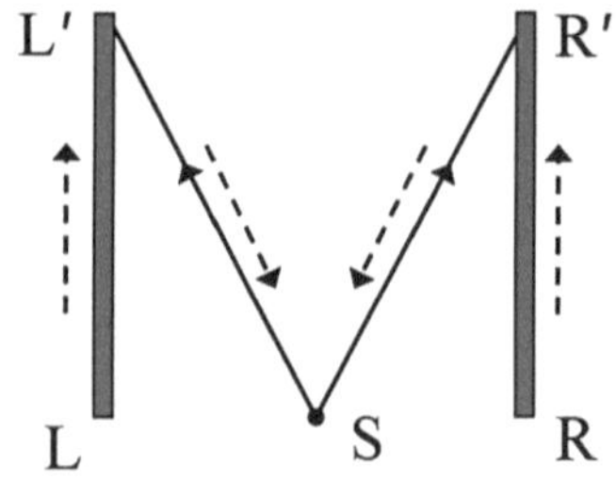

Fig. 3.2 Bell experiment with retrocausation

(2014a, 22). Bell concurs: "this way of arranging quantum mechanical correlations would be even more mind-boggling than one in which causal chains go faster than light" (2004, 154). It is hard to take such a theory seriously.[13]

But there is another option here, namely the possibility of a *retrocausal* mechanism correlating the measuring device settings with the particle properties (Price 1994). For the Bell experiment, the proposal is outlined in Fig. 3.2. The basic idea is that the choices of measurement settings at L and R cause the actual measurements at L' and R', and these measurements causally influence the *earlier* particle emission event S. By this means there is no need for the vast conspiracy of the common cause approach: the particles have to interact with the devices that measure them anyway, so there are no new causal links, just an unexpected *direction* for some of the links. Admittedly, backwards causation is potentially conceptually problematic, and there is no well-developed theory along these lines in existence yet.[14] But if it can be made to work, the retrocausal model provides a clearly realist and clearly causally local account of quantum phenomena.[15] If it is tenable, the retrocausal approach raises the possibility that the lesson of Bell's theorem is that effects can precede their causes.

So I think it is too soon to say what *the lesson* of Bell's theorem is. All the models of quantum phenomena presented here have their attractions, but also their weaknesses, weaknesses that may prove fatal. At the end of the day, it may be that the lesson of Bell's theorem is that the world is causally non-local. Or it may be that the lesson is that measurements have multiple equally real outcomes. Or it may be that effects can come before their causes. Or it may even be that no description of the quantum world can be given—although this latter conclusion seems to me to be a last resort. In any event, the import of Bell's theorem is far from a settled matter. What Bell did is to demonstrate what quantum mechanics *cannot* be: it cannot be a theory that satisfies *all* the assumptions of his theorem. Something has to give—but *what* precisely has to be given up will have to await future research.

[13]For a more detailed appraisal of this kind of theory, see Lewis (2006).

[14]Some of the potential problems for retrocausal theories are addressed in Price (1996).

[15]That is, each causal link is local, although the sum of a forwards-causal and a backwards-causal link can add up to instantaneous action at a distance. It is the former sense of locality that makes the theory compatible with special relativity.

References

Barrett, J. A. (2003). Are our best physical theories (probably and/or approximately) true? *Philosophy of Science, 70*, 1206–1218.

Bell, J. S. (1964). On the Einstein-Podolsky-Rosen paradox. *Physics, 1*, 195–200. Reprinted in Bell (2004): 14–21.

Bell, J. S. (1982). On the impossible pilot wave. *Foundations of Physics, 12*, 989–999. Reprinted in Bell (2004): 159–168.

Bell, J. S. (2004). *Speakable and unspeakable in quantum mechanics*. Cambridge: Cambridge University Press.

Bohr, N. (1935). Can quantum-mechanical description of physical reality be considered complete? *Physical Review, 48*, 696–702.

Davies, P. C. W., & Brown, J. R. (1986). *The ghost in the atom*. Cambridge: Cambridge University Press.

Einstein, A., Podolsky, B., & Rosen, N. (1935). Can quantum-mechanical description of physical reality be considered complete? *Physical Review, 47*, 777–780.

Laudan, L. (1981). A confutation of convergent realism. *Philosophy of Science, 48*, 19–49.

Lewis, P. J. (2006). Conspiracy theories of quantum mechanics. *British Journal for the Philosophy of Science, 57*, 359–381.

Lewis, P. J. (2010). Probability in Everettian quantum mechanics. *Manuscrito, 33*, 285–306.

Maudlin, T. (1995). Why Bohm's theory solves the measurement problem. *Philosophy of Science, 62*, 479–483.

Maudlin, T. (2014a). What Bell did. *Journal of Physics A, 47*, 424010.

Maudlin, T. (2014b). Reply to comment on 'What Bell did'. *Journal of Physics A, 47*, 424012.

Mermin, N. D. (1981). Quantum mysteries for anyone. *Journal of Philosophy, 78*, 397–408.

Price, H. (1994). A neglected route to realism about quantum mechanics. *Mind, 103*, 303–336.

Price, H. (1996). *Time's arrow and archimedes' point: New directions for the physics of time*. Oxford: Oxford University Press.

van Fraassen, B. C. (1980). *The scientific image*. Oxford: Oxford University Press.

Wallace, D. (2010). *The emergent multiverse: Quantum theory according to the everett interpretation*. Oxford: Oxford University Press.

Wallace, D., & Timpson, C. G. (2010). Quantum mechanics on spacetime I: Spacetime state realism. *British Journal for the Philosophy of Science, 61*, 697–727.

Werner, R. F. (2014). Comment on 'What Bell did'. *Journal of Physics A, 47*, 424011.

Chapter 4
The Universal and the Local in Quantum Theory

Tim Maudlin

Abstract Any empirical physical theory must have implications for observable events at the scale of everyday life, even though that scale plays no special role in the basic ontology of the theory itself. The fundamental physical scales are microscopic for the "local beables" of the theory and universal scale for the non-local beables (if any). This situation creates strong demands for any precise quantum theory. This paper examines those constraints and illustrates some ways in which they can be met.

Keywords Quantum theory · Metaphysics · Local beables · Non-local beables · Conditional wavefunction · Bohmian mechanics

In *Posterior Analytics* Book 1 Chap. 2, Aristotle confronts a methodological puzzle about scientific knowledge of the world. On the one hand, scientific inquiry into the physical world must start from objects "prior and better known to man". In a more recent idiom, scientific inquiry must start from the "manifest image": the world as it appears to us independently of any theoretical postulates. Aristotle calls these objects "closer to sense". The manifest image concerns the universe at mesoscale: objects and their behavior at the scale of everyday life. The microscopic details of these objects form no part of the manifest image. Contra Eddington, a table does not present itself to us in everyday experience as microscopically uniform and homogeneous. Nor does it present itself to us as microscopically atomic. It does not present itself to us microscopically *at all*. Similarly, the universe as a whole at its largest scale has no manifest structure: space as a whole does not appear to the senses as either finite or infinite because it does not appear to the senses as a whole at all.

But the objects that are reliably revealed by human sensory capacities are not the fundamental entities postulated by scientific inquiry. Those objects, which Aristotle

T. Maudlin (✉)
Department of Philosophy, New York University (NYU), New York, NY, USA
e-mail: twm3@nyu.edu

A. Cordero (ed.), *Philosophers Look at Quantum Mechanics*,
Synthese Library 406, https://doi.org/10.1007/978-3-030-15659-6_4

calls "prior and better known without qualification", or "prior and better known absolutely" escape our immediate observation. And just as Aristotle remarked over two millennia ago, so must modern scientific inquiry proceed: from the directly observable at mesoscale to postulated entities that are not directly observed. But since the behavior of the mesoscale objects provides the *empirical evidence* for the physical theory, there must be not merely a heuristic ascent from the manifest image to the theoretical, but also a more logically rigorous return from the scientific image to the manifest. Having postulated the physically fundamental but not-directly-observable, one must be able to derive consequences of the postulates at mesoscale. Were this not possible, the fundamental physical theory would have no empirical consequences and so could not become part of empirical science.

This basic methodological problem arises for quantum theory, but assessing how that theory solves it is problematical. The difficulty arises because, in a sense, there is no such thing as "quantum theory" at all. A rigorously formulated physical theory must contain a clear set of ontological postulates detailing what theory claims to exist. From these postulates follows a *kinematics*: an abstract "space" of all possible physical states, all possible values and arrangements of the physical ontology. Next the theory provides a *dynamics*: specific constraints (deterministic or probabilistic) on how the physical state evolves in time. These are the fundamental laws of the theory. Aristotle's problem, then, arises this way: in a fundamental physical theory the basic ontological postulates of the theory are not immediately observable by the senses. Still, these postulates must be inspired by observations at mesoscale and have logical consequences for observable behavior at mesoscale. But "standard quantum theory" contains no such clear ontological postulates. Solving Aristotle's problem for "standard quantum theory" is therefore impossible.

We should be clear about the problem. It is not that we are asking for too much rigor in the presentation of the theory. Any physical theory under construction will have some gray areas, where the exact ontological commitments of the theory are not clear. Is Newtonian gravitational theory, for example, committed to unmediated action at a distance, or to a gravitational field or to mediating gravitational-force-producing particles of some sort? Research can go forward even when the precise answers to questions like this are hazy. But the *evidence* for Newton's theory, the *data* against which it was tested, was never direct observation of gravitational fields or potentials or forces. The evidence was the observable relative motions of bodies such as apples and planets. So one could calculate observational consequences of the theory without having settled all of the ontological detail. Since the data are determined by the behavior of observable matter, one needs to be clear about how to describe the distribution of matter at mesoscale and how the distribution of matter should change at that scale (according to the theory). Insofar as this can be determined without settling other details of the theory, research and testing can continue.

The problem with "quantum theory" is that not even this much about the basic physical ontology is clear. It is more in the nature of a recipe for making predictions using a certain mathematical formalism. No ontological postulates, either clear or murky, are made. If one takes a standard quantum physics text and asks what *physically exists* according to the theory, no answer is readily forthcoming.

This shortcoming of "quantum theory" is obscured by unfortunate nomenclature. Suppose one endeavors to formulate a rigorous theory with a clear ontology and dynamics that produces the same (or nearly the same) predictions as the textbook predictive recipe. This activity, which is properly speaking the construction of an exact physical theory, is commonly called "providing an interpretation of quantum theory". The bad nomenclature suggests that the activity involved is not theory construction at all—after all, it sounds like one *starts* with a given theory—but rather the "interpretation" of a theory that already exists. A physicist might reasonably wonder both what an "interpretation" of a theory is and why, as a physicist, she should be interested in having such an interpretation. Perhaps "interpretations" are the province of philosophers rather than physicists. Then let the philosophers busy themselves with interpretation and leave the physicists alone!

Those physicists most deeply concerned with physical ontology understood that "quantum theory" as it is usually presented contains no clear ontology and insisted that this constitutes a failure of physics as such. In the last half century, the most eloquent and forceful advocate of this position was John Stewart Bell. Bell was also among the strongest proponents of theories with clearly articulated physical ontologies, such as the pilot wave theory and his own version of the Ghirarid-Rimini-Weber (GRW) collapse theory (Bell 2004, chapters 17 and 22). Bell defended a general approach to connecting theory with the manifest image, which he called the *theory of local beables* (Bell 2004, chapters 7 and 19). This approach aims directly at answering Aristotle's challenge in the context of quantum theory.

4.1 The Theory of Local Beables

If the fundamental ontological postulates of a physical theory are not themselves directly observable, it might at first glance seem to be problematic how the theory can have any observable consequences at all. But in some cases the solution to this puzzle is so direct and simple that the problem never even strikes us.

Consider Democritean atomism. The fundamental ontology of Democritus was clear: atoms and the void; the full and the empty. The void, the empty, was taken to be *empty Euclidean space*. The space has a definite geometrical structure, exposited in the *Elements*. It is three-dimensional, infinite, and (as we can now say) geometrically flat. The empty does not act on the human senses: one cannot see it or hear it or touch it. Therefore our perception is never directly of space itself. Indeed, the passivity and unobservability of the void makes it so remote from direct experience that the atomists sometimes advert to it as "non-being", $\tau o\ \mu\eta\ ov$. What we observe, what acts on our senses, are the atoms.

But the atoms are also individually unobservable. No one can see a single atom or verify by direct observation that matter is atomic rather than infinitely divisible. So, one might wonder: if the two basic ontological postulates of Democritean atomism are both unobservable, how can the theory be an empirical theory?

This never seemed to puzzle the ancients and similarly does not puzzle us now. Individual atoms may be unobservable, but large collections of atoms can easily constitute observable collectives. If a table is nothing but a very large collection of Democritean atoms, then the shape, location and orientation of the individual atoms determine the gross macroscopic geometrical structure of the table in an obvious, ineluctable and conceptually transparent way. The observable motion of the table is nothing but the collective motion of the individual atoms described in a coarse-grained vocabulary. If all the atoms that constitute the table move to the right, then the table as a whole automatically does. And if we see the table move to the right, then we know in a general way how the individual atoms are moving, although many distinct precise individual motions are consistent with the observable gross behavior.

The intrinsically unobservable geometrical structure of space also manifests itself. Since the atoms are moving in a three-dimensional Euclidean space, their geometrical relations are always the geometrical relations among some points or regions in that space. The table appears to us as a three-dimensional (approximately) Euclidean shape because the atoms occupy some region in an (at least approximately) Euclidean arena. Space alone cannot act on our senses, but the atoms can, and the atoms are constrained in their configurations by the geometry of the space in which they move.

Just as the fine, exact details of the Democritean atoms would escape our direct observation, so too does the precise geometry of the space in which they move. Our everyday experience of the world is consistent with space being Euclidean, but also consistent with deviations from Euclidean geometry that are small at mesoscale. It has been obvious from antiquity (to Zeno, for example) that for all we can tell space might be either continuous or discrete microscopically. Similarly, it might deviate significantly from a Euclidean structure at cosmological scale. All of these possibilities could be consistent with the everyday structure of the manifest image.

In sum, neither the exact microscopic character of space (or space-time) nor the microscopic character of matter is evident to the senses. Nonetheless, a physical theory that makes precise postulates about these things can have straightforward empirical consequences via *coarse-graining*. Given a precise disposition of matter in a precise space-time structure described at microscale, the theory has implications about the approximate shapes, locations and motions of mesoscopic objects at mesoscale, which can be tested against sensory observation.

This basic idea, so simple and transparent as to be easily overlooked, is the central idea of Bell's theory of local beables. Bell invented the term "beable" to refer to the things that a physical theory postulates to exist or to *be*:

> In particular we will exclude the notion of 'observable' in favor of that of '*be*able'. The beables of a theory are those elements that might correspond to elements of reality, to things which exist. Their existence does not depend on 'observation'. Indeed, observation and observers must be made out of beables. (Bell 2004, 174)

The requirement that beables exist independently of being observed must be made explicit because of the peculiar history of quantum theory. There, many suprising claims have been made about observation "bringing reality into existence", and about quantum entities not having positions and momenta antecedently to being observed. But although there is some chance of making sense of this sort of talk in a restricted domain, the universal application of such a principle is immediately self-undermining. If *nothing* exists until it is observed, then there can be no observers to do the necessary observing, and nothing will ever exist.

The "local" requirement for local beables requires a little unpacking. The locality is in space-time, and refers to the beables themselves being located at particular, small, bounded regions of space-time. Individual local beables must occupy particular locations if collections of them are to determine the shapes and motions of perceptible things. Some beables fail this test because they are global in nature: Bell gives the example of the total energy of a system (Bell 2004, 53). But even some mathematically definable quantities that are associated with precise space-time locations fail to be local beables. The center of mass of the Earth-Moon system, for example, has a reasonably precise location, but still there need not be anything of physical interest that exists *at* that location: it may well just be a point in empty space.

There are many examples of possible local beables in the relevant sense. Particles—either point particles or microscopic particles with geometrical shapes like Democritean atoms—clearly qualify. So do classical fields such at the electromagnetic field. If one reifies a classical gravitational potential it also would be a local beable. Microscopic vibrating strings would do. Bell himself suggested some possible local beables. For quantum field theory, he suggested the fermion number density: how many fermions are located in each small bounded region of space-time (Bell 2004, 175). And in his presentation of the GRW collapse theory, he introduced a novel proposal for the local beables of the theory: point events in space-time, which have come to be called the "flash ontology".

It is useful to pause on the flash ontology because it is both unfamiliar and *prima facie* quite shocking. The quantity of local beables in this ontology is vastly less than one might have thought possible in an empirically adequate theory. According to the flash ontology, at most times there are no local beables at all: the whole universe is just empty space. The only exception to this complete vacuity occurs when there is a spontaneous GRW collapse of the quantum state. When such a collapse occurs, the only local beable that comes into existence is a single point-event with no spatial or temporal extension. The point event has a precise location in space-time. So in this theory there are exactly as many distinct local beables as there are GRW collapses, each being one physical point event.

Given the dynamics of the theory, we can make back-of-the-envelope calculations for the number of such flashes in a region of space-time that we regard as occupied by matter such as a table or human body or human cell. Let's do a strand of DNA since we are accustomed to think of DNA as having a characteristic geometrical structure, a double-helix, which it retains at all times irrespective of being "observed" or "measured". To what extent, in this theory, is this geometrical structure realized by the distribution of local beables at microscopic scale?

There are about 200 billion atoms in a strand of DNA, mostly carbon, oxygen, hydrogen and nitrogen. Since carbon is the heaviest, we overestimate by assigning each atom 8 electrons and 16 nucleons (i.e. 48 quarks). So there are about 10^{13} elementary particles in a strand. In the GRW dynamics, each elementary particle suffers a GRW collapse once every 10^{15} s. So on average, there would be less than a single GRW collapse associated with a complete strand of DNA every minute! One single, solitary, dimensionless point in space-time per minute to form the basis of the geometrical shape of the strand. In such a theory, it is misleading to say that DNA actually has a double-helix structure at all times.

But if the local beables are so scarce at microscopic scale, how can the theory be empirically adequate at mesoscopic scale? The same calculation reveals the answer: in a whole human body, there will be something like 10^{14} flashes per second. This many points in space-time, appropriately configured, could straightforwardly correspond to a shaped object indicated in much more detail than is apparent to simple observation. At *mesoscopic* scale, this collection of points is quite sufficient to constitute the positions, shapes and motions of familiar observable bodies in more detail than we can directly apprehend. The microscopic local beables, shockingly sparse at microscopic scale, yield a coarse-grained distribution of matter at mesoscopic scale that corresponds (or fails to correspond) to what we take ourselves to know by direct experience about the behavior of matter.

One might take the sparseness of the local beables in this theory as good grounds to dismiss it. Surely, one thinks, our understanding of DNA requires that there actually *be*, at all times, double-helix-shaped configurations of matter in the nuclei of our cells. The flash ontology correctly predicts the observable output of microscopes, resonance imaging, etc. Those technologies all produce output at mesoscopic scale (so we can read it!) and the GRW flash theory will get the mesoscopics right. But in an obvious sense all of these scanning outputs are, according to the theory, highly misleading. They suggest the existence of microscopic local structure that isn't really there at all.

All of this shows how modest the demand for empirical adequacy of the theory is: it is enough to get the mesoscopic aspects of things right to render the theory empirically unassailable. The range of possible microscopic local beables that could serve this purpose is vast, and includes proposals that one might, for non-empirical reasons, find incredible.

Bell not only suggested the flash ontology as the local beables of the GRW theory, but also an ontology of fermion number density for quantum field theory. His comment on the methodological adequacy of this choice sums up the situation:

> Not all 'observables' can be given beable status, for they do not all have simultaneous eigenvalues, i.e. do not all commute. It is important to realize therefore that most of these 'observables' are entirely redundant. What is essential is to be able to define the positions of things, including the positions of instrument pointers or (the modern equivalent) of ink on computer output... [Bell considers and rejects energy density as a choice of local beable].

> We fall back then on a second choice—fermion number density. The distribution of fermion number in the world certainly includes the positions of instruments, instrument pointers, ink on paper, . . . and much much more. (Bell 2004, 175)

Half of our discussion is now done. The objects "prior and better known to man", localized physical objects at mesoscopic scale, can be accommodated by the postulation of local beables in a space-time structure whose coarse-grained description matches the manifest image. This can be accomplished in myriad ways. Both the exact nature of the local beables and the exact microscopic (and cosmological) geometry of the space-time can differ wildly from our naive guesses. The flash ontology illustrates an unexpected choice of local beable, and the 11-dimensional space-time of string theory, with 7 "compactified" dimensions, illustrates an unexpected choice of precise space-time geometry. These choices, and many others, can coarse-grain to correspond to what is prior and better known to us.

It is at least logically possible for a physical theory without either a space-time structure or a choice of local beables that coarse-grains in this way to nonetheless be empirically adequate. Some physicists and philosophers regard this as the most likely possibility: the manifest image somehow "emerges" from a fundamentally non-spatio-temporal physical reality in a way quite unlike coarse-graining. I cannot review the proposals or prospects for such "emergence" here. I do insist, though, that any such proposal for recovering the manifest image owes us an account as precise and as clear as the one arising from coarse graining of local beables in a space-time.

4.2 Non-local Beables

A physical theory can posit only local beables. Einstein advocated this in a letter he wrote to Max Born, detailing the progress of physical theories toward locality in both ontology and dynamical law:

> If one asks what, irrespective of quantum mechanics, is characteristic of the world of ideas of physics, one is first of all struck by the following: the concepts of physics relate to a real outside world, that is, ideas are established relating to things such as bodies, fields, etc., which claim a 'real existence' that is independent of the perceiving subject - ideas which, on the other hand, have been brought into as secure a relationship as possible with the sense-data. It is further characteristic of these physical objects that they are thought of as arranged in a space-time continuum. An essential aspect of this arrangement of things in physics is that they lay claim, at a certain time, to an existence independent of one another, provided these objects 'are situated in different parts of space' . . . .
>
> This principle has been carried to extremes in the field theory by localizing the elementary objects on which it is based and which exist independently of each other, as well as the elementary laws which have been postulated for it, in the infinitely small (four-dimensional) elements of space. (Born 1971, 170)

Classical electro-magnetic field theory is local in both respects: the fields themselves are local beables and the laws that govern the fields depend only on the local field configuration. Events that take place far from a system can only influence it via a spatio-temporally continuous sequence of local changes in the field.

This sort of locality, which Einstein sought in all of physics, turns out to be incompatible with the predictions of quantum theory (provided that the spatio-temporally continuous sequences of local disturbances propagate no faster than light). That is the main consequence of Bell's theorem. Therefore, any precise physical theory capable of recovering the predictions of quantum theory must fail to be local in Einstein's sense (for more detail, see Maudlin 2014).

One way to implement the required non-locality would be to retain a completely local ontology but provide it with a spatio-temporally continuous but superluminal dynamics. That is, a perturbation of the local physical state at one location could give rise to a continuous sequence of local perturbations whose trajectory is space-like. But this has not been the sort of resolution implemented in any precise theory based on the standard quantum formalism. Why is that?

The quantum recipe, as we have called it, is a mathematical technique for generating probabilistic predictions about the outcomes of experiments. The central mathematical object used in the recipe is the "wavefunction" of the system of interest. For simplicity, we will here consider the wavefunction used in non-relativistic theory. The main points about non-locality are already apparent here.

One begins by characterizing the system of interest as an "N-particle system". The scare quotes are important. Naively, one would expect an N-particle system to be a system containing N particles, N local beables that follow continuous trajectories in space-time. On this naïve understanding, just using the phrase "N-particle system" already commits one to some local ontology, viz. particles in the classical sense. But in the quantum domain the phrase "N-particle system" is commonly used by physicists who would stoutly deny the existence of any such particles. Indeed, the phrase is used by physicists who maintain that quantum theory requires abandoning all hope for such an ontology. Classical particles following definite trajectories through space-time automatically have both definite positions and definite velocities (and hence momenta) at all times. But, they claim, the Heisenberg uncertainty relations preclude "particles" from having definite positions and definite momenta simultaneously. Hence (they conclude), "particles" in quantum theory cannot possibly mean particles in the classical sense. Nonetheless, the term "N-particle system" persists.

Why call something an "N-particle system" while simultaneously denying that it contains any particles at all? One answer lies in the mathematics of the wavefunction. A classical N-particle system determines a *configuration space*. The configuration of a collection of particles is nothing but the set of their locations. So one can define a mathematical space, each point of which corresponds to a unique possible configuration of the system. If the N particles inhabit an M-dimensional space, the configuration space will be (N × M)-dimensional. This classical abstract configuration space inherits its geometry in a natural way from the geometry of the space that the particles move around in.

In non-relativistic quantum theory, the wavefunction assigned to a system is a mathematical function defined over a mathematical space. And the reason one calls the system an "N-particle system" is simply that this base space has the mathematical form of the configuration space for N classical particles in physical space. If the "particles" have no spin, then the wavefunction assigns a complex number to each point in the configuration space; if the "particles" have spin then the wavefunction assigns a spinor. For simplicity, we will discuss the spinless case.

Suppose that we have a "19-particle" system. Taking physical space as 3-dimensional, the classical configuration space for such a system is 57-dimensional. A corresponding spinless quantum wavefunction is therefore a complex function over a 57-dimensional space. We can label these dimensions just as if they were variables for the positions of 19 classical particles, so the wavefunction is a complex function: $\psi(x_1, y_1, z_1, x_2, y_2, z_2, \ldots x_{19}, y_{19}, z_{19})$.

This mathematical function is ascribed to the system as a whole, since the space over which it is defined reflects all the "particles". So the situation with respect to part/whole relation is exactly the reverse of that for the local beables. If we consider a system as a collection of local beables, then the local beables of the whole are nothing but the collective local beables of the parts. The whole is the aggregation of the parts, which is why the motion and geometrical characteristics of the larger collectives follow directly from the geometrical disposition and motion of the parts. But the wavefunction of a system is assigned to the system as a whole rather than being derived from wavefunctions assigned to the parts. This raises the question of how wavefunctions can be assigned to subsystems of the large system at all.

Suppose, for example, we want to treat our "19-particle" system as composed of two subsystems: the first 3 particles and the remaining 16. Given the wavefunction $\psi(x_1, y_1, z_1, x_2, y_2, z_2, \ldots x_{19}, y_{19}, z_{19})$ for the whole, is it possible to specify what might be meant by the wavefunction of just "the first three particles"?

For one particular sort of wavefunction this is simple. Suppose it happens that $\psi(x_1, y_1, z_1, x_2, y_2, z_2, \ldots x_{19}, y_{19}, z_{19})$ can be written as the product of two other functions, each defined over the configuration space of a subsystem. That is, suppose that there exist two complex functions $\chi(x_1, y_1, z_1, x_2, y_2, z_2, x_3, y_3, z_3)$ and $\xi(x_4, y_4, z_4,, \ldots x_{19}, y_{19}, z_{19})$ such that $\psi(x_1, \ldots, z_{19}) = \chi(x_1, \ldots z_3)\xi(x_4, \ldots z_{19})$. ψ is then called a *product state*, and χ and ξ are obvious wavefunctions to assign to the two subsystems. In this case, the wavefunction of the whole system can be recovered from the wavefunctions assigned to the parts: just multiply them together.

But—and this underlies the radical departure of quantum from classical physics—not every possible wavefunction of the large system has this mathematical feature. Most wavefunctions of our 19-particle system cannot be written as the product of a wavefunction of the 3-particle subsystem and a wavefunction of the 16-particle subsystem. We say that the two such subsystems are *entangled.* In such cases, it is not clear what "the wavefunction of the 3-particle subsystem" might mean.

This situation can obtain even though, intuitively, the 3 particles have been isolated in space and spatially separated from the 16 particles. We might, for example, have a box on one side of the laboratory that we would say contains the 3-particle system and a box on the other side that contains the other 16. If all of the beables postulated by the physical theory were local, then the 3-particle subsystem would have its own local physical state, the 16-particle system would have its own local state, and the physical state of the complete 19-particle system would be nothing but the aggregation of these two. But if we take wavefunctions seriously as somehow representing a real physical characteristic of a system, then such a local ontology cannot be maintained. When entangled, the physical state of the whole is not determined by the physical states of the spatially separated parts.

Let's dig a little deeper into the mathematics. We have already seen that if the wavefunction of a total system is entangled, then it cannot be recovered from any pair of wavefunctions assigned to its subsystems. There is, however, a particular mathematical item (not a wavefunction as defined above) that one can assign to each subsystem. This mathematical item is called a *density operator* or *reduced state*, and the quantum recipe provides a means to derive probabilistic predictions for experiments carried out on the subsystem from this. From the density operator ascribed to the 3-particle subsystem one can make accurate probabilistic predictions for experiments carried out on it, and from the density operator ascribed to the 16-particle subsystem one can make accurate probabilistic predictions for experiments carried out on it. Why not take the complete physical state of the 19-particle system to be exhaustively described by this pair of density operators?

One thing that gets left out here are predictions about *correlations* between the outcomes of experiments carried out on the two subsystems. So, for example, the density operator for the 3-particle subsystem may assign a 50% chance for an experiment to have an outcome A, and the density operator of the 16-particle subsystem may assign a 50% chance for an experiment on that subsystem to have outcome B. But these probabilities alone have no implications about whether these results will be correlated. If they are uncorrelated, then in the long run the pair of experiments will yield the results (A, B), (not-A, B), (A, not-B) and (not-A, not-B) each 25% of the time. Given these statistics, knowing the result of one experiment provides no information about the outcome of the other: one would still bet on the other at even odds. But it is also possible that the outcomes be correlated. In the most extreme case, they might be perfectly correlated: an A outcome on the 3-particle system *always* accompanied by a B outcome on the 16-particle system, or an A outcome *never* accompanied by a B outcome. In this scenario, an experiment on one system provides perfect information about the other: one goes from complete uncertainty about how the other experiment will come out to complete certainty. There are also intermediate cases, with weaker but non-zero correlations, in which conditionalizing on the outcome of one experiment improves predictions with respect to the other.

The full wavefunction of the 19-particle system provides not just the probabilistic predictions for the subsystems, but predictions for the correlations as well. Thus one loses information in passing from the full wavefunction to the density operators of the parts, and hence cannot reconstruct the full wavefunction from the density operators. This is the irreducible holism implicit in the wavefunction. And if one takes the wavefunction to represent a real physical feature of the system, this mathematical holism suggests some sort of *ontological* holism. Such a physical feature would be a *non-local* beable.

As the passage cited above shows, Einstein thought that physics progresses by the successive elimination of non-local elements from its ontology. He was therefore intent on interpreting the wavefunction as something other than the mathematical representation of an objective physical feature of an individual system. Einstein inclined instead to a *statistical* account of the wavefunction: wavefunctions provide information about the statistical properties of *collections* of systems rather than information about the states of *individual* systems. According to this approach, there is no physical feature of an individual system that the wavefunction represents.

The most obvious advantage of this approach, in Einstein's eyes, is that it appears to provide the means to account for correlations between distant subsystems in a boring, commonplace way. Consider, again, the example mentioned above. Suppose the wavefunction of our 19-particle systems is entangled in such a way that the quantum recipe yields the following predictions: some experiment on the 3-particle subsystem has a 50% chance of yielding outcome A, some experiment on the 16-particle subsystem has a 50% chance of yielding outcome B, and these outcomes are perfectly correlated: whenever A occurs, B does as well. This is an example of an EPR correlation: exactly the sort of thing that Einstein, Podolosky and Rosen discuss in their classic paper (Einstein et al. 1935). If one regards these probabilities as irreducible physical chances associated with the subsystems in a single experiment, then a puzzle arises. Since the subsystems might be separated arbitrarily far from one another in space, if the 3-particle subsystem has a real, irreducible, non-zero physical chance of yielding the outcome A and also of the outcome not A, *how can the distant 16-particle subsystem always manage to yield the correctly corresponding outcome*? To Einstein, this was the "spooky action-at-a-distance" inherent in the standard understanding of quantum theory.

But on the statistical view, this problem seems to disappear. All one has to imagine is that each individual 19-particle system is in one of two distinct physical states. In State 1, the 3-particle subsystem is disposed, with certainty, to yield outcome A and the 16-particle subsystem is disposed to yield outcome B. In State 2, the 3-particle subsystem is disposed to yield outcome not-A and the 16-particle subsystem is disposed to yield outcome not-B. If half of a large collection of 19-particle systems are in State 1 and half are in State 2, then the statistics mentioned above follow immediately and without anything spooky at all.

Einstein's notion of locality—no spooky action-at-a-distance—does not itself imply that the dynamical laws must be deterministic. Irreducibly chancy outcomes are permitted. But if such chancy outcomes occur for widely separated systems, then Einstein-locality requires that they be uncorrelated: conditionalizing on the outcome of one should not improve predictions for the other. If there happen to be perfect correlations for the outcomes, then Einstein-locality does require the physics to be deterministic. But even in the absence of perfect correlations, Einstein-locality implies constraints on what the observed statistics between distant systems can be.

These constraints were discovered by Bell in 1964. The constraints are violated by the predictions of the quantum recipe, which have since been confirmed in the lab. So no Einstein-local theory can recover the predictions of the quantum recipe; Einstein's hope for a statistical understanding of the wavefunction has been dashed.

The recent no-go theorem of Pusey, Barrett and Rudolf (2011) put yet another nail in that coffin. In sum, every viable precise quantum theory on offer today is ψ-*ontic* in the sense that the wavefunction is taken to represent some real physical characteristic of the individual system to which it is assigned.

Our problematic is now complete. For, on the one hand, we have been more or less forced to accept some element of our physical ontology that is represented by the wavefunction. But on the other, the wavefunction is irreducibly holistic: the wavefunction of a system cannot be regarded as determined by either wavefunctions or density operators assigned to its parts. The *fundamental* or *basic* ontological object represented by the wavefunction must be ascribed to the *largest* and *most inclusive* system there is, the system of which all other systems are parts. And that universal system is, of course, nothing less than the entire universe. So we have been led to posit, as part of the fundamental ontology of the physical universe, an irreducibly holistic universal quantum state, represented by a universal wavefunction Ψ.

From the magisterial perspective of fundamental metaphysics, then, our precise quantum theories have a tripartite ontology: a space-time structure that assumes a familiar approximate form at mesoscopic scale; some sort of local beables (particles, fields, matter densities, strings, flashes) in that space-time; and a single universal non-local beable represented by the universal wavefunction Ψ. Any other ontology that we wish to accept must be somehow *derived from* these.

The derivation of the local aspects (shape, size and motion) of mesoscopic localized objects such as tables and chairs and cats and people and pointers has already been covered: that is nothing but the collective behavior of the fundamental microscopic local beables. But we are still left with a problem in the other direction. The universal wavefunction Ψ, which we posit to represent the quantum state of the universe, is something that we cannot observe, cannot know, cannot write down, cannot calculate with. What we do write down, and gets fed into the quantum predictive recipe, is always the wavefunction of some small subsystem of the entire universe. And by "small' here, I do not mean "mesoscopic" (small with respect to universal scale). The systems actually treated by quantum theory are much smaller than that, typically only small numbers of particles. So our puzzle is this: if at the level of ontology, the universal wavefunction represents the fundamental non-local item, how are these *derivative* wavefunctions ascribed to small subsystems derived from it? And why should the quantum predictive recipe, which always makes use of wavefunctions of small subsystems, work as well as it does?

4.3 Bohmian Mechanics and the Conditional Wavefunction

To review: if the fundamental local beables are microscopic, we have to solve the problem of how to use these to define the local characteristics of non-microscopic (and particularly mesoscopic) systems. This is easily and transparently solved by simple aggregation of microscopic parts. But if the fundamental non-local beable is of universal scale, represented by a universal wavefunction Ψ, then we have to solve the "top down" problem of defining some ontologically derivative wavefunctions for small subsystems of the universe, the wavefunctions we actually use to make predictions. It might seem at first that these two problems are completely unconnected. But, in fact, one clean solution to the latter problem depends on the solution to the former problem, viz. the choice of local beables.

From a purely mathematical point of view, the situation is simple. We are given a universal wavefunction $\Psi(x_1, y_1, \ldots z_{19})$ (imagining that the whole universe is just a "19 particle system") and a specification of a subsystem such as "just the first three particles". (We treat the "particles" here as distinguishable from one another by, e.g., mass and charge; the treatment of qualitatively identical particles is a little more complicated.) Ascribing a wavefunction to the subsystem would require defining one for it: $\Phi(x_1, y_1, z_1, x_2, y_2, z_2, x_3, y_3, z_3)$. How, as a purely mathematical matter, are we to go from the big Ψ defined over the 57-dimensional space to the little Φ defined over the 9-dimensional space?

It is here that the *local* beables postulated by the theory can come to the rescue. Suppose that according to the theory the "19-particle system" actually contains 19 classical particles with definite positions that follow definite trajectories. (This is *not*, therefore, anything like the ontology of "standard quantum mechanics".) Each of these particles will, at every moment, have a precise location in space-time. We refer to these actual locations using capital letters: the x-location of particle 1 at a given time will be X_1, its y-location Y_1, etc. So while the little x_1, y_1, etc. are all variables, the capital X_1, Y_1, etc. are values for these variables.

As noted above, "standard quantum theory" (whatever that is) does not postulate actual particles with locations, and so has no room for our Xs, Ys and Zs. But the *pilot wave* theory, A.K.A. *Bohmian mechanics*, does postulate such particles. In Bohmian mechanics it makes sense to go from the universal wavefunction $\Psi(x_1, y_1, \ldots z_{19})$ to a subsystem wavefunction $\Phi(x_1, y_1, z_1, x_2, y_2, z_2, x_3, y_3, z_3)$ by plugging in the actual particle positions for the remaining variables. In Bohmian mechanics one defines the *conditional wavefunction* of the three-particle subsystem as.

$$\Phi\,(x_1, y_1, z_1, x_2, y_2, z_2, x_3, y_3, z_3) =_{\mathrm{df}} \Psi\,(x_1, y_1, z_1, \ldots X_4, Y_4, Z_4, \ldots X_{19}, Y_{19}, Z_{19})\,.$$

The definition illustrates how the conditional wavefunction is a *derivative* entity: the items used on the right side—the universal wavefunction and the actual particle positions—are all fundamental physical posits of the theory. So defining the conditional wavefunction does not require us to expand the fundamental ontology of the theory. This sort of situation is called *grounding* in the metaphysical literature: the particles and the universal quantum state (represented by the universal wavefunction) are the fundamental ontology, and the definition shows how the conditional wavefunction is grounded in that ontology.

Because the conditional wavefunction derives from the fundamental ontology, its dynamical behavior is also derivative: the dynamics of the conditional wavefunction follows from the dynamics of the universal wavefunction (which never collapses) and the motions of particles 4–19. This latter is determined in the theory by the "guidance equation", which fixes the evolution of the complete particle configuration at all times. So the dynamics of the conditional wavefunction follows from the fundamental dynamics by analysis. How does it behave?

This is a complicated matter (see Dürr and Teufel 2009, 213 ff.), but the first pass answer is this: if our subsystem does not interact with the rest of the universe ("the environment") in a significant way, then the conditional wavefunction evolves according to the same dynamical equation as the universal wavefunction and hence does not collapse. But if the subsystem interacts with its environment in the sort of way required to make a "measurement" (which requires entangling the subsystem with the environment), then the conditional wavefunction will "collapse" in just the way and with just the probabilities that appear in the textbook quantum recipe. This is so *even though the universal wavefunction never collapses*. So the derivative dynamics that describes the derivative ontology can differ substantially from the dynamics that governs the fundamental ontology. Note how clear metaphysical analysis into fundamental and derivative can play a central role in explaining the practical success of our predictive techniques. Metaphysics meets the nuts-and-bolts explanation of the predictive success of science.

Our example also illustrates how clarity about the fundamental ontological postulates can introduce some subtlety into the relation between the everyday practice of physics and the basic metaphysics. The predictive success of the quantum recipe is a plain fact about the world and must be susceptible to physical explanation. One naïve way of trying to do is to reify the mathematical elements of the quantum recipe in the most literal and direct way possible. This sort of route has led some philosophers to declare that being a "wavefunction realist", i.e. thinking that the wavefunction represents some real physical characteristic of a system, requires also being a realist about "configuration space", i.e. thinking that fundamental physical reality must include some high-dimensional physical space (Albert 1996). It also leads to the suspicion that the "wavefunction collapse" in the quantum predictive recipe is best accounted for by the *physical* collapse of some part of the fundamental ontology. It is instructive to see how neither of these claims need be true in a theory that nonetheless accounts for the success of the standard predictive techniques.

But our main moral cuts deeper than this, or at least in a different direction. As metaphysicians, we care first and foremost about the fundamental ontology of the world. And the fundamental ontology is naturally associated with a fundamental scale, which is not the scale of everyday life. In the case of the local beables, that scale ought to be microscopic. If we accept the existence of localized mesoscopic items, these ought to be nothing more than collections of microscopic beables. And in the case of the quantum-mechanical non-local beable, the piece of physical reality represented by the wavefunction, the only natural scale is universal scale. The fundamental quantum state is that of the whole universe. Insofar as we make reference to the quantum states of small parts of the universe, that must be reference to derivative ontology, not fundamental ontology. The conditional wavefunction gives an example of how this can be done. Other theories, which do not postulate actual particles, cannot tell this story. But they must come up with some other story to tell.

John Bell articulated this situation with his usual incisiveness. The "Copenhagen interpretation" of the quantum formalism was, he notes, committed to the existence of local beables in the form of the disposition of apparatus that characterize an experimental situation. These "classical" everyday facts were not, themselves, "brought into existence by measurement". These everyday facts were just there. It is only by reference to these mesoscopic local beables that the Copenhagen approach could make sense of the quantum-mechanical treatment of microscopic systems. But ultimately laboratory equipment is *nothing but* some very complicated sort of quantum-mechanical system. So there is a conceptual incoherence in the standard interpretation.

> The kinematics of the world, in this orthodox picture, is given by a wavefunction (maybe more than one?) for the quantum part, and classical variables—variables which *have* values—for the classical part: $(\Psi(t,q\ldots), X(t)\ldots)$. The Xs are somehow macroscopic. This is not spelled out very explicitly. The dynamics is not very precisely formulated either. It includes a Schrödinger equation for the quantum part, and some sort of classical dynamics for the classical part, and 'collapse' recipes for their interaction.
>
> It seems to me that the only hope of precision with the dual (Ψ,x) kinematics is to omit completely the shifty split [between classical and quantum], and let both Ψ and x refer to the world as a whole. Then the xs must not be confined to some vague macroscopic scale, but must extend to all scales. In the picture of de Broglie and Bohm, every particle is attributed a position $x(t)$. Then instrument pointers—assemblies of particles—*have* positions, and experiments *have* results. (Bell 2004, 228)

As Bell notes, if the local beables are to refer to "the world as a whole" (i.e. to both the "classical" apparatus *and* to the "quantum system", then they should be defined at the microscopic scale of the quantum system. But equally, the non-local beable, represented by the wavefunction, referring to the world as a whole requires that it be fundamentally defined at universal scale. There should be only one fundamental quantum state, and any reference to the wavefunctions of subsystems must somehow be derivative.

The microscopic local beables must aggregate together to provide local characteristics of the mesoscopic objects that populate the manifest image. Getting the behavior of these right is exactly what it takes to make the theory empirically adequate. And in the other direction, there must be some way to define wavefunctions not just at the universal scale but also at the scale of the systems we actually treat using quantum theory. These derivative wavefunctions must somehow provide information about how the microscopic local beables will behave, if we are to understand how the theory as a whole is empirically successful. This downward connection of fundamental to derivative ontology is both mathematically and conceptually more difficult to achieve than the upward path from microscale to mesoscale.

Neither of these basic metaphysical problems has gotten much attention in the standard physics literature. There, it is clear neither what local beables are being postulated nor how to relate the wavefunctions of small systems to the wavefunction of the larger system they are part of. Both of these problems must be solved if the ontology of a quantum theory is to be made clear. There may be other ways to achieve these goals, but the example of Bohmian mechanics provides an undisputable proof-of-concept for one sort of solution.

References

Albert, D. (1996). Elementary quantum metaphysics. In J. Cushing et al. (Eds.), *Bohmian mechanics and quantum theory: An appraisal*. Dordrecht: Kluwer.

Bell, J. (2004). *Speakable and unspeakable in quantum mechanics.*, 2. Cambridge: Cambridge University Press.

Born, M. (1971). *The Born-Einstein letters* (I. Born, Trans.). New York: Walker.

Dürr, D., & Teufel, S. (2009). *Bohmian mechanics*. Dordrecht: Springer.

Einstein, A., et al. (1935). Can quantum mechanical description of reality be considered complete? *Physical Review, 47*, 777–780.

Maudlin, T. (2014). What Bell did. *Journal of Physics A: Mathematical and Theoretical, 47*, 424010.

Pusey, M., et al. (2011). *The quantum state cannot be interpreted statistically*. arXiv:1111.3328v1 [quant.ph] 14 Nov 2011.

Part II
Ontological Explorations of QM

Chapter 5
The Reality of the Wavefunction: Old Arguments and New

Harvey R. Brown

Abstract The recent philosophy of Quantum Bayesianism, or QBism, represents an attempt to solve the traditional puzzles in the foundations of quantum theory by denying the objective reality of the quantum state. Einstein had hoped to remove the spectre of nonlocality in the theory by also assigning an epistemic status to the quantum state, but his version of this doctrine was recently proved to be inconsistent with the predictions of quantum mechanics. In this essay, I present plausibility arguments, old and new, for the reality of the quantum state, and expose what I think are weaknesses in QBism as a philosophy of science.

5.1 Non-realist Interpretations of the Wavefunction

Whatever the quantum mechanical wavefunction is, *it is not fundamental*. The wavefunction ψ and its unitary dynamics are emergent elements within relativistic quantum field theory (RQFT), associated with the non-relativistic, low energy regime.[1] This state of affairs is no impediment in principle to the reality of ψ, or more generally of the statistical (density) operator, if a realist stance is taken for

---

We show that not only individual atoms but matter in bulk would [in the absence of the Pauli exclusion principle] collapse into a condensed high-density phase. The assembly of any two macroscopic objects would release energy comparable to that of an atomic bomb (Freeman Dyson 1967).

Thus our daily experience that 2 l of gasoline contain only twice as much energy as 1 l is a pathological property of small clumps of matter containing fermions. ...For fermi-matter only objects somewhat heavier than our sun are doomed to gravitational collapse but if mountains were made of bose-matter they would crush under their own weight (Walter Thirring 1986, p. 345).

---

[1]See, e.g., Wallace and Timpson (2010) and Myrvold (2015).

H. R. Brown (✉)
Faculty of Philosophy, University of Oxford, Oxford, UK
e-mail: harvey.brown@philosophy.ox.ac.uk

A. Cordero (ed.), *Philosophers Look at Quantum Mechanics*,
Synthese Library 406, https://doi.org/10.1007/978-3-030-15659-6_5

states in RQFT.[2] But if a non-realist stance is taken for ψ, then it is hard to see how a realist reading of states in RQFT is tenable – a point we return to in Sect. 5.3 below.

Arguments for the non-reality of the wavefunction take various forms. Some prominent advocates of the de Broglie-Bohm pilot wave theory argue that the quantum state can be thought of as part of the laws of nature, with a status akin to that of the Hamiltonian. Adoption of such a *nomic* view is critical in rebutting the criticism that the theory is essentially Everettian quantum theory in denial. I will not repeat here arguments which Wallace and collaborators have advanced which question the Hamiltonian analogy.[3] The considerations in Sect. 5.2 below complement these arguments.

A prominent advocate of the alternative *epistemic* view of the quantum state is Christopher Fuchs.

> ... the quantum state represents a collection of subjective degrees of belief about something to do with that system (even if only in connection with our experimental kicks to it) ...
>
> Our foremost task should be to go to each and every axiom of quantum theory and give it an information theoretic justification if we can ...
>
> Quantum states are states of information, knowledge, belief, pragmatic gambling commitments, not states of nature.[4]

Such a view has *prime facie* a lot going for it. If it is right, then it would seem that the notorious collapse of the wavefunction in the act of measurement is innocuous: it corresponds to nothing other than Bayesian updating.[5] As a consequence, the threat of instantaneous action-at-a-distance in the 1935 Einstein-Podolsky-Rosen (EPR) scenario involving entangled systems is also removed.[6]

If only things were so simple!

Let us start with the well-known, and surely most obvious, articulation of the ψ-epistemic view which I shall call the Einstein version. To borrow Fuch's words, Einstein suggested from at least as early as 1929 that "the quantum state represents a collection of subjective degrees of belief about something to do with that system". The "something" in Einstein's understanding was the hidden, ontological state of the system. Einstein, unlike Fuchs, was proposing a deterministic hidden variable theory of a certain kind, precisely in the hope that it would remove not only what he saw as the spectre of non-locality in othodox quantum mechanics (QM).[7] but

---

[2] I will bypass here the debate between realists about the quantum state regarding whether the state should be defined on configuration space (see e.g. Ney 2015) or (nonseparably) on space (see Wallace and Timpson *op.cit.*).

[3] See Wallace and Timpson *op.cit.* and Brown and Wallace (2005). The strongest arguments for the nomic reading of the wavefunction in my opinion are found in Callender (2017), which build on the case made by Dürr et al. (1997), and address the criticism in Brown and Wallace *ibid.* In this connection see also Maudlin (2010).

[4] Fuchs (2002a).

[5] See Fuchs et al. (2014) and Leifer (2014), p. 68.

[6] See Fuchs et al. (2014) and Timpson (2008).

[7] See Harrigan and Spekkens (2010).

also the "paradox" involved in obtaining definite outcomes in generic measurement procedures[8] – essentially what is known today as the measurement problem.

The prospects of the Einstein version of the ψ-epistemic view look very bleak. Starting with the work of Pusey, Barrett and Rudolph (PBR) in 2012, a series of no-go proofs have appeared in the literature, which show, on the basis of plausible auxiliary assumptions, that the Einstein version is inconsistent with the predictions of quantum mechanics.[9] But even before these recent dramatic results were obtained, there were grounds for doubting the success of the Einstein version as a solution of both the measurement and nonlocality problems. It has long been known that the process of measurement must, in general, disturb the hidden state (if any) of the system in question, whatever view is taken on the status of the wavefunction in the theory.[10] Whether this disturbance is compatible with the inter-measurement dynamics would depend on the details of the theory and cannot be guaranteed a priori.[11] In relation to the EPR challenge, I refer of course to the many non-locality theorems inspired by the 1964 work of J. S. Bell, and to a great deal of subsequent experimentation, which together show that any deterministic hidden variable theory must incorporate action-at-a-distance if it is consistent with the proven predictions of QM.[12]

The so-called Copenhagen interpretation is widely understood to deny a realist status to the quantum state, which is nonetheless taken to be a complete description of the system. The state is a mathematical tool within the quantum algorithm, allowing for probabilistic predictions to be made concerning the outcome of measurements involving macroscopic instruments which themselves can and must be described "classically". I have no intention of rehearsing all the well-known challenges facing this interpretation, in so far as it can be regarded as a single thing. But it will be useful to remind ourselves of the stinging criticism John Bell raised against it in 1990:

> To restrict quantum mechanics to be exclusively about piddling laboratory operations is to betray the great enterprise. A serious formulation will not exclude the big world outside the laboratory.[13]

Need this exhortation have as its target all versions of the ψ-epistemic view? Not according to Fuchs and collaborators: their relatively recent philosophy of Quantum Bayesianism, or QBism, is, they claim, an exception. I use the word philosophy advisedly. In its attempt to resolve the puzzles of quantum mechanics, QBism makes the jaw-dropping claim to "liberate us from the grip of an ancient Greek maneuver

[8]This is particularly clear in Einstein (1970), pp. 670 and 683.

[9]A detailed review of these recent results is found in Leifer (2014).

[10]For details see Squires et al. (1994), p. 429.

[11]The de Broglie-Bohm theory suffers from no such incompatibility, but it is not a ψ-epistemic theory of the Einstein version.

[12]For a recent comprehensive collection of essays on this matter, see Bell and Gao (2016).

[13]Bell (1990).

that worked for over two millennia",[14] to overturn the allegedly dominant natural philosophy in which science has an "object" but not a "subject". These heady matters deserve special attention, and I will return to QBism in Sect. 5.3 of this essay. For the moment I note that since QBism denies that the "something" the quantum state refers to probabilistically is itself an element of observer-independent reality, the theory not only claims to solve both the measurement and nonlocality problems,[15] it also survives the recent PBR-type no-go results.[16] But at what cost?[17]

5.2 Wider Concerns

When John Bell in 1966, and Simon Kochen and Ernst Specker in 1967, independently proved that non-contextual hidden variable theories are inconsistent, there was little to indicate that such a result was likely within the prior literature on the foundations of quantum mechanics.[18] The post-2012 no-go results concerning Einstein's ψ-epistemic position, impressive as they are, surely are not as surprising. Powerful plausibility arguments have long been available, some since the birth of QM, to the effect that the quantum state is something real. They almost all have to do, in one way or another, with quantum phase, with the fact that the wavefunction, in its relation to probability, is strictly a (generally complex) probability *amplitude*: it has more structure than a probability distribution does.

5.2.1 Interference

Arguably the oldest and most striking of these plausibility arguments is based on interference effects. Whether it be the old chestnut, the two-slit experiment (in both its spatial, and less well-known temporal variants[19]) or the Mach-Zender

[14]Fuchs et al. (2014).

[15]See Fuchs et al. (2014) and Timpson (2008).

[16]See Pusey et al. (2012) and Leifer (2014), section 14.4. For details of advocates of such ψ-epistemic views other than the authors of QBism, see *ibid* p. 72, and Healey (2016), which also contains a useful review of QBism and its history. Healey's own "pragmatist" approach of the wavefunction (for details see *ibid*) has much in common with QBism but important differences as well.

[17]The following section of this paper is an attempt to make the case for the realist interpretation of the wavefunction; a more elaborate discussion is found in Gao (2017).

[18]David Bohm's 1952 hidden variable theory had already shown that von Neumann's 1932 no-go result was inconclusive.

[19]For a striking experimental version of the latter involving atomic interferometry, see Szriftgiser et al. (1996). For an experimental proposal involving neutrons, with references to earlier optical variants, see Brown et al. (1992).

interferometer for photons, or neutron or atomic interferometers, such displays of of single-system interference effects cry out for a realist interpretation of the wavefunction. Of course the case is not completely water-tight, as Leifer has recently stressed.

> Interference phenomena also occur in [certain toy models] simply because they reproduce fragments of quantum theory exactly and those fragments contain coherent superpositions. It is arguable whether the mechanisms explaining interference in all these models are plausible, but the main point is that the direct inference from interference to the reality of the wavefunction is blocked by them. If there is an argument from interference to be made then it will need to employ further assumptions.[20]

Although not what Leifer had in mind, a particularly intriguing recent example is a fluid mechanical ("walking droplet") model of diffraction, tunneling, quantisation and other quantum-like effects.[21]

Neither this nor the toy models Leifer is referring to capture all of the quantum predictions, and a striking omission is entanglement and its manifold manifestations. (Of particular interest for our purposes is the antisymmetric nature of the many-electron wavefunction, of crucial importance in accounting for the stability of bulk matter; see below.) So perhaps an analogy will help in addressing Leifer's skepticism. Consider the explanation of the gravitational redshift phenomenon in general relativity. Although in the actual experimental confirmations of this phenomenon tidal effects are negligible, the explanation refers to a metric field with curvature, a solution of Einstein's field equations. Would it not be odd to cast doubt on this explanation just because the experimental redshift phenomenon can also be explained in flat spacetime?[22] It is common scientific practice that an explanation for a given physical phenomenon is provisionally accepted when the theory behind it is uniquely capable of accounting for a wide gamut of diverse phenomena, even when in relation to the given phenomenon it may not provide the only explanation.[23]

I would particularly like to mention the case of partial absorption experiments in single neutron interferometry which were performed by Helmut Rauch and his collaborators in Vienna in the 1980s. In one experiment, a rotating toothed wheel, or "chopper", constructed out of fully absorbing material (cadmium), is placed in one of the two beams inside the interferometer; it deterministically absorbs a certain percentage of the successive neutrons "in" that beam, and in doing so changes (weakens) the interference pattern recorded in the beams of (unabsorbed) neutrons emerging from the interferometer. In the other experiment, a static piece of gold foil replaces the chopper; the nuclei in the new absorber will likewise absorb a certain

[20] Leifer *op. cit.*, p. 79.

[21] See Bush (2015) and further references therein.

[22] That (first order) redshift is consistent with flat Minkowski spacetime has long been known, but it is not always acknowledged; for details see Brown and Read (2016).

[23] Attempts to describe all known gravitational effects in a theory based on flat spacetime generally turn out to be awkward reformulations of general relativity, and I suspect that any future "toy" model that accounted for more than a fragment of quantum theory would likewise be an awkward reformulation of that theory.

percentage of the neutrons inside the interferometer, but this process is intrinsically probabilistic. The experiments corroborate the prediction in quantum mechanics that even when the absorption coefficients are the same in both cases, and therefore *so is the Shannon information* concerning which beam the neutron is "in", the interference effects are different – there is a greater degree of interference in the case of the gold foil. (A third hypothetical example involves a slit in partially absorbing material; again the Shannon information can be arranged to be the same as in the previous examples, but the loss of interference will be intermediate, depending on the penetration of the neutron wavepacket in the slit material.[24])

Finally, certain quantum interference experiments involving electrons and neutrons provide strong, if not conclusive, grounds for supposing that the properties of mass (inertial and gravitational), charge and magnetic moment adhere to the wavefunction itself[25] – if this is not already seen to follow from the simple fact that such properties appear in its equation of motion. It would seem to follow that in both the nomic version of de Broglie-Bohm theory and QBism, such properties have no describable observer-independent physical entities in which to reside.

5.2.2 Phase Matters

In an important paper of 1962, Merzbacher investigated the conditions in quantum mechanics required for the quantization of angular momentum for a spinless particle.[26] In particular, he was concerned to show that the single-valuedness of the wavefunction is one of the conditions, as it is in the derivation of the original Aharonov-Bohm effect. Merzbacher demonstrated that single-valuedness itself is motivated when the background space (whether 2 or 3-dimensional) is simply connected. In the case of a system of identical particles, where the wavefunction is defined on configuration space, or rather the reduced space obtained by identifying the configuration points related by particle permutations, the topology of the reduced space is again important, as Leinaas and Myrheim demonstrated in 1977.[27] If points corresponding to two or more particles coinciding spatially are excised from the space, so that it becomes non-simply connected, the wavefunction is no longer single-valued. The upshot is that if the physical space in which the particles live has three or more dimensions, then the wavefunction can be shown to be either symmetric or antisymmetric under permutations of particle labels. This constraint is widely regarded as a postulate in standard quantum mechanics, but here it is derived on topological grounds. Even more remarkably, if the physical space is two dimensional, intermediate phase factors between 1 and -1 are possible under

[24]For further details on all these cases, see Kaloyerou and Brown (1992).

[25]See Brown et al. (1995).

[26]Merzbacher (1962).

[27]Leinaas and Myrheim (1977).

permutations, and this leads to the possibility of 'fractional' or 'braid' statistics ranging between Bose-Einstein and Fermi-Dirac. This is not a mere theoretical oddity. It is apparently displayed in two-dimensional electron gases in a transversal external magnetic field exhibiting the fractional quantum Hall effect.[28] Certain systems exhibiting the fractional quantum Hall effect are being investigated with a view to application in quantum computation.

I do not claim that such considerations are outright inconsistent with the ψ-epistemic position. But it is again unclear to me how the topology of physical space in the case of single particles, and the topology of the reduced configuration space as well as the dimensionality of physical space in the case of the many (identical) particles system, can be understood to play such important roles in determining critical properties of the wavefunction within this interpretation.

5.2.3 *The Stability of Matter*

In his systematic 2014 review of no-go theorems for ψ-epistemic theories, Leifer referred to what he called the neo-Copenhagen views which, like QBism, reject the notion that the wavefunction is a probability distribution over ontic states. He wrote in this connection:

> For my part, I think that if one denies the existence of an observer-independent reality then it becomes very difficult to maintain a clear notion of explanation at all. Closing explanatory gaps by denying the need for any explanation at all does not seem that appealing to me.[29]

These remarks arguably do not do justice to the role of the external world in QBism (see Sect. 5.3(vi) below), but the point is well taken. In his earlier detailed 2008 study of QBism, Christopher Timpson had also raised the issue of an "explanatory deficit" in the theory.[30] He questioned whether, for example, the standard explanation in quantum theory of the thermal and electrical conductivity properties of solid matter, can be incorporated into QBism. Timpson's core point was that the QBist can explain why someone would believe that, for example, matter conducts but cannot explain why matter does conduct. He also mentioned in this context the explanation of the stability of matter, but did not provide details. I intend in this section to provide some of these details, in the spirit of Timpson's critique.[31]

(i) It is a remarkable fact that a satisfactory quantum mechanical explanation of the stability of bulk matter emerged only in 1967. But let us consider the single-electron atom/ion first. Here, the proof of stability is older, but the full story is still

[28]See, e.g., Prange and Girvin (1990). It is notable that space reflections and time reversal are not symmetries of such electron gases. See Frohlich (2009), p. 56.

[29]See Leifer (2014), p. 139.

[30]Timpson (2008).

[31]I will restrict myself to non-relativistic quantum mechanics; the relativistic version of the story of stability can be found in Lieb and Seiringer (2010).

often omitted from textbooks. Quantum mechanics explains the stability of discrete spectral lines (modulo a satisfactory solution to the "measurement problem"!), which were of course mysterious from a classical perspective. But it is of greater significance that the theory accounts for the fact that the energy of the electron is bounded from below. The key challenge is the nature of the $1/r$ Coulomb electrostatic potential, as Jeans had noted in 1915.[32] Bound electrons have negative potential energy. What is to prevent the electron from getting arbitrarily close to the nucleus, so that its potential energy approaches negative infinity, while its kinetic energy remained arbitrarily small? Were this to happen, in the words of Elliott Lieb,

> ...the hydrogen atom would be physically unstable; in a gas of many atoms another particle or atom could collide with our atom and absorb energy from it. After many such collisions our electron could find itself in a tiny orbit around the nucleus and our atom would no longer be recognizable as an object whose radius is supposed to be 10^8 cm. Each atom would be an infinite source of energy which could be transmitted to other atoms or to radiation of electromagnetic waves.[33]

One can solve the time-independent Schrödinger equation for the hydrogen atom to show that the ground state energy is finite, but this procedure is unfeasible for large atoms and a simpler, generalisable one is desirable. To this end, a variant of the Heisenberg uncertainty relation is often employed. Consider the kinetic energy $T = p^2/2m = -\hbar^2\Delta/2m$ and its expectation value for any particle of mass m and wavefunction ψ:

$$\langle T\rangle_\psi = \frac{\hbar^2}{2m}(\psi, -\Delta\psi) = \frac{\hbar^2}{2m}\int_{\mathbb{R}^3} |\nabla\psi(\mathbf{x})|^2 d\mathbf{x}. \tag{5.1}$$

The Heisenberg uncertainty relation is, then, for any ψ of unit norm,

$$\langle T\rangle_\psi \langle x^2\rangle_\psi \geq \frac{9\hbar^2}{8m}, \tag{5.2}$$

where

$$\langle x^2\rangle_\psi = \int_{\mathbb{R}^3} x^2|\psi(\mathbf{x})|^2 d\mathbf{x}. \tag{5.3}$$

The inequality (2) means in this case that increasing localisation of ψ around the origin (the nucleus) is associated with a correspondingly large value of the kinetic energy, so stability of the atom is secured. But the argument fails if, for example, ψ has two "bumps", one localised around the nucleus and containing most of the mass, and the other localised at, say, the moon. In this case, $\langle x^2\rangle_\psi$ is large, so $\langle T\rangle_\psi$ can be small, while the average potential energy decreases without bound.

---

[32] See Lieb (1990), p. 7.

[33] Lieb (1990).

Fortunately help is at hand. In 1938 Sobolev proved the following inequality[34]:

$$\langle T\rangle_\psi \geq \frac{3\hbar^2}{2m}\left(\frac{\pi}{2}\right)^{\frac{4}{3}}\left\{\int_{\mathbb{R}^3}\rho_\psi(\mathbf{x})^3 d\mathbf{x}\right\}^{\frac{1}{3}}, \tag{5.4}$$

where $\rho_\psi(\mathbf{x}) = |\psi(\mathbf{x})|^2$. It can be shown that when ψ is of unit norm, it follows from the Sobolev inequality that the mean value of the ground state energy of the Hydrogen atom is bounded from below.[35]

Now a special case of the Hölder inequality[36] states

$$\int_{\mathbb{R}^3}\rho_\psi(\mathbf{x})^{\frac{5}{3}} d\mathbf{x} \leq \left\{\int_{\mathbb{R}^3}\rho_\psi(\mathbf{x})^3 d\mathbf{x}\right\}^{\frac{1}{3}}\left\{\int_{\mathbb{R}^3}\rho_\psi(\mathbf{x}) d\mathbf{x}\right\}^{\frac{2}{3}}, \tag{5.5}$$

so assuming as before that ψ has unit norm (so the second term on the RHS of (5.5) is unity), applying (5.5) to the Sobolev inequality yields

$$\langle T\rangle_\psi \geq \frac{3\hbar^2}{2m}\left(\frac{\pi}{2}\right)^{\frac{4}{3}}\int_{\mathbb{R}^3}\rho_\psi(\mathbf{x})^{\frac{5}{3}} d\mathbf{x}. \tag{5.6}$$

Elliott Lieb expresses the content of this inequality "poetically" as follows:

> An electron is like a rubber ball, or a fluid, with an energy density proportional to $\rho_\psi^{5/3}$. It costs energy to squeeze it and this accounts for the stability of atoms.[37]

This is also the fundamental reason why dynamical collapse models of QM involve non-conservation of energy and momentum. For the QBist, however,

> The notorious "collapse of the wave-function" is nothing but the updating of an agent's state assignment on the basis of her experience.[38]

It is not clear to me how easy it is to reconcile this claim with the fact that in the case of a localisation measurement, collapse is accompanied by a change in the expected energy of the system.

(ii) An even more profound analogue of the inequality (5.6) holds in the case of bulk matter, containing many electrons, protons and neutrons. Not surprisingly the details in this case are far more complicated; I shall do no more than sketch the main results.

---

[34]Sobolev (1938). We are concerned here with the three-dimensional version of the original inequality. For further details see Seiringer (1990) section 1.3.

[35]See Lieb (1976), section 1, Lieb (1990) Part III, and Seiringer (1990), section 1.4. It should not be concluded however that a proof of this kind of the stability of the hydrogen atom was only possible in 1938, with the appearance of the Sobolev inequality. A weaker, but less useful inequality due to Hardy (1920) suffices; see, e.g., Seiringer (1990) and particularly Frank (2011).

[36]For further details see Lieb (1976), p. 555, or Seiringer (1990), p. 9.

[37]Lieb (1990). Note that none of the considerations here require that the wavefunction be complex.

[38]Fuchs et al. (2014).

Consider the ground state energy E_0 of a system comprised of N electrons and M nuclei, defined by

$$E_0 = \inf\{(\Psi, H\Psi) : ||\Psi|| = 1, \Psi \in \mathcal{H}\}, \tag{5.7}$$

where H is the Hamiltonian associated with the system and $\mathcal{H}$ is the Hilbert space of possible states Ψ. The Hamiltonian contains Coulombic terms describing the attraction of the nuclei and electrons, the repulsion between the electrons and the repulsion between the nuclei. Because the proton mass is three orders of magnitude greater than that of the electron, the nuclei can be treated as classical objects at fixed locations, and it is the many-electron wavefunction that is the object of study:

$$\Psi = \Psi(\mathbf{x}_1, \sigma_1; \ldots; \mathbf{x}_N, \sigma_N), \tag{5.8}$$

where the space variables $\mathbf{x}_i$ range over $\mathbb{R}^3$, and the spin variables can take q values. (For electrons the σ_i take values in $\{-1/2, 1/2\}$, so $q = 2$.)

The first issue associated with the stability of bulk matter is, again, how to avoid of the possibility of *implosion*. As with the case of the individual atom, the ground state energy E_0 must be bounded from below: $E_0 > -\infty$. This is called *stability of the first kind*. But we also require that E_0 satisfy another inequality: $E_0 \geq -C(N + M)$, where C is non-negative and independent of N and K; it depends on the maximum positive charge on the nuclei. This is called *stability of the second kind*. The reason for this requirement needs to be spelt out.

When we mix two equal quantities of (say) water together, we expect to the quantity of water to double, without the release of any significant amount of energy. But the terms in the Coulomb interaction quadruple, and the electrostatic energy grows with the square of the number $N + M$, not linearly.[39] Now the total ground state energy is $2E_0(N + M)$ before mixing. After mixing, the ground state energy becomes $E_0(2(N + M))$ so the energy released will be $\Delta E_0 = 2E_0(N + M) - E_0(2(N + M))$. So suppose that the energy content of matter is proportional to (minus) the square of the number of particles $N + M$. Then on mixing the water, an energy proportional to $2(N + M)^2$ would be released, where $N + M$ is of the order 10^{26}. As Elliott Lieb remarked, a chunk of any such matter "would be very unpleasant stuff to have hanging around the house.".[40]

Some mechanism must exist to offset the quadratic dependence of the Coulomb energy on $N + M$. The first conclusive proof of stability of the second kind was due to Dyson and Lenard in the late 1960s,[41] and it relied critically on a fact that *electrons are fermions*: the many-electron wavefunction (5.8) must be antisymmetric under the interchange of $(\mathbf{x}_i, \sigma_i)$ and $(\mathbf{x}_j, \sigma_j)$ for any $i \neq j$. Dyson

[39]See Loss (2005) p. 53.

[40]Lieb (1990), p. 23.

[41]Dyson and Lenard (1967, 1968).

was also able to show that bosonic matter is not stable[42]: The ground state energy E_0 of $2N$ charged bosons, N with charge $+1$ and N with charge -1 satisfies

$$E_0 \leq -CN^{7/5}. \tag{5.9}$$

For such matter, its volume would decrease with N; more particles would take up less space. Again, in Elliott Liebs' words:

> ... *the imposition of the Pauli exclusion principle raises* $[E_0]$. The miracle is that it raises $[E_0]$ enough so that the stability of the second kind holds. While it is easy to say that ψ must be antisymmetric ... it is not easy to quantify the effect of antisymmetry. Even the experts have difficulty, for it is not easy to think of an antisymmetric function of a large number of variables.[43]

An alternative, and relatively simple proof of stability of the second kind for fermionic matter was provided by Lieb and Thirring in 1975. This proof exploited features of the Thomas-Fermi theory of the electronic structure of many-body systems,[44] which puts emphasis on the single particle density function ρ_Ψ rather than the wavefunction:

$$\rho_\Psi(\mathbf{x}_1) = \sum_{i-1}^{N} \sum_{\sigma_1,\ldots,\sigma_N} \int_{\mathbb{R}^{3N-1}} |\Psi(\mathbf{x}_1, \sigma_1; \ldots; \mathbf{x}_N, \sigma_N)|^2 d\mathbf{x}_2 \ldots d\mathbf{x}_N. \tag{5.10}$$

Lieb and Thirring showed first that there is a many-body analogue of (5.6) for wavefunctions of unit norm (so $\int_{\mathbb{R}^3} \rho_\psi(\mathbf{x})d\mathbf{x} = N$):

$$\langle T \rangle_\psi \geq \frac{\hbar^2}{2m} \frac{K}{q^{2/3}} \int_{\mathbb{R}^3} \rho_\Psi(\mathbf{x})^{\frac{5}{3}} d\mathbf{x}, \tag{5.11}$$

where

$$\langle T \rangle_\psi = \frac{\hbar^2}{2m} \sum_{\sigma_1,\ldots,\sigma_N} \sum_{i=1}^{N} \int_{\mathbb{R}^{3N}} |\nabla_{\mathbf{x}_i} \Psi(\mathbf{x}_1, \sigma_1; \ldots; \mathbf{x}_N, \sigma_N)|^2 d\mathbf{x}_1 \ldots d\mathbf{x}_N. \tag{5.12}$$

It is speculated that the best constant in the Lieb-Thirring inequality (5.11) is $K = (3/5)(6\pi^2)^{2/3}$. Note that if the wave function is such that the single particle density is distributed in N equal disjoint bumps across space then the right side of (5.11) is proportional to N.

More generally, Lieb and Thirring went on to prove stability of the second kind, $E_0 \geq -C(N+M)$, with a much improved value of the constant C in relation to that

[42]Dyson (1967). For further details see Loss (2005), p. 7.

[43]Lieb (1990), p. 15.

[44]One such feature is the important result originally due Teller that atoms do not bind: the energy of a system of electrons and nuclei is minimised if the atoms are infinitely far apart and neutral.

of Dyson and Lenard.[45] (In fact, this inequality has been shown to hold with $N+M$ replaced by M.[46]) A further comforting consequence of this result is that fermionic matter in its ground state is indeed bulky: its volume is proportional to N.[47]

So far we have been discussing the problem of avoiding *implosion* of bulk matter associated with the near-range singularity in the Coulomb potential. But in the treatment of macroscopic systems which purport to have typical thermodynamic behaviour, it is also necessary to account for the non-trivial fact that such systems don't *explode*! Here we are concerned with the long-range behaviour of the Coulomb potential, and the demonstration that E_0/N has a limit as $N \to \infty$. Happily, a proof of the existence of a thermodynamic limit in this sense was provided by Lieb and Lebowitz in 1972.[48] It is another interesting chapter in the story of the stability of matter, but once stability of the second kind is established, it turns out that little further quantum mechanics is needed to complete it.[49]

Let's go back to 1931, when Ehrenfest raised the question as to why an atom of lead, for example, doesn't pack more of its 82 electrons into the orbits close to the nucleus, and so be smaller than it appears to be. He realized the size of the atom, and the bulky nature of matter generally, must have something to do with the Pauli exclusion principle. He addressed the following point to its originator:

> You must admit, Pauli, that if you would only partially repeal your prohibitions, you could relieve many of our practical worries, for example the traffic problem on our streets.[50]

Thanks to the hard work of later quantum physicists, we know why Ehrenfest was right. Matter is both stable and bulky because the many-electron wavefunction has a key property when the electrons are not confined to two dimensions: it is antisymmetric under exchange of particle indices.

5.3 Remarks on QBism

(i) QBism is nothing if not ambitious. It "corrects a profound misconception in our general view of science, which led us into major confusion in the twentieth century."[51] This misconception is that science is about an external reality that can and should be described without introducing the human agent – the "subject". QBism regards the root cause of this misconception to be the failure to fully appreciate that, in the words of David Mermin, a convert to QBism, "scientific

45 Lieb and Thirring (1975).
46 See Lieb and Seiringer (2010).
47 Lieb and Thirring (1976).
48 Lieb and Lebowitz (1972).
49 See Lieb (1976), section V.
50 Quoted in Dyson (1967); see also Lieb (1990), p. 25.
51 Fuchs et al. (2014).

pictures of the world rest on the private experiences of individual scientists", and "each of us has a view of our world that rests entirely on our private personal experience."[52] QBism puts the "subject" alongside the "object" (the world) in scientific discourse:

> According to QBism, quantum mechanics is a tool anyone can use to evaluate, on the basis of one's past experience, one's probabilistic expectations for one's subsequent experience. ... [Q]uantum mechanics itself does not deal directly with the objective world; it deals with the experiences of that objective world that belong to whatever particular agent is making use of the quantum theory.[53]

Now I cannot think that the ultimate grounding for this view is the innocuous notion that science is an attempt by humans to make sense of the world given to us through our senses, and that science is a human construct. On the contrary, it seems that the basis of a variant of Berkeleyian idealism which suffuses QBism (an admittedly provocative claim, but see (v) below) may be more directly linked to the subjectivist or "personalist" interpretation of probability that is central to the theory.

> Since probabilities are the personal judgments of an agent, it follows that a quantum state assignment is also a personal judgment of the agent assigning that state.[54]

QBism adopts a subjectivist stance on probability in physics, inspired principally by the writings of Bruno de Finetti. For the purposes of this essay, I have no objections to it; indeed I largely share it.[55] So suppose we accept the premiss that probabilities are, loosely speaking, related to betting quotients that rational agents place on chance events. No agents, no probabilities. What I question is the further inference in QBism that our scientific reasoning should primarily be about our personal experiences, our "beliefs", and not the objective world. E. T. Jaynes was perhaps the most prominent and astute defender of a Gibbsian approach to classical statistical mechanics based on a subjectivist interpretation of probability. Jaynes was also a fan of de Finetti. It did not lead him to say that statistical mechanics is essentially about his and other agents' personal expectations; he never concluded that theory "does not deal directly with" the world of molecules in gases, and stars in galaxies, etc., for which it provides dynamics. There is more to statistical mechanics than just the probabilities, and arguably it is no different in quantum theory.[56]

(ii) There are *two* principles of probabilistic updating in QBism. Besides the Bayesian updating associated with the registration of measurement outcomes, the wavefunction is also updated between measurements: it evolves according to the

[52] Mermin (2016).

[53] Fuchs et al. (2014).

[54] Fuchs et al. (2014).

[55] My own views on probability are partly spelt out in Brown (2011). But for a critique of the subjectivist interpretation of probability in the context of QBism, see Timpson (2008).

[56] For a clear account of why Jaynes thought equilibrium statistical mechanics works, which has little to do with the choice of probability assignments, see Jaynes (1957). Fuchs (2016) himself states that "there is more to quantum mechanics than just three isolated terms (states, evolution, and measurement)", but he has something quite different in mind; see (vi) below.

Schrödinger equation, whether there are external forces or not. Now it is a recurring theme in QBism, to which we return in (v) below, that our beliefs about likely perceived events in the future are a result of our interacting with the world.

> In QBism the outcome of a measurement is the experience the world induces back in the user who acts on the world.[57]

Suppose then that the quantum system in question evolves freely over a finite interval of time, in which there are no measurements of the system taking place, and so no "experimental kicks". The notion that an agent's subjective quantum probabilities related to the system undergo a non-trivial change in this interval – determined by a specific Hamiltonian that carries no information about previous or future measurements – seems mysterious to me. The agent might even be asleep! Quantum process tomography, involving initial and final measurements, confirms that time evolution in such cases exists, but does not account for its happening. According to QBism there is no ontic state objectively evolving and dragging the probabilities along with it, in analogy with the Liouville evolution of the probability distribution in classical statistical mechanics.[58] It is as if von Neumann's two motions in quantum mechanics have reappeared in a different guise! The difference now is that the mystery lies with the unitary evolution.

(iii) QBists make a point of distinguishing between information and belief; they argue that it is the progression of the latter that quantum theory describes. I suppose a typical agent's past experience will, if the agent is sufficiently clued up, believe that the quantum state will evolve according to the Schrödinger equation between measurements. But this possible response to the problem posed in (ii) is unconvincing – it cuts no ice in relation to the key question as to whether the wavefunction itself is "belief".

Mermin accepts that

> My reification of the concepts I invent, to make my immediate sense of [*sic*] data more intelligible, is a useful tool of day-to-day living. But when subtle conceptual issues are at stake, related to certain notoriously murky scientific concepts like quantum states, then we can no longer refuse to acknowledge that our scientific pictures of the world rest on the private experiences of individual scientists.[59]

The arguments given in Sect. 5.2 above are attempts to show, following Timpson (2008) lead, that this unevenness is unwarranted, even for a proponent of Bayesianism. Wavefunctions and their properties allow us to make sense of our

---

[57]Mermin (2016).

[58]Consider the claim made recently by Leifer (*op. cit.*, p. 71) that in the epistemic view of the state in quantum mechanics "the appropriate analogies are between quantum states and probability distributions, and between the Schrödinger equation and Liouville's equation." This holds for the Einstein version of the epistemic state, but not for QBism. Timpson (2008), section 2.2, is also concerned with the issue of objective evolution of the state in QBism, but to different ends.

[59]Mermin (2016).

experiences not only in the laboratory but in certain day-to-day phenomena, just as the concept of other agents does. As Elliott Lieb wrote in 1990:

> But we also see the effects of quantum mechanics, without realizing it, in such mundane facts about stability as that a stone is solid and has a volume which is proportional to its mass, and that bringing two stones together produces nothing more exciting than a bigger stone.[60]

(iv) Mermin writes:

> Some claim, for example, that quantum states were evolving (and even collapsing) in the early universe, long before anybody existed to assign such states. But the models of the early universe to which we assign quantum states are models that we construct to account for contemporary astrophysical data.[61]

Yet it is hard to avoid the question: *what* was evolving in the early universe, if not quantum states? *It is not that the question is ill-posed in QBism; it is rather that it leads nowhere.* Fuchs tells us that the universe is "made of something else than quantum states",[62] but details are not thick on the ground.

An analogous scenario suggests itself in the spatial, rather than temporal domain. Using the Pauli exclusion principle, Chandrasekhar famously explained in 1931 the gravitational stability and instability of stars in their late evolutionary phase as white dwarfs.[63] I take it that the QBist is committed to saying that such stellar models are only constructed to account for what humans see in their telescopes; so the quantum states of stars are no more than the figments of the highly trained imaginations of astrophysicists. Again, we seem to be left with an explanatory gap.

It is worth noting at this point the reason QBists consider Bell's criticism of the Copenhagen interpretation – recall Sect. 5.2 above – not to apply to their theory. It is that what QBism encompasses are not just the agent's experiences of the "piddling" results of measurements in Earth-bound scientific laboratories. The theory allows

> each of us to take the scope of physics to be any of the manifold aspects of our own experience …[64]

And

> Users are making measurements more or less all the time more or less everywhere. Every action on her world by every user constitutes a measurement, and her experience of the world's reaction is its outcome.[65]

I cannot help but find this response to Bell unconvincing. Generalising the notion of measurement to include the myriad experiences of agents gained outside the laboratory will in large part lead to theories, or "personal modes of thought" that

[60]Lieb (1990), p. 1. See also the two first epigraphs at the start of the present paper.

[61]Mermin (2016).

[62]Fuchs (2016), footnote 5.

[63]Chandrasekhar (1931).

[64]Fuchs et al. (2014).

[65]Mermin (2016).

have very little to do with quantum physics. More importantly, I suspect Bell – the inventor of the word beable – would have thought that to circumscribe physics to what is going on in the minds of human beings, even when outside laboratories, would still be "to betray the great enterprise". I suspect he wanted science to try to tell us, amongst many other things, what actually happened in the early universe, and what has actually gone on inside stars since then.

(v) In Sect. 5.2.1 above it was mentioned that in QBism there is no physical entity describable in the theory which is the seat of such properties of the quantum system as mass. This raises the question as to how one is to understand the sense in which matter interacts with the metric field in Einstein's general theory of relativity (GR). Of course, in its standard form this theory represents matter, at least insofar as it appears in the Einstein field equations, by way of a stress energy tensor field that is associated with classical fields/particles. But applications of quantum field theory in curved spacetime has led to important results in GR, arguably the most famous being Hawking's 1975 prediction concerning the evaporation of black holes. Here, the treatment is semi-classical, the metric field being considered classical and the matter fields quantum mechanical, and the (weak) back reaction of the matter fields on the metric is ignored. There are well-known problems associated with the semi-classical approach in other contexts, but even where it is regarded as successful, as in the case of Hawking radiation, there is arguably something odd going on from the perspective of QBism. Gravitational degrees of freedom are coupled with specific features of quantum fields the literal reality of which is, presumably, in question. The only way that I see to surmount this conundrum within QBism is to appeal to the yet-to-be-developed, more fundamental theory of quantum gravity. Now one expects the QBist to desist from assigning reality to the states of the gravitational field as well. Parity is restored, but at the price of adding gravity to what for QBists is the shadow world of quantum physics, where nothing is truly what it seems.

(vi) Earlier, I used the description "a variant of Berkeleyian idealism" in relation to QBism. This may well seem inappropriate. Bishop Berkeley did not believe in a reality external to human perceptions, apart from God. QBists do. Indeed, Fuchs says that QBism and related views "should be regarded as attempts to make a deep statement about the nature of reality."[66] So it would seem that QBism is not strictly idealism in Berkeley's sense; it does not "deny the existence of an observer-independent reality", as Leifer claimed (see Sect. 5.2.3 above). But I find the ineffable nature of the external world in QBism troubling, and it is this concern that leads me to make the analogy with George Berkeley's metaphysics.

A key notion in the theory is, as was mentioned in (i) above, that of the interaction between the agent and the world. Here is the way Fuchs and Schack put the point:

> [O]ne. . .might say of quantum theory, that in those cases where it is not just Bayesian probability theory full stop, it is a theory of stimulation and response . . . . The agent, through the process of quantum measurement stimulates the world external to himself. The world, in return, stimulates a response in the agent that is quantified by a change in his beliefs –

[66] Fuchs (2016), p. 1.

> i.e., by a change from a prior to a posterior quantum state. Somewhere in the structure of those belief changes lies quantum theory's most direct statement about what we believe of the world as it is without agents.[67]

In Mermin's words:

> Science is about the interface between the experience of any particular person and the subset of the world that is external to that particular user.[68]

Let us remind ourselves why the QBist needs to postulate such an interface, since after all we cannot be certain such a world external to our subjective experiences exists. Here is Fuchs' reason:

> I would say all our evidence for the reality of the world comes from without us, i.e., not from within us. We do not hold evidence for an independent world by holding some kind of transcendental knowledge. ... We believe in a world external to ourselves precisely because we find ourselves getting unpredictable kicks (from the world) all the time.[69]

Not quite. What we find ourselves getting is forever changing subjective experiences. Berkeley, and many other thinkers over the ages, have not been content to leave it at that. They have looked for an explanation of these more-or-less structured experiences (predictable or otherwise), and in particular, an explanation of the correlations between the experiences of different agents. Berkeley chose the intervention of God; QBists (and scientists generally) choose that of the world. I am willing to grant they are not the same thing, but in the case of QBism there are, I think, analogies that are striking. I will try to spell this out.

Berkeley's famous dictum *esse est percipi (aut percipere)* – to be is to be perceived (or to perceive) – was based on the claim that the action of matter on mind is inexplicable, implying that to postulate the existence of matter is pointless:

> ...though we give the materialists their external bodies, they by their own confession are never the nearer knowing how our ideas are produced: since they own themselves unable to comprehend in what manner body can act upon spirit, or how it is possible it should imprint any idea in the mind. Hence it is evident the production of ideas or sensations in our minds, can be no reason why we should suppose matter or corporeal substances, since that is acknowledged to remain equally inexplicable with, or without this supposition.[70]

Modern philosophers of mind tend not to follow Berkeley in rejecting the material world, but they are certainly divided on how to make sense of the relationship between conscious states and the underlying neurophysiological states of the brain of the agent in question. The problem of how "qualia" – the introspectively accessible, phenomenal aspects of our mental experiences – relate to the physical world both in the brain and in its environment is a central contentious issue in the philosophy of mind.[71] But it is widely accepted that *some* important kind of

[67] Fuchs and Schack (2004).

[68] Mermin (2016).

[69] Fuchs (2002b), also quoted in Fuchs (2016).

[70] Berkeley (1710).

[71] For an introduction to the problem of qualia, see Tye (2016).

connection exists between mental experiences and physical brain states, and that physics has something to say at least in principle about how the latter are affected by the external environment. Is there space in QBism for something like this picture?

I take it that the things "external" to the agent (call her Alice), such as atoms, tables and chairs and other agents (such as Bob), are part of what QBists call the world. Consider then the following way QBists distinguish their position from that of Bohr:

> Acting as an agent, Alice can use the formalism of quantum mechanics to model any physical system external to herself. QBism directs her to treat all such external systems on the same footing, whether they be atoms, enormous molecules, macroscopic crystals, beam splitters, Stern-Gerlach magnets, or even agents other than Alice. ...
>
> ... But because Alice can treat Bob as an external physical system, according to QBism she can assign him a quantum state that encodes her probabilities for the possible answers to any question she puts to him.[72]

Quantum mechanics is seen then to "model" physical systems like atoms and agents, but the notion is a subtle one. To repeat, "... quantum mechanics itself does not deal directly with the objective world; it deals with the experiences of that objective world ...". The modelling is done purely by way of specifying quantum states and their dynamical behaviour, the states themselves being "beliefs" belonging to a single agent. The external physical systems float free of the quantum formalism. No *describable* objective attributes can be assigned to these systems in QBism, because, as we have seen, the universe is made of something other than quantum states, and quantum states are all we have in the formalism of quantum mechanics.

So how are we to understand the nature of the interface between agents and the world that plays such an important role in QBism and specifically its claims to be "realist"? Think first of Alice concerning herself with Bob's interaction with things in the world around him. Insofar as she is equipped with knowledge of quantum mechanics, and provides a formal model for this interaction, it will not strictly be between agents like Bob and other parts of the external world but between quantum states associated with these systems. These are all part of Alice's "personal mode of thought". What about Alice's own interface with the world? I quote Fuchs, Mermin and Schack:

> In QBism the only phenomenon accessible to Alice which she does not model with quantum mechanics is her own direct internal awareness of her own private experience. ... Her awareness of her past experience forms the basis for the beliefs on which her state assignments rest. And her probability assignments express her expectations for her future experience.[73]

This seems to go beyond acknowledgment of the mystery of qualia (if mystery is the right word); there seems to be an indication that there is no model based on the quantum formalism that Alice can construct to account even for the states of the physical brain substrate underlying her own personal experiences. Indeed, it is not

[72]Fuchs et al. (2014).

[73]Fuchs et al. (2014).

clear to me how such a thing could exist in QBism, in which assignment of quantum states rests on beliefs, which in turn rest on subjective experience. But even if Alice can somehow assign a quantum state to her own brain, the nature of its interaction with the external world is opaque because, again, for the QBist the world is itself not made of wavefunctions. What does it mean then to say that we find ourselves getting kicks from the world? Here is a variation on the QBism theme:

> When an experimentalist reaches out and touches a quantum system – the process usually called quantum 'measurement' – that process gives rise to a birth. It gives rise to a little act of creation. And it is how those births or acts of creation impact the agent's expectations for other such births that is the subject matter of quantum theory.[74]

The language is colourful, and the recent term "participatory realism" Fuchs has used to describe QBism[75] is alluring, but the nature of the agent-world interface in QBism seems to be entirely obscure from a physics perspective for that agent. How, in particular, does Alice know that she interacts only with a "subset" of the world? What does subset in this context even mean? Why does the world react to and act on Bob in a way similar to its interaction with Alice? If Alice applies physical notions like locality (no action-at-a-distance), and divisibility into subsystems, to the external world that is purportedly acting on her, she does so with no clear justification. That part of QBism which relates to "a theory of stimulation and response" between the agent and the world is not grounded in known physics.

For Berkeley, the nature of God's action in creating living minds is a mysterious affair. That's the way it is with God. I fail to see how the action of the external world on human agents in QBism is any less mysterious.

(vii) It would arguably be a step in the right direction if QBists, in their zeal for realism, were to conclude that understanding the universe is the true aim of physics, and that current quantum theory, as they see it, is a stop-gap. If I have understood him correctly, Fuchs has gone some way to adopting this stance.

> Ultimately, as physicists, it is the quantum world for which we would like to say as much as we can, but that is not our starting point. Quantum theory rests at a level higher than that. To put it starkly, quantum theory is just the start of our adventure. The quantum world is still ahead of us.

But recall the self-proclaimed revolutionary nature of QBism and its philosophy of nature:

> We bring QBism to the readers attention because it corrects a profound misconception in our general view of science, which led us into major confusion in the twentieth century. Now that we are well into the 21st and we all agree that quantum mechanics works spectacularly well for every practical purpose, surely it is time to expand our ancient view of the nature of science, to dispel the murkiness that has obscured the foundations of the theory for too long.[76]

---

[74]Taken from the introduction to a 2004 lecture by Christopher Fuchs, and reproduced in Fuchs (2016).

[75]Fuchs (2016).

[76]Fuchs et al. (2014).

Whether this is consistent with the view that quantum theory is merely a stop-gap is surely debatable, as is the question as to how revolutionary QBism really is. Given the nature of modern philosophy of science at least in the Anglophone tradition, it is easy to overlook the existence of a major idealist trend in natural philosophy that might be said to have started with Leibniz, Berkeley and Kant, and which gives, in varying ways, the "subject" a prominent role in the understanding of scientific thinking.

What I find startling is Fuchs' recent comparison of QBism with Einstein's philosophy of science, in which he "cannot see any way in which the program of QBism has ever contradicted what Einstein calls the program of "the real". ..."[77] This remarkable claim is worth examining.

If the QBist is truly to treat atoms and laboratory equipment and human agents on the same footing, then elements of the familiar macroworld, as much as elements of the microworld, are to be treated as "concepts" in "a personal mode of thought", i.e. theory, "that any agent can use to organize her own experience". As we have seen, the QBist is resolutely silent on the precise nature of the external world. Better to say it is *as if* the world is populated by such entities as atoms, tables and chairs, according to our best theories. As I understand Einstein's brand of scientific realism, it is indeed not far from this "as-if" reconstruction of QBism. Einstein's view was that in explaining the structure of human experience (what he called the "subjective factor"), the scientist is charged with coming up with coherent models of an external reality involving mind-independent elements (the "objective factor"), including presumably the constituents of the brains of sentient beings. For Einstein, who had a life-long interest in philosophy, this realist commitment is tentative. As he said,

> ...the "real" in physics is to be taken as a type of program, to which we are, however, not forced to cling a priori. ...[78]

Einstein was aware that the program could fail in principle; it was a dogma about which he was not dogmatic, though he recognised of course that so far the historical record has been encouraging. But were he asked if the fundamental objects in a successful model of some domain in physics *actually correspond* to the relevant part of the actual world, Einstein's answer would be a smile. He too would not be drawn into a discussion of what transcends the "as if" world.[79] In philosophical jargon, Einstein was an advocate of a deflationary theory of truth, not a correspondence theory.

[77]Fuchs (2016).

[78]Einstein (1970), p. 674. This is part of a longer Einstein quotation found in Fuchs (2016).

[79]See Einstein (1970), p. 680, where Einstein attributes this position to the influence of Immanuel Kant. For further references to Einstein's realist philosophy, and to that of commentators, see Brown and Lehmkuhl (2016), footnote 4.

But I think there is an essential difference between Einstein's position and "as if" QBism, and it has to do with the scope of the "objective factor":

> The ...objective factor is the totality of such concepts and conceptual relations as are thought of as independent of experience, viz., of perceptions. So long as we move within the thus programmatically fixed sphere of thought we are thinking physically.[80]

I find it hard to reconcile this reasoning with the notion that quantum mechanics is a complete theory and, according to Fuchs,

> ...the best understanding of quantum theory is obtained by recognizing that quantum states, quantum time-evolution maps, and the outcomes of quantum measurements all live within what Einstein calls the subjective factor.[81]

Of course, for the post-1927 Einstein the wavefunction is, as we have seen, essentially a probability distribution over hidden ontic states; it is (at least) these ontic states that correspond to a "concept" that is "independent of experience", if we are "thinking physically". According to Einstein, orthodox quantum mechanics is incomplete precisely because it does not specify what such ontic states are. The idea that quantum physics can do without them altogether seems to me to be antithetical to Einstein's program, metaphysically shy though it is.

Fuchs strongly resists criticisms to the effect that QBism is non-realist. In a 2016 paper, he addresses the authors of such criticisms and accuses them of indulging in a *non sequitur*:

> This is because, if any of these cads were to take a moment to think about it, they would recognize that there is more to quantum mechanics than just three isolated terms (states, evolution, and measurement) – there's the full-blown theory that glues these notions together in a very particular way, and in a way that would have never been discovered without empirical science.

I am not sure I entirely understand what is meant here, but presumably the glue in QBism has something to do with the role of the external world in underpinning the notion of experience itself. But this world is not the "as if" world populated by well-defined mind-independent concepts in physical theory, as Einstein understood it. The ineffable world of QBism would, I submit, have held little interest for Einstein.

Acknowledgements I thank the organisers of the XII International Ontology Congress for the kind invitation to contribute to these proceedings. I am also grateful to David Wallace for useful remarks, and to Rhys Borchert, James Read and particularly Christopher Fuchs and Christopher Timpson for invaluable critical comments on the first draft of this paper. None should be taken to endorse the arguments presented here.

---

[80] Einstein (1970), pp. 673–4; again this is part of the longer quotation given in Fuchs (2016).

[81] Fuchs (2016).

References

Bell, J. S. (1990). Against 'measurement'. *Physics World, 3*(8), 33–40.

Bell, M., & Gao, S. (Eds.) (2016). *Quantum nonlocality and reality*. Cambridge: Cambridge University Press.

Berkeley, G. (1710). Of the principles of human knowledge: Part 1, republished In A. A. Luce & T. E. Jessop (Eds.), *The works of George Berkeley, bishop of cloyne*. London: Thomas Nelson and Sons.

Brown, H. R. (2011). Curious and sublime: The connection between uncertainty and probability in physics. *Philosophical Transactions of the Royal Society A, 369*, 1–15. http://philsci-archive.pitt.edu/8571/

Brown, H. R., & Lehmkuhl, D. (2016). Einstein, the reality of space, and the action-reaction principle. In P. Ghose (Ed.), *Einstein, Tagore and the nature of reality* (pp. 9–36). London/New York: Routledge. arXiv:1306.4902v1.

Brown, H. R., & Read, J. (2016). Clarifying possible misconceptions in the foundations of general relativity. *American Journal of Physics, 84*(5), 327–334.

Brown, H. R., & Wallace, D. (2005). Solving the measurement problem: de Broglie-Bohm loses out to Everett. *Foundations of Physics, 35*, 517–540. quant-ph/0403094; PITT-PHIL-SCI 1659.

Brown, H. R., Summhammer, J., Callaghan, R., & Kaloyerou, P. (1992). Neutron interferometry with antiphase modulation. *Physics Letters A, 163*, 21–25.

Brown, H. R., Dewdney, C., & Horton, G. (1995). Bohm particles and their detection in the light of neutron interferometry. *Foundations of Physics, 25*, 329–334.

Bush, J. W. M. (2015). The new wave of pilot-wave theory. *Physics Today, 68*(8), 47–53. https://doi.org/10.1063/PT.3.2882

Callender, C. (2017). The Redundancy Argument Against Bohm's Theory, manuscript available at http://philosophyfaculty.ucsd.edu/faculty/ccallender/publications.shtml

Chandrasekhar, S. (1931). The density of white dwarfstars. *Philosophical Magazine, 11*, 592–596.

Dürr, D., Goldstein, S., & Zanghi, N. (1997). Bohmian mechanics and the meaning of the wave function. In R. S. Cohen, M. Horne, & J. Stachel (Eds.), *Experimental metaphysics: Quantum mechanical studies for Abner Shimony*, Vol. 1 (Boston studies in the philosophy of science, Vol. 193, pp. 25–38). Dordrecht: Kluwer Academic Publishers

Dyson, F. J. (1967). Ground-state energy of a finite system of charged particles. *Journal of Mathematical Physics, 8*, 1538–1545.

Dyson, F. J., & Lenard, A. (1967). Stability of matter. I. *Journal of Mathematical Physics, 8*, 423–434.

Dyson, F. J., & Lenard, A. (1968). Stability of matter. I. *Journal of Mathematical Physics, 9*, 698–711.

Einstein, A. (1970). Remarks concerning the essays brought together in this co-operative volume. In P. A. Schilpp (Ed.), *Albert Einstein: Philosopher-scientist* (3rd ed., Vol. 2). La Salle: Open Court.

Frank, R. L. (2011). Sobolev inequalities and uncertainty principles in mathematical physics: Part 1. Lecture notes (LMU Munich). Available at https://web.math.princeton.edu/?rlfrank/sobweb1.pdf

Frohlich, J. (2009). Spin, or actually: Spin and quantum statistics. In B. Duplantier, J.-M. Raimond, & V. Rivasseau (Eds.), *The spin* (pp. 1–60). Basel/Switzerland: Birkhäuser Verlag.

Fuchs, C. A. (2002a). Quantum mechanics as quantum information (and only a little more). arXiv:quant-ph/0205039v1.

Fuchs, C. A. (2002b). The anti-vaxjo interpretation of quantum mechanics. http://arxiv.org/abs/quant-ph/0204146

Fuchs, C. A. (2016). On participatory realism. arXiv:1601.04360v3 [quant-ph].

Fuchs, C. A. & Schack, R. (2004). *Unknown quantum states and operations, a Bayesian view* arXiv:quant-ph/0404156v1.

Fuchs, C. A., Mermin, N. D., & Schack, R. (2014). An introduction to QBism with an application to the locality of quantum mechanics. *American Journal of Physics, 82*(8), 749–754.
Gao, S. (2017). *Meaning of the wavefunction. In search of the ontology of quantum mechanics*. Cambridge: Cambridge University Press. arXiv:1611.02738v1 [quant-ph].
Hardy, G. H. (1920). Note on a theorem of Hilbert. *Mathematische Zeitschrift, 6*(3–4), 314–317.
Harrigan, N., & Spekkens, R. W. (2010). Einstein, incompleteness, and the epistemic view of quantum states. *Foundations of Physics, 40*(2), 125–157. arXiv:0706.2661. https://doi.org/10.1007/s10701-009-9347-0
Healey, R. (2016). Quantum-Bayesian and pragmatist views of quantum theory. https://plato.stanford.edu/entries/quantum-bayesian/
Jaynes, E. T. (1957). Information theory and statistical mechanics. *Physical Review, 106*(4), 620–630.
Kaloyerou, P., & Brown, H. R. (1992). On neutron interferometer partial absorption experiments. *Physica B, 176*, 78–92.
Leifer, M. S. (2014). Is the quantum state real? An extended review of ψ-ontology theorems. *Quanta, 3*, 67–155.
Leinaas, J. M., & Myrheim, J. (1977). On the theory of identical particles. *Il Nuovo Cimento B, 37*(1), 1–23.
Lieb, E. H. (1976). The stability of matter. *Reviews of Modern Physics, 48*, 553–569.
Lieb, E. H. (1990). The stability of matter: From atoms to stars. *Bulletin of the American Mathematical Society, 22*(1), 1–49.
Lieb, E. H., & Lebowitz, J. L. (1972). The constitution of matter: Existence of thermodynamics for systems composed of electrons and nuclei. *Advances in Mathematics, 9*, 316–398.
Lieb, E. H., & Seiringer, R. (2010). *Stability of matter in quantum mechanics*. Cambridge: Cambridge University Press.
Lieb, E. H., & Thirring, W. (1975). Bound for the kinetic energy of fermions which proves the stability of matter. *Physical Review Letters, 35*(11), 687–689.
Lieb, E. H., & Thirring, W. (1976). Inequalities for the moments of the eigenvalues of the Schrödinger Hamiltonian and their relation to Sobolev inequalities. In E. Lieb, B. Simon, & A. Wightman (Eds.), *Studies in mathematical physics* (pp. 269–303). Princeton: Princeton University Press.
Loss, M. (2005). *Stability of matter*. http://www.math.lmu.de/~lerdos/WS08/QM/lossstabmath.pdf
Maudlin, T. (2010). Can the world be only wavefunction? In S. Saunders, J. Barrett, A. Kent, & D. Wallace (Eds.), *Many worlds? Everett, quantum theory, & reality* (pp. 121–143). Oxford: Oxford University Press.
Mermin, D. (2016). Why QBism is not the copenhagen interpretation and what John Bell might have thought of it. In *Quantum [Un]Speakables II*. Part of the series the frontiers collection (pp. 83–93). Springer. arXiv:1409.2454V1 [quant-ph].
Merzbacher, E. (1962). Single valuedness of wave functions. *American Journal of Physics, 30*(4), 237–247.
Myrvold, W. C. (2015). What is a wavefunction? *Synthese, 192*(10), 3247–3274.
Ney, A. (2015). Fundamental physical ontologies and the constraint of empirical coherence: A defense of wave function realism. *Synthese, 192*(10), 3105–3124.
Prange, R. E., & Girvin, S. M. (Eds.) (1990). *The quantum hall effect* (Graduate texts in contemporary physics). Berlin/Heidelberg: Springer.
Pusey, M. F., Barrett, J., & Rudolph, T. (2012). On the reality of the quantum state. *Nature Physics, 8*, 475–478. arXiv:1111.3328. https://doi.org/10.1038/nphys2309
Seiringer, R. (1990). Inequalities for Schrödinger operators and applications to the stability of matter problem. http://www.ueltschi.org/AZschool/notes/RobertSeiringer.pdf
Sobolev, S. L. (1938). On a theorem of functional analysis (in Russian). *Mat. Sb., 46*, 471–497. English transl.: American Mathematical Society, Transl., II. Ser. 34, 39 68 (1963)
Squires, E., Hardy, L., & Brown, H. R. (1994). Non-locality from an analogue of the quantum Zeno effect. *Studies in History and Philosophy of Science, 25*(3), 425–435.

Szriftgiser, P., Guéry-Odelin, D., Arndt, M., & Dalibard, J. (1996). Atomic wave diffraction and interference using temporal slits. *Physical Review Letters, 77*(1), 4–7.
Thirring, W. (1986). Stability of matter. In V. Gorini & A. Frigerio (Eds.), *Fundamental aspects of quantum theory* (pp. 343–354). New York: Plenum Press.
Timpson, C. (2008). Quantum Bayesianism: A study. arXiv:0804.2047v1 [quant-ph].
Tye, M. (2016). Qualia. In E. N. Zalta (Ed.), *The Stanford encyclopedia of philosophy* (Winter 2016 Edition). https://plato.stanford.edu/archives/win2016/entries/qualia/
Wallace, D., & Timpson, C. (2010). Quantum mechanics on spacetime I: Spacetime state realism. *British Journal for the Philosophy of Science, 61*, 697–727.

Chapter 6
Preliminary Considerations on the Emergence of Space and Time

David Albert

Abstract This paper explores the idea that the wave-function is the unique fundamental concrete physical stuff of the world *itself*. The paper focuses on two suggestions: (a) First-quantized non-relativistic quantum mechanics is a not a theory of the 3-dimensional motions of *particles*, but of the 3N-dimensional *undulations* of a concrete physical *field* –the wave-function itself – where N is a very large number that corresponds, on the *old* way of thinking, to the number of elementary particles in the universe. (b) This particularly radical coming-apart of the geometry (on the one hand) and the fundamental arena (on the other) is what's at the bottom of everything that's exceedingly and paradigmatically *strange* about quantum mechanics.

All I mean to do here is to assemble a few pretty simple observations that seem to me to suggest that space and time, as we are usually accustomed to thinking of them, are not fundamental features of the world. And let me mention at the outset that there are lots of other people who seem to me to have had thoughts that are very much in this ballpark, and whose work this is very much built upon – people like D. C. Williams and Brad Skow and Harvey Brown and Carlo Rovelli and any number of others besides.

One sometimes hears talk about the emergence of space and time in the context of various attempts at constructing a quantum theory of gravity – but I'm going to be coming at it from an entirely different angle. My own interest in all this initially had to do with thinking about the direction of time, and about the ontology of quantum-mechanical wave-functions – and I want to try, in the space I have here, to tell you something about how it bears on those questions.

Let me start out with some very simple considerations about the classical mechanics of particles.

D. Albert (✉)
Department of Philosophy, Columbia University, New York City, NY, USA
e-mail: da5@columbia.edu

A. Cordero (ed.), *Philosophers Look at Quantum Mechanics*,
Synthese Library 406, https://doi.org/10.1007/978-3-030-15659-6_6

Consider a world that consists of N classical particles floating around in a 3-dimensional Euclidian space – with Cartesian co-ordinates x, y, z – under the influence of a Hamiltonian[1] of the form:

$$H = \left\{\Sigma_i{}^N(((dx_i/dt)^2 + (dy_i/dt)^2 + (dz_i/dt)^2)/2m_i)\right\} + \left\{\Sigma_{k\neq j}{}^N \; V_{kj}((x_k - x_j)^2 + (y_k - y_j)^2 + (z_k - z_j)^2\right\}. \quad (6.1)$$

And imagine that the functions V_{jk} are structured in such a way as to accommodate the existence of tables and chairs and measuring-devices and information-processing systems and so on.[2] Then we can ask questions about how a world like this would *appear* to its *inhabitants*.

The specifically *geometrical* appearances of a world like this are (one supposes) going to be Euclidian and 3-dimensional – but it will be worth thinking for a minute about exactly how it is that those appearances arise. The fact that this world happens to *be* Euclidian and 3-dimensional certainly does not explain, in and of itself, why it *appears to its inhabitants* to be that way. Maybe the explanation has to do (rather) with the way that the fundamental geometrical properties of the world (on the one hand) and the mathematical structure of the fundamental dynamical laws (on the other) *fit together*. The fact that V_{kj} is a function of the distance between particle k and particle j in the background 3-dimensional Euclidian space makes it possible to think of the interaction between those particles as *inscribing* that distance, as making it *manifest*, as making it *visible*, in the *motions* of those particles. The fact that the V_{kj} are functions of the distances between pairs of particles is what makes it possible (for example) for there to be *collections* of such particles whose *stable configurations* have a characteristic *length* in the background 3-dimensional Euclidian space – and can therefore be put to work as *measuring-rods*.

But now consider *another* world, one that consists of N classical particles floating around in an arbitrarily *curved* 3-dimensional space, with the topology of R^3, and with some *generalized curvilinear* co-ordinates x, y, z. And suppose that those particles happen to float around in that space under the influence of a Hamiltonian which depends on the co-ordinates x, y, z in *exactly* the way that the earlier one – the one (that is) in Eq. (6.1) – does.

[1]Maybe a quick disclaimer is in order, at this point, about the metaphysics of lawhood: Nothing that I'm going to say here has any implications whatever – in so far as I am aware – about the dispute between Humean and Necessatarian ideas about the nature of laws. My own experience (mind you) is that many of the topics we will be talking about here, and many of the important questions about the foundations of physics in general, turn out to be a little easier to get one's head around if one has a Humean conception of laws in the back of one's mind – but none what I say here is going to in any way *require* or *entail* or *depend on* a conception like that.

[2]This (of course) is almost certainly impossible – that's why we need quantum mechanics! But imagining otherwise will do no harm – for the moment – in so far as our purposes here are concerned.

Note that this superficial similarity masks very profound differences. In *this* world – unlike in the previous one – $(x_k - x_j)^2 + (y_k - y_j)^2 + (z_k - z_j)^2$ is emphatically *not* the distance between particles j and k in the background 3-dimensional space. Indeed, in worlds like this, there is in general going to be *no* simple and uniform expression for the distance between two particles in terms of their generalized co-ordinates, and Hamiltonians like the one in (1) will *not* accommodate the existence of collections of particles that have stable configurations that have any particular characteristic *length* in the background curved space.

But note, as well, that a Hamiltonian like this *could* support the existence of collections of particles that have stable configurations of particular characteristic $((\Delta x)^2 + (\Delta y)^2 + (\Delta z)^2)$-values – and (more generally) the fact that V_{kj} is a function of $(x_k - x_j)^2 + (y_k - y_j)^2 + (z_k - z_j)^2$ makes it possible to think of the interactions between those particles as *inscribing* that quantity, as making *it* manifest, as making *it* visible, in the evolutions of the co-ordinates of those particles.

And this is worth pausing over, and thinking about. We were talking a few paragraphs back (remember) about the fundamental *dynamical laws* of the world (on the one hand) and the fundamental *geometrical properties* of the world (on the other) as somehow working *together* with one another in the production of geometrical *appearances*. But what seems to be emerging here is that the "actual" geometry of the background space has nothing to do with the production of geometrical appearances *at all*. A Hamiltonian like the one in (1) can apparently produce flat, Euclidian, three-dimensional geometrical appearances in a way that is *entirely independent* of what the "actual" geometry of the background 3-dinmendional manifold happens to be. And (in the light of all this) we might begin to wonder what the *point* is, we might begin to wonder what useful scientific *purpose* can be served, by supposing that that manifold has any "actual" geometry *at all*.[3]

This second example is an inverted and somewhat better-dressed variant of Poincare's old parable of the finite two-dimensional Euclidian disk-world – the one that contrives (by means of the effects of spatial variations in temperature on the lengths of measuring-rods) to appear to its inhabitants as an infinite Lobachevskian plane. That parable led (of course) to a famous debate between conventionalist, verificationist, and scientific realist accounts of the epistemology of geometry – but the lesson I want to draw from it here is something at right angles to all that, something (in particular) about metaphysics. What this second example seems to me to suggest is something about the structure of what you might call the *fundamental arena* in which the history of the world unfolds.

Think of the fundamental arena as a set of points that amounts to something like *the totality of opportunities for things, at any particular time, to be one way or another*. Or you could put it this way: what we have in mind, what we mean to say, when we refer to some set of points as the *fundamental arena* of the world, is that

[3]The business of thinking about geometrical appearances as dynamical effects has been discussed by (among others) John Foster (1982) and Harvey Brown (2005), and Marco Dees, in a very interesting unpublished manuscript called *The Causal Theory of Space-Time*.

a specification of what is physically going on, at each one of those points, at any particular time, amounts to *a complete specification of the physical situation of the world* at that time.

And what this second example seems to me to suggest is that this fundamental arena doesn't need to be thought of as having any particular affine or metrical structure *at all*. The *geometrical* properties of the world might turn out (rather) to be *emergent* things, *mechanical phenomena of nature* – things (that is) that have to do not with the structure of the arena itself, but (like tables and chairs and universities) with the action of the *dynamics*.

Let's suppose (unless or until we find that this goes too far, or that something might be gained by going farther) that the fundamental arena has the structure of a *differentiable manifold*.[4]

Good. Let's recap: The conception of the 'fundamental space' of the world that all of us grew up with includes *both* the idea of a fundamental arena *and* the idea of a fundamental *geometry* – but what we've learned here suggests that these two ideas are worth prying apart. The fundamental *arena* of the world may not need to be thought of as *having* any geometry – and what geometry the world *does* have may turn out not to be any part of it's fundamental structure, but (again) a *mechanical* phenomenon, a by-product of the *dynamical laws*.

What we are ordinarily in the habit of referring to as 'space' (then) can perhaps be thought of as emerging from something very different, and less structured, and more fundamental. But – since that emergence involves *dynamics* – that more fundamental stuff must apparently include *time*. And this (in turn) raises questions of whether time and dynamics can *themselves* be understood in terms of still *more* fundamental things. And it turns out that they *can*. It turns out (that is) that one can play a game with time (in the manner of D.C. Williams and Brad Scow and others) which is analogous to the one we have just now been playing with space.

Consider (to that end) the manifold of the totality of opportunities for things to be one way or another. This is not (mind you) the manifold of the totality of opportunities for things to be one way or another *at this or that particular time* – which is what we have been calling the 'fundamental' arena – but the manifold of

[4]Carlo Rovelli (as we will see in section 3 of this essay) has been thinking along lines like these, for some years now, about the General Theory of Relativity. Carlo sometimes speaks as if he aims to do away, at the fundamental level, with even so much as a differentiable *manifold*. But what I think he actually means to deny is not that there *is* a fundamental differential manifold, but (rather) that that manifold has a certain particular *metaphysical status*. What I think he actually means to deny (that is) is not that the world has some ultimate and fundamental set of topological *possibilities*, but (rather) that those possibilities *inhere* in, and are *parasitic* on, some ultimate and fundamental *substance*. Marco Dees' unpublished doctoral dissertation *The Causal Structure of Space-Time* (on the other hand) aims to go genuinely further. Dees aims to treat the fundamental arena as a *completely* unstructured set of points, and proposes that not only geometrical and affine structure, but topological and differential structure as well, be understood as by-products of the dynamics. My own suspicion – for reasons that should become clear in section 2 of this essay – is that Dees' very imaginative and ambitious program is likely to prove very difficult to actually carry through.

the totality of opportunities of things to be one way or another *simpliciter*. The *total* totality of opportunities for things to be one way or another. Call that the *Ur*-arena.

Imagine (then) some manifold of points, with the topology of R^M, and with some co-ordinization $x_1 \ldots x_M$, over which local properties (the presence or absence of particles, say, or the values of fields, or what have you) are distributed. It may happen (or, of course, it may not) that this distribution has the special mathematical feature that there is some relatively simple set of *rules* which relate the various *sub*-distributions of those properties over the various (M-1)-dimensional *sub-manifolds*, $\{x_1 = \alpha\}$, for different values of α, to one another. And rules like that are called *dynamical laws*, and the various sub-manifolds $\{x_1 = \alpha\}$ that those rules *link* are called *temporal instants*, and the co-ordinate x_1 – the one that *indexes* those sub-manifolds – is called *time*.

So, just as the fundamental arena lacks any intrinsic *geometry* the Ur-arena apparently lacks anything which is either intrinsically spatial (even in the more primitive sense of the fundamental arena) or intrinsically temporal. *All* of that, apparently, is something *emergent* – something which is built out of the pattern in which this total totality of opportunities to be one way or another actually ends up getting *taken*.

There is (needless to say) much more to the phenomenon of temporality than a mere foliation of the Ur-arena into times. And this particular way of *understanding* that foliation ought itself to be understood as a sort of prologue to a much larger and more ambitious enterprise aimed at understanding temporality *as a whole* along very similar lines. What we've done here is to treat the existence of time not as a feature of the fundamental metaphysical structure of the world, but (again) as something more akin to a *mechanical phenomenon of nature* – something (that is) that bottoms out in facts about how the fundamental physical properties of the world happen to be *distributed* over the *Ur-arena.* And it will be very natural to go on from there by attempting to understand the various familiar temporal *asymmetries* – the second law of thermodynamics (for example) and the fact that we have a very different kind of epistemic access to the past than we have to the future, and the fact that by acting now we can apparently influence the future but not the past, and the fact that time seems to us to *pass*, and that the history of the world seems to us to *unfold*, in a particular *direction*, and so on – as mechanical phenomena of nature, as somehow bottoming out in those contingent distributions of local properties, or (more particularly) in *asymmetries* of those contingent distributions of those local properties, as well. But that (as I said) is a whole other project.[5]

Anyway, the business of attending carefully to this distinction between the *arena* of the world (on the one hand) and it's *geometry* (on the other) seems to me to throw a good deal of useful light on a broad range of issues at the foundations of physics – and I want to use the few minutes I have left to tell you a little about one of those.

---

[5]And it happens that a project like that has in fact been *underway* for something on the order of 20 years now. For progress reports, see my *Time and Chance*, and chapters 1 and 2 of my *After Physics*.

Let me start by rehearsing one particular train of thought about the quantum-mechanical measurement problem – the one (in particular) that leads from the measurement problem to speculations about, and theories of, the collapse of the wave-function. And let me ask you to forget (for the time being) all of the forgoing discussion – and think of the world as unfolding, in the old familiar way, in some fundamental, pre-dynamical, geometrical, *space*.

The measurement problem was put in its clearest and most urgent and most ineluctable form, in the first half of the twentieth century, by figures like Schrodinger and Von Neumann and (especially and particularly) Wigner. They thought of quantum mechanics – at least in its first-quantized, non-relativistic version – as a theory of fundamental material particles, moving around in a fundamental three-dimensional space. And they supposed that those particles were the sorts of things to which one could coherently attribute dynamical properties like position and momentum. And they treated quantum-mechanical wave-functions as complete and exact and realistic representations of the states of systems of those particles – the wave-function of such a system was thought of (more particularly) as fixing the values of the dynamical properties of that system, and the dynamical properties of all of its sub-systems, by means of the eigenstate-eigenvalue link. And that (of course) brought with it the principle of superposition, and the phenomenon of non-separability. And what Schrodinger and Von Neumann and Wigner were able to show was that all of *that*, together with the linearity of the fundamental laws of the evolution of quantum-mechanical wave-functions in time, led directly to a puzzle about how it is that measurements ever manage to have outcomes.

Now, what people like Wigner and Von Neumann had to say about the business of actually *coming to terms* with that problem – which famously involved distinctions between 'measurements' and 'ordinary physical processes', or between 'micro' and 'macro', or between 'conscious' systems and merely 'physical' ones, or between 'subject' and 'object', and so on – was silly. But the very inadequacy of those proposals helped to clear a space, and to produce a demand, for the decisive advances in our understanding of these matters which are now associated with names like Bell, and Pearle, and Ghirardi and Rimini and Weber. *Their* innovation was to approach the question of measurement as if it were a traditionally *scientific* sort of a problem, and to look for precise and explicit and unambiguous and traditionally scientific sorts of modifications of the fundamental quantum-mechanical equations of motion that were aimed at actually *solving* it.

And this new approach very naturally brought with it a new and more straightforward and more flat-footed and more traditionally scientific way of thinking about the wave-function *itself*. This *new* way of thinking turns everything about the foregoing tradition elegantly inside out: The wave-function is not an abstract mathematical *representation* of the states of concrete physical systems, but (rather) the unique fundamental concrete physical stuff of the world *itself*. First-quantized non-relativistic quantum mechanics is a not a theory of the 3-dimensional motions of *particles*, but (rather) of the 3 N-dimensional *undulations* of a concrete physical *field* – which is nothing other that the wave-function itself – where N is a very large number that corresponds, on the *old* way of thinking, to the number of elementary particles in the universe. And once this new picture is fully taken in, there are no longer any such metaphysical conundrums in the world as indeterminacy or superposition or non-separability: On the GRW theory,[6] the complete fundamental physical condition of the world, at any particular time t, is just the 3 N-dimensional configuration of this *field* at t, and there is a perfectly definite fact of the matter about the *value* of that field, at every single time t, at every single point in the 3 N-dimensional space in which it undulates, and everything is exactly as crisp and as sharp and as concrete and as straightforwardly intelligible as it was (say) for Newton, or for Maxwell.

But unlike in the case of Maxwell, what we are confronted with here is a field not in *3*-dimensional space, but (rather) in a space of 3 *N* dimensions – where N (remember) is a very large number. And the question that immediately arises is *why* – if that's the way space actually *is* – it *appears to us* to be *3*-dimensional.

And the *answer* to that question, in a nutshell, turns out to be very much like the answer to the question of why the *curved* 3-dimensional space that we considered in the skeptical scenario above appears to be a *flat* one. The answer (that is) is that dimensional appearances, like geometrical ones, are products of the *dynamics* – and that dynamics laws like the ones that actually hold sway in worlds like ours are going to produce 3-dimensional appearances even if the real, fundamental, pre-dynamical space of our world has 3 N dimensions.

But the idea that the fundamental space of the world has as many dimensions and as unfamiliar a structure as the one we have been playing with here has nevertheless struck many investigators as preposterous, or grotesque, or insane. The offence against our ordinary ways of thinking is so enormous, and so outrageous (so these investigators say) that it is simply not possible, it is simply not *scientific*, to take such a picture seriously.[7]

And at this point – it seems to me – it will be useful to remind ourselves that (as a matter of fact) the affine and metrical structure of the background, fundamental, pre-dynamical space does *no explanatory work whatsoever* in this story – and that once that structure is *dismantled*, the very *idea* of anything like a 'Fundamental Space of the World' disappears along with it. And it is much less clear how serious the offence might be, or whether (indeed) there should be any offence at all, in supposing that the fundamental *arena* of the world (as opposed to its fundamental *space*) has 3 N dimensions.[8]

---

[6]Or rather, on the particular version of the GRW theory that I am thinking about here – the original version, in which the wave-function is not supplemented with any further "primitive ontology", the one which is usually referred to in the literature nowadays as GRW_0.

[7]Various strategies have been proposed – strategies that go under the collective name of *Primitive Ontology* – for somehow *hanging on* to the claim that the fundamental space of the world is (notwithstanding everything) 3-dimensional. The interested reader can find detailed accounts of these strategies a number of the essays in Albert & Ney (2013); arguments against these strategies can be found in chapter 7 my book *After Physics* (Albert 2016).

[8]Note that earlier on, when we were dealing with a classical point-like 'item', we needed a fundamental, pre-dynamical, geometrical structure in order to even *write down* our dynamical

Note (to begin with) that the *geometry* of the world, on a theory like GRW, is thoroughly Euclidian and 3-dimensional. The only conception of *distance* that does any dynamical or predictive or explanatory *work* in that theory, the only conception of a distance (that is) that plays any *mathematical role* in the quantum-mechanical version of the Hamiltonian in (1), is the 3-dimensional Pythagorean distance $(x_k - x_j)^2 + (y_k - y_j)^2 + (z_k - z_j)^2$.[9] And it deserves to be emphasized that there is absolutely nothing approximate or defective or misleading or illusory or otherwise second-class – on a picture like this one – about our everyday experience of the geometry of the world as Euclidian and 3-dimensional. On the sort of picture I have been sketching here, the Euclidian 3-dimensional geometry of our everyday experience – notwithstanding that it is something *emergent* – is (again, and indeed, and on the contrary) the *true* and *unique* and *authentic* and *exact* and *complete* geometry of the world. Period. Case closed. End of story.[10]

And there can certainly not be any dispute about the question of the dimensionality of the *fundamental arena* of a world like that, since (after all) it follows immediately from the *definition* of the fundamental arena of the world, and from the mathematical structure of the quantum-mechanical wave-function, that the fundamental arena of a non-relativistic quantum-mechanical N-particle world is 3 N-dimensional.

Let me just pursue this one small step further. Here's where we are: The conception of the 'fundamental space' of the world that all of us grew up with seems to include both the idea of an *arena* and the idea of a *geometry*, and those two ideas are (in fact) importantly *distinct* from one another. But the full *force* of that distinction, the full *importance* of that distinction, only really begins to announce itself when we move from classical theories to quantum-mechanical ones.

laws. That's what footnote 18 was about. But now that we are dealing with a quantum-mechanical, field-like *wave-function*, the thought is that a fundamental differential manifold, with no affine or geometrical structure at all, will suffice. That's what section 2 was about.

[9] It might be thought that the transition to Quantum Mechanics introduces new and potentially worrisome issues here. It might be thought (in particular) that the *non-local* influences that we encounter in a theory like GRW – the ones (that is) associated with Bell's Theorem – will bring *other* or *additional* or *conflicting* geometrical structure into the picture with them. But a little reflection will show that what's *non-local* about those influences is not that they depend on some other or additional or conflicting conception of distance, but (rather, and precisely) that they do *not* depend on any conception of distance *at all*.

[10] Once upon a time (in papers like "Elementary Quantum Metaphysics," which dates back to the late 1990's) I used to say that the world of a theory like GRW, or Bohmian Mechanics, was only *approximately* 3-dimensional and Euclidian – that it was only 3-dimensional and Euclidian in so far as one was careful *not to look too closely*. And this now strikes me as a very bad way to have put it. What one *does* discover – if one looks at the world closely enough to see that it is quantum-mechanical rather than classical – is that the topology of the fundamental arena (on the one hand) and the topology induced by the emergent geometry (on the other) *come apart*. But it is no less the case in a quantum-mechanical world than it was in any classical one that *the only conception of distance that has any physically significant role to play is the three-dimensional Euclidian one*. The geometry of a non-relativistic first-quantized quantum-mechanical world (then) is not in any sense, and not by any measure, one *whit* less Euclidian and 3-dimensional than the geometry of a Newtonian world is.

In classical worlds (and maybe this is what it *is* – or some important *part* of what it is – to be a classical world) the arena and the emergent geometry tend to fit smoothly and obviously together. But in quantum-mechanical worlds, as we have just begun to see, they literally, and radically, come apart.

What happens in classical worlds is (more particularly) that the dynamics merely *decorates* the fundamental arena with a *metric*. What the dynamics does, in the classical case, is simply to establish some determinate *fact of the matter* about how *far away* any two points in that arena *are* from one another – and it does so in a way that *respects*, it does so in a way that takes entirely *on board*, the topological and smoothness properties of the arena that were already, intrinsically, pre-dynamically, *there*.

In the quantum-mechanical case (on the other hand) the topology which the emergent geometry brings with it is *different* – the topology which the emergent geometry brings with it is (as we have seen) of a different *dimension* – than the topology of the fundamental arena. And that's not even the half of it. The sort of thing that we *usually* have in mind when we imagine (in science fiction, or in string theory, or what have you) that the world has more than 3 dimensions is (after all) that the 3 dimensions of our everyday *experience* of the world are the dimensions of some *sub*-manifold of the world's real fundamental *arena*. But what's going on *here* is something else entirely. In the quantum-mechanical case, the 3-dimensional manifold within which my hand and the table in front of me occupy different and non-overlapping regions – and (indeed) the very 'points' between which the emergent 3-dimensional geometry *fixes a distance* – are simply *nowhere at all* in fundamental pre-dynamical structure of the world. It would be nearer the mark to say (instead) that in the quantum-mechanical case it's the three-dimensional manifold *itself*, and not merely it's *geometry*, that turns out to be a product of the dynamics.

And I think a case might be made that this particularly radical coming-apart of the geometry (on the one hand) and the fundamental arena (on the other) is what's at the bottom of everything that's exceedingly and paradigmatically *strange* about quantum mechanics.

But that's a topic for another talk.

References

Albert, D. Z. (2016). *After physics*. Cambridge, MA: Harvard University Press.
Brown, H. (2005). *Physical relativity*. Oxford: Oxford University Press.
Dees, M. (unpublished manuscript). *The causal theory of space time*.
Foster, J. (1982). *The case for idealism*. New York: Routledge & Keegan Paul.
Ney, A., & Albert, D. Z. (Eds.). (2013). *The wave function*. New York: Oxford University Press.

Chapter 7
Decoherence and Ontology

Roland Omnès

Abstract This paper discusses the consequences of quantum mechanics for our understanding of physical reality, particularly regarding how classical concepts emerge from quantum laws; how common sense logic stands out as a special case of quantum logic applied to macroscopic objects; how causality and locality are found to be "provincial" consequences of quanta; how tiny probabilities that would seem to turn reality into an appearance are so small that unreality does not matter; how quantum theory agrees with everything observed, except for a uniqueness that (alas) is the very essence of reality.

There is now a trend among philosophers to refuse the drastic revision in the philosophy of knowledge that was thought necessary after the work by Bohr, Heisenberg and Pauli (and still is in my opinion). Together with a few physicists, many philosophers today nurture the hope of seeing a new realistic theory come out, although there is no sign of a complete or consistent one yet. A very different trend of research existing among physicists might be described on the other hand as being both pragmatic and theoretical, two characteristics which are apparently opposite and require therefore an explanation. When I say that our understanding of quantum mechanics is now completely pragmatic, I mean that every concept in it, either important or tiny, is not only appreciated as a building block in a grand construction, but as an individual piece of knowledge expressing directly the results of some specially dedicated experiments. The present experimental techniques are so powerful that the investigation of an aspect or another of the quantum world has become a most enjoyable testing ground for the tools at our disposal. The paradigm of this approach goes back presumably to the experiment that was performed in 1982 by Alain Aspect for checking Bell's inequalities and resulting in an evidence for

R. Omnès (✉)
University of Paris, Paris Sud XI, Orsay, France

A. Cordero (ed.), *Philosophers Look at Quantum Mechanics*,
Synthese Library 406, https://doi.org/10.1007/978-3-030-15659-6_7

entangled states.[1] The emphasis has slightly changed however. Whereas John Bell asked a deep question, more akin to the structure of our understanding than anything else (such as locality or a causality hidden in hidden variables),[2] the experimental result is now more or less taken as meaning: "Never mind philosophical issues. Just think as Nature lets you to do it by giving you plain facts". The number of plain facts confirming quantum concepts and resulting from smart experiments has now tremendously increased and all of them, up to now, are in essential agreement with the Copenhagen rules.

When one tries to catch more of the spirit of this time (l'esprit du temps) among the majority of physicists, one does not always find them fully adept of the Copenhagen philosophy, in spite of their acceptance of the practical Copenhagen rules. The real existence of wave packet reduction, for instance, is often considered as remaining more or less an open question; the gap between classical and quantum physics and between determinism and probabilism, is considered as somewhat less important than their correspondence, particularly in the case of quantum-behaving macroscopic systems or mesoscopic ones.[3]

So, clearly, the present episteme is pragmatic. But it is also systematically theoretical. Theoreticians accept readily the basic principles of quantum mechanics, with no reservation, because these principles have withstood a tremendous widening of their range when extended from atoms to the standard model of particles, and without a hint of weakness. These basic laws and concepts consist of the Hilbert space framework incorporating non-commuting physical quantities, of a law for dynamics resulting from the Schrödinger equation and a few more assumptions among which relativistic invariance and the existence of identical particles are the main ones, since they are essential for the existence of quantum fields. Similarly, since a few decades, interpretation has also become more and more a topic of theoretical research, where one tries to extract from the basic principles themselves some new consequences allowing to deepen their understanding. Theoretical work or, essentially, mathematical investigation becomes accordingly some sort of a required preliminary before any further philosophical reflection.

I thought that such a brief description of the state of research would be useful when addressing a majority of philosophically-trained people as in the present case, if only because it is so different from the approach you may be used to. We shall see together however that philosophical questions are like the heads of the famous Hydra, always poking them from new unexpected directions.

The main pioneer of the present spirit among the founding fathers of quantum mechanics was certainly John von Neumann. Although he worked at a time when speculations, research programmes and personal convictions were not considered as suitable for publication, there is little doubt that he did not accept the double talk of using sometimes a classical language and sometimes a quantum one in the

[1] A. Aspect et al. (1981).

[2] J. S. Bell (1964).

[3] J. Clarke et al. (1988).

statement of physical laws. He does not seem either to have had much respect for the complementarity principle, probably too Hegelian for a logician such as him. As a matter of fact, von Neumann began his career by doing research in logic, with the foundations of set theory, and his later work on the logic of computers is well known. It may be mentioned that he belonged to the school of thought headed by Hilbert, according to whom theoretical physics should rely on well-formulated basic axioms and then consist of their rigorously derived consequences. The famous book by von Neumann, *Mathematische Grundlagen der Quantenmechanik*, was a powerful attempt at satisfying these exacting requirements and one may remember how he pointed out there Hilbert spaces as the right framework for quantum concepts, while improving considerably for that purpose the theory of operators.

Another important idea in that book was to define a proposition, a statement expressing a physical property of a quantum system, by "elementary predicates" according to which "the value of some observable A lies within a certain range of real numbers at a time t". Indeed, it turns out that every statement concerning a quantum event can always be expressed by such a predicate. Furthermore, one can always translate the statement into the language of mathematics by associating it with a definite "projection operator", namely the mathematical operator projecting a state vector on a Hilbert subspace (consisting of the eigenstates of the operator A(t) with an eigenvalue in the said range). An important consequence of this construction is to subject the logic of quantum properties to the rigor of mathematics, a paradigm to which von Neumann later gave another expression with the logic of computers. In both cases, there is emphasis on the Aristotelian principle according to which a proposition can only have two possible values: "true" or "false". In a computer, this is obtained by assigning the value of a proposition to some memory, which can either withstand a standard voltage for "true", or a zero voltage for "false". In quantum mechanics, a projection operator has the same two-valuedness since it can only have two eigenvalues, 1 or 0.

With his projection operators, von Neumann had discovered something essential for the understanding of quantum mechanics, namely a language that could bridge the gap between the intuition of a pragmatic physicist (and also, why not, of a philosopher) and the formal requirements of a mathematician, since the concepts and laws of quantum physics can only be fully stated in mathematical terms. When however his book was published, the readers were much more impressed by three failures he had met, all of them of course duly acknowledged. One of these great difficulties was the prediction of macroscopic quantum interferences in the final state of a quantum measurement device, a result appearing in the last two pages of the book offering a quantum model for a measurement. This remark was later to become famous when explained by Schrödinger with the example of his unforgettable Cat.

There were two other difficulties, one of them with classical physics. Classical properties do not only state for instance that the position of a particle lies in some range of values, but that the position and the velocity of some macroscopic part of an apparatus can be assigned simultaneously some values, even with large enough errors allowing a minimal violation of the uncertainty relations. Although he made a nice try at it, von Neumann was not able to associate a projection operator with a classical property and his predicates could not be therefore considered at that time as providing a universal language for physics.

The last difficulty had to do with logic. If all the possible predicates, or all the possible projectors, express so many sensible propositions, then it is impossible to define the basic logical operations (not, and, or) and the corresponding relations (equivalence and inference) while satisfying the standard rules of logic. For a long time, this difficulty generated a trend of thought according to which the logic of quantum mechanics might be non-standard. Finally, it seemed that von Neumann's language for quantum mechanics was neither universal nor convincingly sensible, which is why it did not much influence interpretation for a long time.

Though rather old, this story remains certainly the best introduction to more recent research. It was first seen negatively: three outstanding difficulties stood on the way of a deductive interpretation of quantum mechanics. They looked so insuperable that von Neumann himself proposed almost incredible solutions, such as leaving to consciousness the burden of removing macroscopic superpositions. He also considered seriously the possibility of non-standard logic as a key to the understanding of quanta.[4] With hindsight, these somewhat desperate attempts show that the difficulties were really non-trivial but, from a positive point of view, it can also be said that they were well-defined problems, which held the key for a deeper interpretation.

The theoretical approach to interpretation has led to answers for the three von Neumann problems in the last two decades or so. For macroscopic interferences, the name of the answer is "decoherence". The precise derivation of classical physics from the quantum principles was obtained by three different methods using either "coarse graining"[5], "coherent states"[6] or a newcomer in mathematics known as "microlocal analysis"[7]; it shows explicitly how classical determinism is a consequence of probabilistic quantum laws. Finally the problem of logic was solved by introducing "consistent histories",[8] in which the propositions describing physical properties are not single predicates but so-called histories. When put together, the three answers lead easily to a completely deductive interpretation of quantum mechanics, in which the basic principles are enough for generating their own interpretation.[9] The usual rules of measurement theory for instance become so many theorems.

My purpose here is not however to describe this interpretation in detail, because it would be too long, but to identify the new philosophical problems one is left with at the end. The most important new item is certainly decoherence. I shall not try

---

[4]G. Birkhoff and J. von Neumann (1936).

[5]M. Gell-Mann and J. B. Hartle (1991).

[6]K. Hepp (1974).

[7]R. Omnès (1989, 1997a, b).

[8]R. G. Griffiths (1984).

[9]R. Omnès (1999).

to describe it technically, but only mention the essential ideas. When considering a macroscopic object, we perceive only some obvious collective degrees of freedom describing for instance the position and the shape of its various parts. We should not forget however that the object itself (for instance a detector) contains typically billions of billions of billions other degrees of freedom for all the atoms in it. The degrees of freedom of an external environment (atmospheric molecules around the object or photons in the surrounding light) can also play a role at the quantum level and, globally, it has become conventional to call the formal subsystem containing all these degrees of freedom (internal and external) the environment. The accessible (collective) degrees of freedom one can directly perceive and measure label, from a formal standpoint, another subsystem, both abstract mathematically and empirically. The whole object is therefore considered formally as made of two systems, one we can see and one we cannot control in detail.

You may remember that Heisenberg already considered the environment (without the name) as opening a possible way out of the cat problem. This possibility has now been investigated in some detail and, basically, the following schedule is found: The two systems (collective and environment) are coupled. They can exchange energy as we know from the existence of friction and dissipation, so that a part of the total Hamiltonian must connect them, couple them. Now I suppose that you do not easily envision a complex wave function of the environment depending on so many billions of variables any more than I do, but let us say that it is very complicated and, most importantly, extremely sensitive to the external coupling. When a wheel turns even so slightly in a clock, what it provokes in the environment wave function is a cataclysm: atoms move, electrons are shaken and phases, which are the most sensitive and delicate features of a wave function, change practically at random.

Suppose now that the wheel belongs to a measuring device, guiding for instance a voltmeter pointer whose position will indicate the actual result of a measurement. What happens? According to Schrödinger and his cat or von Neumann and his mathematics, the final wave function of the measuring device is a sum of two terms, one indicating, say, that the pointer did not move and another according to which the final position of the pointer has turned by 90 degrees, indicating that something has been detected. But in fact, these two parts of the wave function are very different. Already when the wheel was beginning to move, the environment wave function, with its many billions of variables, was behaving very differently for the static wheel than it did for the moving one. The corresponding phases (i.e. a phase for every value of every variable in the crowd), I say, had soon lost any hint of coherence: they decohered, according to a useful neologism stamped out by Gell-Mann.

In the example of the cat, one would say that the wave functions of the set of atoms inside a live cat and a dead one differ so much in their multitudinous phases that they cannot be anything but orthogonal: they do not allow any visible interference at the level of the cat body. The non-existence of macroscopic interferences which looked so troublesome has now an obvious origin: destructive interferences in the environment wave functions suppress constructive ones at the atomic level.

The theory of decoherence is of course more precise than the sketch I just gave.[10,11,12] As it turns out, decoherence is in fact a special kind of irreversible process.[13] It is moreover an extraordinarily effective effect, so quick in action that it completely suppressed any interference before it could be spotted by an observation. And so, for many years, theorists have lived with a solution that no experiment could establish. Who would then believe equations when so much is at stake! But fortunately, 4 years ago, the effect was seen at last in an experiment of quantum optics, where the number of degrees of freedom in an environment could be made to vary from 0 to 10. Interferences were then seen to disappear, gradually, in exact agreement with theoretical predictions.[14] Something essential is therefore now established. Decoherence exists and it is as much effective as we did expect. So much then with physics. The next question should be to consider the kind of consequence it has for our understanding of physical reality.[15,16]

It is certainly not a surprise if various people have very different reactions to the experimental discovery of decoherence. People believing in Bohmian mechanics do not care: they live in a world much above any experimental reach. People who are fond of actual reduction mechanisms recognise that decoherence does the job more rapidly and completely than any unconventional effect they had proposed for the same purpose. Some of them say that two effects are still better than one.

What about people who accept quantum mechanics as complete? It is very instructive for instance to put together the answers for the three von Neumann problems and draw the consequences. There are in fact so many of them that only a book can give their list but I can mention one: It is found that a definite direction of time comes out from three different origins: there is dissipation (with the second principle), decoherence (!) and also logic (leaving aside the cosmological direction of time). As far as logic is concerned, some histories for a quantum system make sense with standard logic with one direction of time but they do not with the opposite direction, when the film of events is run back. These three directions are furthermore necessarily identical.[17] There could be a nice Kantian echo when this result is expressed as follows: the quantum thing-in-itself has no specific direction of time but pure reason, i.e. logic, can only give an account of it by selecting once and for all a unique direction. Think of it: the direction of time as a categorical a priori judgement!

Decoherence is closely related with wave packet reduction, but they are not identical. Decoherence is a genuine quantum effect occurring inside a measurement

[10] W. H. Zurek (1982).

[11] A. O. Caldeira and A. J. Leggett (1983).

[12] E. Joos and H. D. Zeh (1985).

[13] R. Omnès (1997a, b).

[14] M. Brune et al. (1996).

[15] R. Omnès (1999).

[16] R. Omnès (1999).

[17] R. Omnès (1999).

device, whereas reduction was supposed to affect directly the measured quantum object and was at variance with the Schrödinger equation. Their statistical consequences are however identical, because decoherence implies that one can compute the probability for the results of a second measurement as if there had been reduction in a previous measurement (this result being most easily shown by using histories). But there was another aspect to reduction. It was also supposed to insure the uniqueness of the measurement result, by selecting a single outcome among various possible ones; it explained, or at least it preserved the uniqueness of reality by an actualising one possibility among many. Decoherence on the other hand performs only the first step of the process. When acting on the environment wave functions (or rather on the state operator), it removes superpositions, entanglements, and leaves only ordinary, classical, probabilities for the various possible results. Being however a quantum effect, it cannot go further and cannot explain how a specific result is selected as the actual one.

It seems at this juncture that we are left again with a very old problem, though now it is rather differently stated, a problem one may call that of actualisation or "objectification". When naively stated, it amounts to the question: is there a genuine effect enforcing actuality? Less naively, it becomes much more subtle because one cannot even state it as a problem in the framework of quantum logic. This logic implies, indeed, that the only logically consistent histories are the ones referring to a unique result, whatever it may be, and the problem of actuality asking "which one?" has no content in the theory.

The question (if not the problem) is made deeper because of some new powerful results. Not only is the theory of measurement becoming a collection of theorems, but classical logic and even common sense can be deduced from the quantum principles, in a macroscopic situation. One may then confidently assert that everything observed has been proved to be a direct consequence of these principles! Everything? Well, there is still this question about uniqueness for which a genuine probabilistic theory cannot obviously provide any cause or mechanism. But if one asks philosophers what is the most essential feature of reality, they say: uniqueness. This is at least what comes out of Wittgenstein's games of language when the apprentice does not understand the word "stone" or "brick" and the sole resource of the master mason is to point out a real stone and say: "that". "That" is meaningful because and only because "that" is unique. We are thus left with a theory agreeing with every feature of reality, except one, but it is essential. My own belief is that we can learn much more about ontology by studying this question deeper, since it is a pure case of the relation between thought and reality, theory and actuality, mathematics and physics. One might try of course to get out of it through Everett's many-sheeted reality, but it means that one believes quantum theory above the unique wonder of a reality we can contemplate every day. It looks to me as the extreme of ideology and I would rather prefer bishop Berkeley's unique dream of reality by God, if things have to go that far.

Let us go back however to less elevated questions. I did not yet mention that decoherence is a dynamical effect that is never perfectly exact. Entangled states of a measured quantum object and a measuring device are disentangled, but a tiny amount of entanglement (or superposition) always survives. The probability for observing a macroscopic interference effect between a dead and a live cat is never exactly zero, but extremely small and becoming exponentially smaller with larger values of time. As a matter of fact, very small probabilities pop up everywhere in the new interpretation: in determinism, which is a logical equivalence between two classical properties holding at different times, and which has always a tiny probability for being wrong (because of gigantic quantum fluctuations); in the expression of classical properties, which are always slightly spoiled by the uncertainty relations and in other places we can leave out. Borrowing a famous expression from ancient philosophy, we might say with Simplicius and St Augustine that quantum theory preserves every appearance of reality, except for extremely small probabilities for having things spoiled. This is again a question about the exact meaning of a physical theory, and certainly the oldest question of that sort.

One may look more carefully at the question of very small probabilities. It is often said that they are negligible, but what is the exact meaning or the precise evaluation of "very small"? We might say: a probability that cannot be checked experimentally, even if the measuring device contained all the matter in the Universe (though excessive, this is at least a "pragmatic" definition and not so crazy when considering that this kind of probabilities always involves an exponential). The next question is then: if our theory agrees with a primary, intuitive, classical experience of the world through our senses (phenomena), except for very small probabilities of error or misconception, can we neglect safely these probabilities and on which ontological grounds?

The first person to ask this question was Emile Borel, the famous mathematician and probabilist. He asked it when thinking of quantum theory in the late thirties and early forties and his answer, which I endorse, can be summarised as such[18]: An interpretation of probability calculus must be decided before any interpretation of quantum mechanics, since the second theory relies conceptually on the first one. The existence of too small probabilities cannot be falsified by any experiment (in the sense of Popper's notion); the corresponding "strange events" or "miracles" (like the Earth leaving the neighbourhood of the Sun to go revolving around Sirius after a tunnel effect or a dead cat coming back to life) are not of course reproducible and therefore, again, their probability has no scientific meaning. Borel went even as far as stating as an "Axiom Zero" of probability calculus that events with too small a probability should be considered as never occurring.

I do not wish to conclude hastily on the fascinating ontological questions we are now discovering with the new data and theoretical results on the foundations of quantum mechanics. Consider how classical concepts are found to emerge from quantum laws; how common sense stands out as a special case of quantum logic, when applied to macroscopic objects and beings; how causality and locality are found as standard consequences of quanta, although they are not universal principles; how very small probabilities would seem to turn reality into an appearance,

[18] E. Borel (1937, 1941).

and nevertheless are so small that unreality does not matter; how quantum theory agrees with everything observed, except for a uniqueness that is the very essence of reality; how we must therefore reconsider the meaning of the Cartesian project in which all of Nature is supposed to be mathematically expressed; what are then the consequences for the ontological status of mathematics; what could be changed and what should remain if a breakthrough occurred on the frontiers of quantum theory and, most probably, of general relativity. We need bold and careful philosophers for helping us to see more clearly and surely through that wonderful maze. But it cannot be done by cooking again the old meal in old pans.

References

Aspect, A., Grangier, P., & Roger, G. (1981). Experimental tests of realistic local theories via Bell's theorem. *Physical Review Letters, 47*, 460–463.

Bell, J. S. (1964). On the Einstein Podolsky Rosen paradox. *Physics, 1*, 195–200.

Birkhoff, G., & J. von Neumann. (1936). The logic of quantum mechanics. *The Annals of Mathematics*, 2nd Series, *37*, 823–843.

Borel, E. (1937). *Valeur pratique et philosophie des probabilités*. Paris: Gauthier-Villars.

Borel, E. (1941). *Le jeu, la chance et eta théories scientifiques modernes*. Paris: Gallimard.

Brune, M., et al. (1996). Observing the progressive decoherence of the "meter" in a quantum measurement. *Physical Review Letters, 77*, 4887.

Caldeira, A. O., & Leggett, A. J. (1983). Path integral approach to quantum Brownian motion. *Physica A, 121*, 587.

Clarke, J., Cleland, A. N., Devoret, M., Esteve, D., & Martinis, J. M. (1988). Quantum mechanics of a macroscopic variable: The phase difference of a Josephson junction. *Science, 239*(4843), 992–997.

Gell-Mann, M., & Hartle, J. B. (1991). Quantum mechanics in the light of quantum cosmology. In W. H. Zurek (Ed.), *Complexity, entropy and the physics of information*. Redwood City: Addison-Wesley.

Griffiths, R. G. (1984). Consistent histories and the interpretation of quantum mechanics. *Journal of Statistical Physics, 36*, 219–272.

Hepp, K. (1974). The classical limit for quantum mechanical correlation functions. *Communications in Mathematical Physics, 35*, 265–277.

Joos, E., & Zeh, H. D. (1985). The emergence of classical properties through interaction with the environment. *Zeitschrift für Physik B Condensed Matter, 59*, 223–243.

Omnès, R. (1989). Logical reformulation of quantum mechanics. IV. Projectors in semiclassical physics. *Journal of Statistical Physics, 57*, 357.

Omnès, R. (1997a). General theory of the decoherence effect in quantum mechanics. *Physical Review A, 56*, 3383.

Omnès, R. (1997b). Quantum-classical correspondence using projection operators. *Journal of Mathematical Physics, 38*, 697.

Omnès, R. (1999). *Understanding quantum mechanics*. Princeton: Princeton University Press.

Zurek, W. H. (1982). Environment-induced superselection rules. *Physical Review D, 26*, 1862.

Chapter 8
Bohmian Mechanics and Its Ontological Commitments

James T. Cushing

Abstract One of the putative lessons from quantum mechanics is that the mathematical structure of that theory and empirical evidence demand that we accept a view of our physical world in which fundamental physical processes at the microlevel are irreducibly and ineliminably indeterministic and even that there cannot exist an objective, observer-independent reality (or "truth of the matter"). This is certainly a world view that is consonant with the standard, or "Copenhagen", interpretation of quantum mechanics, often associated with some of the founding fathers of quantum theory, such as Niels Bohr, Max Born and Werner Heisenberg. I first substantiate this representation of the Copenhagen interpretation by examining typical claims made by these founders and succinctly summarize those positions. I then argue that this common acceptance of the necessity of indeterminism is unfounded, since there exists an alternative version of quantum mechanics, one due to David Bohm, that can be in principle empirically indistinguishable from standard quantum mechanics. Moreover, in Bohmian mechanics (BM), fundamental physical processes at the microlevel are irreducibly and ineliminably deterministic and there exists an objective, observer-independent reality. While this alternative formulation of quantum mechanics **does** allow one to have an ontology that is much closer to that of classical physics than is usually associated with quantum phenomena, it does at the same time raise foundational questions about the status of the special theory of relativity and about the ontology of spacetime.

James T. Cushing was deceased on 2002.
Some of the mathematical and historical illustrations used in this paper were also used in my presentation "The Quantum-Mechanical World View: Deterministic or Indeterministic?" at the David Bohm Symposium held in São Paulo, Brazil, September 21–25, 1998.

J. T. Cushing (✉)
Department of Physics and The John J. Reilly Center, Notre Dame University, Notre Dame, IN, USA

A. Cordero (ed.), *Philosophers Look at Quantum Mechanics*,
Synthese Library 406, https://doi.org/10.1007/978-3-030-15659-6_8

8.1 Introduction

This paper is on ontology, specifically on the ontological import of quantum mechanics (QM). It is conventional wisdom that the virtually complete empirical success of quantum mechanics necessitates our acceptance of an inherent and ineliminable indeterminism, or "chance", in physical processes at the most fundamental level. I see my role in this paper as providing a voice of dissent from such a position. The reason is simple. If this alleged indeterminism were indeed unavoidable, then it would eliminate certain possibilities for the ontological structure of our world. If, on the other hand, it is still in principle legitimate to hold, say, a deterministic view of that world, then the arena of physically possible ontologies is much larger and more interesting. In fact, I shall conclude this paper with the perhaps surprising, but nevertheless defensible, challenge that an insight from quantum mechanics might be that, in a sense, the φύσις ("physics", as "nature" or "physical order") of Aristotle, as opposed to that of Newton, might be quite acceptable after all.

But, first, I must set the stage by outlining the standard, or "Copenhagen", version of quantum mechanics, in order to be able meaningfully to contrast it with a stunningly different formulation of quantum theory.

Even before that, though, two important disclaimers remain to be made at the outset. There is a bit of "truth in advertising" to be done. I am, by training and early professional research interests, a theoretical physicist. However, for about two decades now I have been working in the history and philosophy of modern physics, usually as they relate to the foundations of quantum mechanics. I now kibitz at several tables—physics, history and philosophy. The danger, of course, is that I am now an expert in none of these areas. Therefore, if you disagree with me on something you know about, then beware of what sounds good to you in an area of your ignorance! Furthermore, even though I shall be making a case for the viability of David Bohm's (1952) formulation of quantum mechanics—something I shall henceforth term "Bohmian mechanics" (BM)—I do *not* want to be seen as a defender of that theory as being correct or true. It is, for me, *one* story about how the world might *possibly* be the way that we observe it. Let me paraphrase Quine's graphic and compelling dictum that the world first intrudes as a surface irritation and remains thereafter as a constraint on our imaginations (in constructing scientific theories).[1] For me, that is the status of a successful scientific theory. Such a theory fulfills the function of providing us with a comprehensible, or understandable, (more or less) physically visualizable account of phenomena. "Truth" I would not know about. So, now my cards are on the table.

[1]The sense of this "dictum" is, it seems to me, a central claim of Quine's ***Word and Object*** (1960). This particular sentence is my own recollection of a statement made by Quine during a public lecture at Wittenberg University in late April, 1992.

8.2 A Putative Indeterminism: "Copenhagen"

Let me begin by indicating in thumbnail sketch what I take to be the central tenets of the "orthodox" or "Copenhagen" version of (nonrelativistic) quantum mechanics. There are many, not always mutually compatible, representations of quantum mechanics that are often collectively referred to as the "Copenhagen interpretation." However, in spite of their differences, these orthodox formulations do share some common core commitments. Here I confine myself simply to an illustration of some of these essential tenets and allude to their origins by considering a few specific quotations by the acknowledged founders of orthodox quantum theory. Please bear with me for a few minutes while I go over these to make a point.

For example, in Niels Bohr's early analyses of quantum theory, around the time of his seminal 1927 Como lecture, we read:

> In contrast to ordinary mechanics, the new quantum mechanics does not deal with a space-time description of the motion of atomic particles. ⋯ The difficulties [in constructing pictures of interactions between atoms and radiation] seem to require just that renunciation of mechanical models in space and time which is so characteristic a feature in the new quantum mechanics.⋯[2]
>
> [The quantum] postulate implies a renunciation as regards the causal space-time co-ordination of atomic processes.[3]

Indeed, only by a conscious resignation of our usual demands for visualization and causality was it possible to make Planck's discovery fruitful in explaining the properties of the elements on the basis of our knowledge of the building stones of atoms.[4]

Werner Heisenberg, in his retrospective recollections of these early days, tells us that:

> It should be emphasized, however, that the probability function does not in itself represent a course of events in the course of time, it represents a tendency for events and our knowledge of events. The probability function can be connected with reality only if one essential condition is fulfilled: if a new measurement is made to determine a certain property of the system.[5]
>
> ...
>
> [T]here is no orbit in the ordinary sense.[6]
>
> ...

[2]Bohr 1934, 48–51.

[3]*Ibid.*, 53

[4]*Ibid.*, 108.

[5]Heisenberg 1958, 46.

[6]*Ibid.*, 48.

> [T]he idea of an objective real world whose smallest parts exist objectively in the same sense as stones or trees exist, independently of whether or not we observe them ··· is impossible?[7]

And Max Born weighs in with his own views on the role of causality in physics:

> [I]n quantum theory it is the *principle of causality,* or more accurately that of *determinism,* which must be dropped and replaced by something else. ··· We now have a *new form* of the law of causality. ··· It is as follows: if in a certain process the initial conditions are determined as accurately as the uncertainty relations permit, then the probabilities of all possible subsequent states are governed by exact laws.[8]

...

> [N]o concealed parameters can be introduced with the help of which the indeterministic description could be transformed into a deterministic one. Hence if a future theory should be deterministic, it cannot be a modification of the present one but must be essentially different. How this could be possible without sacrificing a whole treasure of well established results I leave to the determinists to worry about.[9]

Support for similar beliefs can also be found in John von Neumann's work:

> It is therefore not, as is often assumed, a question of a re-interpretation of quantum mechanics, — the present system of quantum mechanics would have to be objectively false, in order that another description of the elementary processes than the statistical one be possible.[10]

My purpose here has not been to set up straw-man positions, but to make an observation. Notice that there is a hallmark shared by all of these claims. While they are actually *consistency* statements about one possible way the world *could* be, they are presented with an air of finality, as though they constitute *impossibility* proofs or arguments that alternative views are logically and in principle untenable.[11] In effect, the Copenhagen school *defined* its interpretation to be true and strengthened its hold on physics, rewriting history so that Einstein, de Broglie and Schrödinger largely fade from view, thus leaving Copenhagen as the only intelligible version of quantum mechanics.[12] Recently the appearance in *Physics Today* of Mara Beller's article on the Sokal hoax and the subsequent voluminous exchange it has triggered on the Internet show that confusion and high passions still abound on these issues.

In spite such pronouncements by the Copenhagen school, Albert Einstein and Erwin Schrödinger remained incredulous at such certitude on these matters.

---

[7] *Ibid.*, 129.

[8] Born 1951, 155, 163–164. (Emphases in original).

[9] Born 1949, 109.

[10] Von Neumann 1955, 325.

[11] Cushing 1994, 26, 112; Beller 1998, 1999.

[12] Heilbron 1988, 219.

> I am, in fact, firmly convinced that the essentially statistical character of contemporary quantum theory is solely to be ascribed to the fact that this (theory) operates with an incomplete description of physical systems.[13]
>
> ...
>
> Bohr's ··· approach to atomic problems ··· is really remarkable. He is completely convinced that any understanding in the usual sense of the word is impossible. (Erwin Schrödinger in a letter to Wilhelm Wien on 10/21/26)[14]

These positions that I have attributed to Bohr, Heisenberg and Born are not merely the products of an earlier era, but are echoed in modern textbooks on quantum mechanics:

> It is clear that [the results of the double-slit experiment] can in no way be reconciled with the idea that electrons move in paths. ···In quantum mechanics there is no such concept as the path of a particle.[15]

Let me characterize these central conceptual commitments of standard versions of quantum mechanics as:

1. No description of quantum phenomena is possible in terms of particle trajectories in a space- time background.
2. Probability enters into quantum theory in an ineliminable and essentially non-classical way (*i.e.*, probabilities are not merely epistemic).
3. It is in principle impossible to give a deterministic account of quantum phenomena (at the level of the events or observables themselves).

I now turn to a discussion of *how* it is in fact possible for there to be two radically different, and in many ways diametrically opposed, world views or ontologies, both of which are consistent with the familiar predictions of quantum mechanics.

8.3 Formalism Versus Interpretation

A scientific theory can be seen as having two distinct components: its formalism and its interpretation. These are conceptually separable, even if they are often entangled in practice. To simplify matters, my remarks will be restricted to theories in modern physics. Here a formalism means a set of equations and a set of calculational rules for making predictions that can be compared with experiment.[16] We shall

[13]Einstein 1949, 666.

[14]Quoted in Moore (1989, 228).

[15]Landau and Lifshitz 1977, 2.

[16]Of course, the correspondence rules between the mathematical symbols that appear in a theory (*e.g.,* the momentum operator $-i\hbar\nabla$ in quantum mechanics) and the physical observables in the world (the momentum $\boldsymbol{p}$ in my example) constitute an interpretation in a sense. However, it is not

see that both standard quantum mechanics and Bohm's version[17] use the same set of rules for predicting the values of observables. The (physical) interpretation refers to what the theory tells us about the underlying structure of these phenomena (*i.e.*, the corresponding story about the furniture of the world—an ontology). Hence, *one* formalism with *two* different interpretations counts as *two* different theories. Such a use of the terms 'formalism' and 'interpretation' is similar to and consistent with, even if a bit technically less explicit than, what is typically done in historical/philosophical analyses of quantum theory. An interpretation is formulated after an only partial examination of a formalism, since one never exhausts *all* of the implications of a (mathematical) formalism.

In briefest outline, the rules and postulates that are usually employed in making quantum-mechanical calculations are the following.

(i) a State vector ***(e.g., Ψ)*** —a vector (in a Hilbert space ***H)*** representing the state of the physical system
(ii) a dynamical equation (*e.g.*, the Schrödinger equation)

$$H\Psi = i\hbar\frac{\partial\Psi}{\partial t},$$

giving the time evolution of the State vector $\boldsymbol{\Psi}$ under the influence of the Hamiltonian H for the physical system

(iii) a correspondence between (Hermitian) operators A on H and physical observables a. These physical observables a can take only the eigenvalues a_j where

$$A\Psi_j = a_j\Psi_j$$

(iv) average values for a series of observations of a given as $\langle\boldsymbol{\Psi}\,|A|\boldsymbol{\Psi}\rangle$
(v) a projection postulate (either explicitly or effectively assumed) upon measurement

$$\Psi = \sum_k a_k\Psi_k \rightarrow \Psi_J$$

Now let me illustrate how a gloss, very different from the standard one, can be put on this formalism.

these correspondences (which I bracket with the formalism) that I am concerned with in discussing various interpretations of the formalism of quantum mechanics.

[17]Bohm 1952.

8.4 A Counter Example: Bohmian Mechanics

Here I cannot go into detail about David Bohm's 1952 program.[18] However, since there may be readers who are not familiar with Bohmian mechanics (BM), please indulge me while I display what I take to be the essential features of that theory, at least as they relate to what I have to say.

First, though, for the benefit of those who are unaware of the various facets of David Bohm's research interests throughout his life and who may have an image of Bohm based on a passing acquaintance with some particular piece written by him, let me point out that there were *three* David Bohms. David Bohm_1 was the true believer who wrote what has been widely acknowledged as one of the best expositions of the standard, or Copenhagen, interpretation of quantum mechanics. His classic *Quantum Theory* (1951) served as the textbook from which at least a generation of physicists learned its quantum mechanics. It went through sixteen printings and has been reissued as a paperback by Dover Publications. Then there was David Bohm_2 who, having written what is arguably one of the best and clearest presentations of orthodox quantum theory, carne to feel that he did not really understand, or, perhaps better put, was not genuinely satisfied with, this orthodoxy and who then published his alternative version of quantum theory.[19] It is this theory, and results subsequently based on it, about which I speak here. Finally, there was David Bohm_3 who wandered off into the quantum quagmire of the (for me) unfathomable implicate order.[20] It is the latter that contributed to Bohm's image as a guru. On my part I must disavow any ability to understand the "implicate order." My remarks below (on the work of Bohm_2), although perhaps radical by the standards of physics and of the philosophy of Science, will be rather unadventuresome if gauged by the new world view of the implicate order.

In his 1952 paper, Bohm showed that, by means of a mathematical transformation alone, the dynamical equations of quantum mechanics *(e.g.,* the Schrödinger equation) can be put into the form

$$\frac{d\mathbf{p}}{dt} = F = -\nabla (V + U)$$

where V is the usual classical potential and U is the so-called quantum potential defined as

$$U = -\frac{\hbar^2}{2m}\frac{\nabla^2 R}{R}, \; R = |\Psi|$$

[18]For varying perspectives on the pilot-wave program, see Dürr et al. (1992a, b), Holland (1993), Cushing (1994), Cushing et al. (1996) and Valentini (1996, 1999).

[19]Bohm 1952.

[20]Bohm 1980.

Here Ψ is the wave function and is the usual solution to the Schrödinger equation. In this interpretation, there are both a particle (with the usual position state variable $\boldsymbol{r}$ and a derived velocity $\mathbf{v}$) and an associated wave function (Ψ). The particle follows definite (if, at times, highly irregular) trajectories (in a space-time continuum), but there are (instantaneous) nonlocal influences produced by the quantum potential. Here Ψ represents the effect of the environment on the microsystem (or particle). There is no collapse of the wave function upon observation. Rather, we (effectively) discover where the particle is. We recover the Heisenberg uncertainty relations (as limitations on the accuracy of our measurements due to the effects of the quantum potential) and all of the statistical predictions of standard quantum mechanics. On this interpretation, a microsystem behaves as a classical chaotic system (of the type now so much discussed in nonlinear dynamics).

In particular, the Bohm theory is calculationally identical to the standard one provided:

(i) $\Psi = Re^{iS/\hbar}$ is a solution to the Schrödinger equation.
(ii) $\mathbf{v} = (1/m)\, \Delta\, S$ is the velocity of the particle ("guidance condition").
(iii) the precise location of a particle is not predicted or controlled but has a statistical (ensemble) distribution according to the probability density $P = |\Psi|^2$.

Notice that these three assumptions are logically independent.

We can characterize the conceptual matrix of Bohm's theory as follows. As far as an ontology is concerned, it provides a picture story in a space-time background (*i.e.,* there are at all times actually existing particles that follow definite trajectories). This theory is completely deterministic and underpins an observer-independent, objective reality. Probabilities are purely epistemic (reflecting our ignorance). The measurement process, which is inherently many-body in nature, is basically an act of discovery —There is *no* quantum-mechanical measurement problem. There is a well-defined criterion for a classical limit (basically $U = 0$),[21] so that there is no *conceptual* mismatch between the classical and quantum domains. Finally, insofar as all measurements are *ultimately position* measurements and quantum equilibrium ($P = |\Psi|^2$) obtains, Bohm's theory gives *complete* empirical equivalence with standard quantum mechanics.[22]

In passing let me mention that the existence of such a viable alternative to the "Copenhagen" version of quantum mechanics is arguably an interesting example of the underdetermination of a scientific theory by its empirical basis.[23] It is a fascinating and important question *how* it happened that the scientific community came to accept the standard, Copenhagen interpretation of quantum mechanics to the virtual exclusion of the equally as empirically adequate, and in some ways more

[21] There is a good deal more that remains to be said about this, in addition to the mere statement that $U = 0$. On this, see Cushing and Bowman (1999).

[22] See Valentini (1996) on the contingent nature of such quantum equilibrium and on the possibility of observing empirical differences from standard quantum mechanics.

[23] See Cushing (1994, especially Chapter 11) on this.

intuitively appealing, version due to Bohm. I have no room to go into that story here,[24] but let me here at least give John Bell the last word on this, since he very effectively reminded us that it was Bohm's 1952 theory that gave life to many of the "orthodox" beliefs about quantum theory by explicitly doing the impossible.

> But in 1952 I saw the impossible done. It was in papers by David Bohm. Bohm showed explicitly how parameters could indeed be introduced, into nonrelativistic wave mechanics, with the help of which the indeterministic description could be transformed into a deterministic one. More importantly, in my opinion, the subjectivity of the orthodox version, the necessary reference to the 'observer,' could be eliminated.
>
> Moreover, the essential idea was one that had been advanced already by de Broglie in 1927, in his 'pilot wave' picture.
>
> But why then had Born not told me of this 'pilot wave'? If **only** to point out what was wrong with it? Why did von Neumann not consider it? More extraordinarily, why did people go on producing 'impossibility' proofs, after 1952, and as recently as 1978? Why even Pauli, Rosenfeld, and Heisenberg, could produce no more devastating criticism of Bohm's version than to brand it as 'metaphysical' and **'ideological'**? Why is the pilot wave picture ignored in the text books? Should it not be taught, not as the only way, but as an antidote to the prevailing complacency? To show that vagueness, subjectivity, and indeterminism, are not forced on us by experimental facts, but by deliberate theoretical choice?[25]

8.5 Ontological Implications: Your Choice

So, here we stand with the Bohr-Heisenberg, or "Copenhagen", interpretation of quantum mechanics that demands a view of our physical world in which fundamental physical processes at the microlevel are irreducibly and ineliminably indeterministic. By some, it is even taken to support the position that there cannot exist an objective, observer-independent reality (or "truth of the matter"). On the other hand, the de Broglie-Bohm, or "causal", quantum theory can be in principle empirically indistinguishable from standard quantum mechanics. In it, fundamental physical processes at the micro- level are irreducibly and ineliminably absolutely deterministic and there exists an objective, observer-independent reality. These two versions of quantum mechanics are both internally logically consistent and are observationally *indistinguishable,* yet they are conceptually incompatible. Therefore, arguably we have a case of genuine underdetermination. Here, sociological factors and the philosophical predilections of the creators of what became the "Copenhagen" hegemony played essential and determinative roles in the creation and selection of one of these theories over the other.

So, ultimately, the choice between determinism and indeterminism in the fundamental laws of quantum mechanics is up to us —your decision! Either way is fine— or is it? Let's look at this a bit more. While Bohm's alternative formulation

[24] See Cushing 1994.

[25] Bell 1987, 160.

of quantum mechanics does allow one to have an ontology that is much closer to that of classical physics than is usually associated with quantum phenomena, it does at the same time raise foundational questions about the status of the special theory of relativity and about the ontology of spacetime. That is, I do not want to leave you with the impression that one is simply back to the ontology of classical physics. One form of Bell's theorem shows that determinism and locality would lead, in our actual world, to a logical contradiction like $1 > 2$.[26]

The essential point here is that there is a tension between an objective, observer-independent version of quantum theory and what are often taken to be the central tenets of the special theory of relativity (**STR**). More specifically, it is usually said that **STR** demands that all of the fundamental equations of physics be manifestly covariant *(i.e.*, that they transform in a specific way under the Lorentz transformations connecting one inertial frame to another). Of course, it is only the predictions (for relations among observables) that must be Lorentz covariant, if one is to avoid contradiction with empirical results. Now it has proven to be enormously fruitful in modern physics to require manifest Lorentz covariance of our equations, since that will insure the necessary covariance of our predictions—and these latter *are* testable experimentally. However, it is well known that there can be theories whose equations are not manifestly Lorentz covariant, yet all of whose predictions are covariant.[27] Recall that Lorentz' theory of electrons was noncovariant, yet empirically identical to Einstein's **STR** in 1905.[28] This is precisely what happens in Bohm's theory when extended to the relativistic domain or to quantum field theory.[29] Bohmian mechanics is a nonlocal theory in which there are instantaneous influences between space-like separated regions —yet there is *no* empirical conflict with **STR**! If we make such a move to **BM** in order to remove the tension (conflict?) between **QM** and **STR**, then we must reexamine the status of relativity —including the questions of a preferred frame and of nonlocality. In the process, Lorentz invariance loses its status as a fundamental symmetry of nature and becomes, instead, a contingent, approximate symmetry at the macrolevel. As Antony Valentini has argued, there might be a subquantum level at which quantum nonlocality is manifest.[30] It is only quantum equilibrium ($P = |\Psi|^2$), which obtains as a contingency, that shields the macroworld from this nonlocality and enforces Lorentz symmetry here. This, I believe, is an exciting possibility for ontology.

There are two obvious ways to go here. One, as I have just indicated, is to countenance a radically different ontology from what one is used to in modem physics. I return to this in the next section. However, one can also effectively prescind from any ontological project —somewhat in the spirit of Bohr. In fact,

[26]Cushing 1994, 193–195.

[27]I am not claiming that ***all*** noncovariant equations make covariant predictions. That would clearly be false.

[28]See, for example, Cushing (1981).

[29]See, for example, Cushing (1994, especially Section 10 4.2).

[30]See Valentini (1992, 1996, 2001).

David Mermin, one of the most creative and clearest expositors of the mysteries of quantum mechanics, has recently suggested precisely this.[31] The basic move is to claim that "Correlations have physical reality; that which they correlate does not."[32] While this is certainly a logically consistent position, I suspect that it will not have much appeal for those specifically interested in ontology —and so here I leave it at that.

8.6 Conclusions

At the beginning of this paper I claimed that an insight from quantum mechanics might be that, in a sense, the φύσις ("physics", as "nature" or "physical order") of Aristotle, as opposed to that of Newton, might be acceptable after all. This was, to my knowledge, first suggested by Antony Valentini.[33] Let me simply cash this out succinctly here as follows. One way to look at Bohmian mechanics is as a first-order theory in terms of the basic dynamical equation (or "guidance" condition) for the velocity

$$\mathbf{v} = \left({}^{1}\!/_{m}\right) \nabla s$$

This is to be contrasted with Newton's second law of motion

$$F = m\, a$$

which is a second-order equation. That is, in BM—as for Aristotle—it is the *velocity* **v** that is the central entity of dynamical interest, not the acceleration ***a***. Perhaps Aristotle was correct in demanding a *cause* for velocity or motion, rather than taking unconstrained motion to be uniform, rectilinear motion (the law of inertia). As they say, "How's them apples?"!

References

Bell, J. S. (1987). *Speakable and unspeakable in quantum mechanics*. Cambridge: Cambridge University Press.

Beller, M. (1998). The Sokal Hoax. At whom are we laughing? *Physics Today, 51*(9), 29–34.

Beller, M. (1999). *The quantum revolution as dialogue and rhetoric*. Chicago: The University of Chicago Press. (to be published).

[31] Mermin 1998.

[32] Mermin 1998, 753.

[33] See Valentini (1992, 1996, 2001).

Bohm, D. (1951). *Quantum theory*. Englewood Cliffs: Prentice-Hall.
Bohm, D. (1952). A suggested interpretation of the quantum theory in terms of 'hidden' variables, I and II. *Physical Review, 85*, 166–179–180–193.
Bohm, D. (1980). *Wholeness and the implicate order*. London: Routledge & Kegan Paul.
Bohr, N. (1934). *Atomic theory and the description of nature*. Cambridge: Cambridge University Press.
Born, M. (1949). *Natural philosophy of cause and chance*. Oxford: Clarendon Press.
Born, M. (1951). *The restless universe*. New York: Dover Publications.
Cushing, J. T. (1981). Electromagnetic mass, relativity, and the Kaufmann experiments. *American Journal of Physics, 49*, 1133–1149.
Cushing, J. T. (1994). *Quantum mechanics: Historical contingency and the Copenhagen hegemony*. Chicago: The University of Chicago Press.
Cushing, J. T., & Bowman, G. (1999). Bohmian mechanics and Chaos. In H. Brown, J. Butterfield, C. Pagonis, et al. (Eds.), *Philosophy and physics*. Cambridge: Cambridge University Press. (to be published).
Cushing, J. T., Fine, A., & Goldstein, S. (Eds.). (1996). *Bohmian mechanics and quantum theory: An appraisal*. Dordrecht: Kluwer Academic Publishers.
Dürr, D., Goldstein, S., & Zanghi, N. (1992a). Quantum equilibrium and the origin of absolute uncertainty. *Journal of Statistical Physics, 67*, 843–907.
Dürr, D., Goldstein, S., & Zanghi, N. (1992b). Quantum chaos, classical randomness, and Bohmian mechanics. *Journal of Statistical Physics, 68*, 259–270.
Einstein, A. (1949). Reply to criticisms. In P. A. Schilpp (Ed.), *Albert Einstein: Philosopher-scientist* (p. 666). La Salle: Open Court.
Heilbron, J. L. (1988). The earliest missionaries of the Copenhagen spirit. In E. Ullmann-Margalit (Ed.), *Science in reflection* (pp. 201–233). Dordrecht: Kluwer Academic Press.
Heisenberg, W. (1958). *Physics and philosophy*. New York: Harper & Row.
Holland, P. R. (1993). *The quantum theory of motion*. Cambridge: Cambridge University Press.
Landau, L. D., & Lifshitz, E. M. (1977). *Quantum mechanics: Non-relativistic theory* (3rd ed.). Oxford: Pergamon Press.
Mermin, N. D. (1998). What is quantum mechanics trying to tell us? *American Journal of Physics, 66*, 753–767.
Moore, W. (1989). *Schrödinger, life and thought*. Cambridge: Cambridge University Press.
Quine, W. V. (1960). *Word and object*. Cambridge, MA: The MIT Press.
Valentini, A. (1992). *On the pilot-wave theory of classical, quantum and subquantum physics*. Unpublished Ph. D. dissertation, ISAS – International School for Advanced Studies, Trieste, Italy.
Valentini, A. (1996). Pilot-wave theory of fields, gravitation and cosmology. In J. T. Cushing, A. Fine, & S. Goldstein (Eds.), *Bohmian mechanics and quantum theory: An appraisal* (pp. 45–66). Dordrecht: Kluwer Academic Publishers.
Valentini, A. (2001). *Pilot-wave theory: An alternative approach to modern physics*. Cambridge: Cambridge University Press.
von Neumann, J. (1955). *Mathematical foundations of quantum mechanics*. Princeton: Princeton University Press.

Chapter 9
The Nomological Interpretation of the Wave Function

Albert Solé and Carl Hoefer

Abstract Friends of the so-called nomological interpretation of the wave function claim that the wave function does not represent a physical substance, nor does it represent a property of physical things; rather, it is law-like in nature. In this paper we critically assess this claim, exploring both its motivations and its drawbacks and reviewing some of the recent debates in the literature concerning such an interpretation.

9.1 Introduction

The wave function is the essential theoretical term of quantum mechanics. It has an obvious instrumental meaning since it codifies measured properties of a system by way of the Born rule. Yet, if we consider that the wave function is not just a device to calculate measurement results, but it rather faithfully represents something out there in the world–some wave function stuff–, then, what is this stuff? In brief and using the philosophical jargon, if we are to be realists about the wave function, what is its nature? In what follows, we will examine one particular answer to this question: according to the so-called *nomological interpretation*, the wave function does not represent a physical substance, nor it is a property (of physical things) but it is rather law-like in nature: it has the status of a law.

This paper corresponds with some few amendments and incorporations to the talk the authors gave at the XII Ontology Congress in San Sebastian, Spain, the 6th of October of 2016.

A. Solé
Departament de Filosofia, Universitat de Barcelona, Barcelona, Spain
e-mail: albert.sole@ub.edu

C. Hoefer (✉)
Departament de Filosofia, Universitat de Barcelona, Barcelona, Spain

ICREA, Barcelona, Spain
e-mail: carl.hoefer@ub.edu

A. Cordero (ed.), *Philosophers Look at Quantum Mechanics*,
Synthese Library 406, https://doi.org/10.1007/978-3-030-15659-6_9

The nomological interpretation of the wave function has been mainly discussed within the context of Bohmian mechanics, a theory empirically equivalent to non-relativistic standard quantum mechanics that postulates particles with well-defined positions at all times. According to this theory, a closed system of N particles (the universe) is completely characterized by specifying both its wave function, Ψ, and $Q \equiv (Q_1, \ldots, Q_N) \in \mathbb{R}^{3N}$, where $Q_k \in \mathbb{R}^3$ is the position of the kth particle of the system in Euclidean 3-dimensional space. Mathematically, the wave function is a field defined in the so-called *configuration space* of the system, that is, it has the form $\Psi(q,t)$, where $q \equiv (q_1, \ldots, q_N) \in \mathbb{R}^{3N}$. The wave function evolves in time according to the *Schrödinger equation*, (in what follows, SE)

$$\text{SE} \qquad i\hbar\frac{\partial\Psi}{\partial t} = \hat{H}\Psi$$

where $\hat{H}$ is the Hamiltonian operator of the system. It should be noticed that, in Bohmian mechanics, this temporal evolution admits no exception. In turn, the velocity of each particle is given by the so-called *guidance equation*, (in what follows, GE)

$$\text{GE} \qquad \frac{dQ_k}{dt} = \frac{\hbar}{\mathrm{m}_k}\,\mathrm{Im}\left(\frac{\vec{\nabla}_{q_k}\Psi(Q,t)}{\Psi(Q,t)}\right)$$

where is m_k the mass of the kth particle and $\vec{\nabla}_{q_k}$ is the gradient with respect to the triple of coordinates q_k.[1] The GE tells us that the velocity of the Bohmian particles depends on the wave function field whereas the SE tells us how this field evolves in time. This situation may be regarded as reminiscent of the way the movement of classical particles depends on the classical fields they interact with. Therefore, many authors have suggested that the wave function in Bohmian mechanics represents a real physical field that guides the particles.[2] Friends of the nomological interpretation, however, deny that the wave function represents a physical substance and consider that it rather has the nature of a law: the wave function is interpreted as a parameter that defines the law of motion of the Bohmian particles. This view was originally proposed and defended by Detlef Dürr, Sheldon

[1]For particles with spin, the GE is slightly more complicated and takes the form:

$$\frac{dQ_k}{dt} = \frac{\hbar}{\mathrm{m}_k}\,\mathrm{Im}\left(\frac{\Psi^*(Q,t)\cdot\vec{\nabla}_{q_k}\Psi(Q,t)}{\Psi^*(Q,t)\cdot\Psi(Q,t)}\right)$$

where the dot represents an appropriate product between spinor wave functions.

[2]This is the view actually endorsed, for instance, by David Bohm in the seminal paper of Bohmian mechanics (see Bohm 1952). Given the peculiarities of the wave function field that we will comment on in the next section, Bohm and Hiley (1993) finally interpret the wave function (or a functional thereof) as a field of *active information*.

Goldstein and Nino Zanghì (in what follows, DGZ) and the purpose of this paper is to provide a critical assessment of it.[3] We consider that this discussion lies at a very nice intersection between physics and philosophy: of course, the nomological interpretation of the wave function is motivated by physical arguments, yet in order to assess the merits and drawbacks of the position, one needs to be clear about what a law (of nature) is and which are the characteristics of a law. And, of course, these latter questions are genuinely philosophical.

We will proceed as follows. In Sect. 9.2, we will motivate the nomological interpretation of the wave function showing the drawbacks of its most natural rival position, namely, the field interpretation of the wave function. Next, we will motivate the nomological interpretation on its own. Here we will focus on an alleged analogy between the wave function and the classical Hamiltonian put forward by DGZ. Then, in Sect. 9.3, we will turn to considering the main problems of the view, as they have been recently discussed in the literature and acknowledged by DGZ themselves. As we will see, these problems have to do with the fact that the wave function is a *time-dependent* and *contingent* solution of the Schrödinger equation. And both the time-dependence and the contingency seem to be at odds with the notion of law. Another problem consists in elucidating the relation of the wave function understood as a law with the other supposed law-like components of the theory such as the temporal evolution sanctioned by the SE. In this section we will specifically focus on the response that DGZ give to these questions. At this point, we will need to introduce the crucial distinction between the universal wave function and the conditional wave function of a given subsystem of the universe, since it should be clear that only the universal wave function is credited with a nomological status. We will see that DGZ's program involves assuming that the universal wave function is a static (and perhaps unique) solution of the so-called Wheeler de-Witt equation of quantum cosmology. However, with this move the authors go beyond non-relativistic Bohmian mechanics. In Sect. 9.4, we will explore whether one can still make sense of the idea of the wave function being nomological within the context of non-relativistic Bohmian mechanics and whether the above-mentioned problems can be overcome. Finally, in Sect. 9.5, we will introduce some notes regarding the relation of the nomological interpretation and another recently discussed interpretation of the wave function: the one that it takes it to represent dispositional properties of the Bohmian particles.

[3]The idea that the wave function has a nomological status is already mentioned in Dürr et al. (1992) but it is fully elaborated in Dürr et al. (1997) and, more recently, in Goldstein and Zanghì (2013). Here, we will closely follow these two later papers.

9.2 The Nomological Interpretation: Motivations and Positive Analogies

As we have already pointed out, the basic idea underlying the nomological interpretation is that the wave function does not represent physical stuff but it is nomological in nature. In words of the proponents of the view:

> "We propose that the wave function belongs to an altogether different category of existence than that of substantive physical entities, and that its existence is nomological rather than material. We propose, in other words, that the wave function is a component of physical law rather than of the reality described by the law." (Dürr et al. 1997, p. 33)

This quote presupposes a duality. On the one hand, we have the "substantive physical entities" that, in the case of non-relativistic Bohmian mechanics, are the particles; on the other hand, we have the "category of existence" of the nomological, this latter decreeing how the substantive physical entities move. This quote also makes clear that the wave function is not itself a law (if it were so, what this law would claim?) but it rather has to be regarded as a component or a *parameter of a law*. It is only in conjunction with the GE that the universal wave function defines a law-like constraint on the movement of the Bohmian particles:

> "[The GE] is an equation of motion, a law of motion, and the whole point of the wave function here is to provide us with the law, i.e., with the right hand side of this equation." (Goldstein and Zanghi 2013, p. 97)

9.2.1 Against the Field Interpretation of the Wave Function

As a first motivation for the nomological view one can consider the many drawbacks of the more natural rival view that, as we have mentioned before, is the idea that the wave function represents an objective physical field. Perhaps the most advertised difficulty with such view is that—as a solution of the SE—the wave function of a system is a field defined in the so-called *configuration space* of the system. But, since each point of the configuration space of a N-particle system represents a possible configuration of the N particles in 3-dimensional space, the configuration space of a N-particle system has $3N$ dimensions. Therefore, only if we consider a system of one particle, its wave function can be straightforwardly interpreted as a field defined in the ordinary 3-dimensional space of our experience. Taking into consideration that the number of baryons of the universe has been estimated to be of the order of magnitude of 10^{80}, the wave function of the universe is defined in a space that has more than 10^{80} dimensions. Now, if one then takes the universal wave function to represent a real physical field, but it is defined in configuration space, the conclusion that configuration space is a real physical space seems inescapable.[4] Despite the notorious revisionist character of this metaphysical posit, there are some that are willing to bite the bullet and defend configuration space realism. Here we have, for instance, David Albert's now famous appraisal:

> "And of course the space the wave functions live in, and (therefore) the space we live in, the space in which any realistic understanding of quantum mechanics is necessarily going to depict the history of the world as playing itself out is configuration-space. And whatever impression we have to the contrary (whatever impression we have, say, of living in a three-dimensional space, or in a four-dimensional space-time) is somehow flatly illusory." (Albert 1996, 277)

We do not assume that this is an untenable metaphysical position *per se*, yet it raises a lot of problems that—in our opinion—have not been adequately resolved despite the many efforts of configuration space realists. First, configuration space realists owe us a story explaining why a world that is really a $3N$-dimensional world nevertheless appears to us as 3-dimensional. At least, a precise account of how the macro-objects of our experience are reduced to the fundamental wave function field ontology should be provided. This is what in in the literature is usually known as *the macro-object problem*.

When it comes to Bohmian mechanics, Albert considers that there exist the wave function field together with what he dubs as the "universal particle" or the "marvelous point" both in configuration space. That is, according to Albert, in Bohmian mechanics both members of the pair (Ψ, Q) represent physical entities inhabiting configuration space. Now, even if there are some Bohmians that follow Albert in considering that the wave function represents a physical field propagating in configuration space, most of them place the Bohmian particles (understood as a plurality) in 3-dimensional space. In this latter case, the macro-object problem can perhaps be alleviated, assuming that macro-objects are composed of particles and that the former are tridimensional because the latter inhabit the 3-dimensional space. However, this "two-space" reading of Bohmian mechanics has problems of its own, since if the wave function field and the particles inhabit altogether different spaces, it is rather mysterious how the former manages to affect the latter.[5] This is what we call the *problem of communication*.

Another difficulty plaguing the idea that configuration space is the *fundamental* arena of the universe has to do with its putative symmetries. Assuming that configuration space has the $3N$-dimensional Euclidean metric, one expects it to naturally

[4]The argument in favour of the reality of configuration space is reinforced by the fact that, in non-relativistic quantum mechanics, the recourse to configuration space (or to spaces of higher dimensionality) is inevitable. By this, we mean that all the information encoded by the wave function in configuration space cannot be encoded by separable properties of points of 3-dimensional space (that is by a finite number of fields all defined in 3-dimensional space). The idea is that some information concerning the correlation of *entangled* systems cannot be represented in a separable way in 3-dimensional space.

[5]It is worth recalling that none of the $3N$ dimensions of configuration space can be identified with any of the 3 dimensions of ordinary space.

have the symmetries of the *Gal*(3*N*, 1) symmetry group, including translations, rotations and boosts along each one of its 3*N* independent axes. On the contrary, the dynamics provided by GE and SE are only covariant under the symmetry group *Gal*(3, 1), this resulting in a patent mismatch among the spatial and the dynamical symmetries if configuration space is taken as fundamental. This is problematic if one considers, along with Earman (1989), that any space-time symmetry should be a dynamical symmetry. A way out of course would be to deny that configuration space has the 3*N*-dimensional Euclidean metric and the corresponding symmetries, assuming that it is an inhomogeneous, anisotropic and highly structured space. However, this posit of unnatural structure would require further explanation.[6]

There are other objections to the view that the wave function represents a physical field that do *not* have to do with it being a field defined in configuration space. It has been noted that one can transform the wave function adding a phase of modulus one without changing the physical situation. In other words, the *quantum state*—what is physically significant—should be better regarded as an equivalence class of wave functions and, therefore, it is not a field.[7,8] We do not see this as a severe objection since one can think that it is a brute metaphysical fact that our universe instantiates a particular wave function field even if it cannot be distinguished from other fields that would have the same physical effects. Another, more serious, objection, has to do with the fact that the wave function does not transform as one would expect a physical scalar field to transform.[9] Finally, one would expect a physical field to carry energy and momentum of its own and not only to act upon the particles but to be acted on by them. But, in Bohmian mechanics, whereas the particles' trajectories depend on the wave function, this latter evolves according to the SE and regardless of the particles' movement. This absence of particle back-reaction has been rightly regarded as one of the main motivations against the field interpretation of the wave function in the context of Bohmian mechanics and as one of the main incentives of the nomological interpretation.

9.2.2 *The Analogy with the Hamiltonian*

As we have already mentioned, in order to motivate the nomological view, DGZ establish a very close analogy between the wave function in Bohmian mechanics and the Hamiltonian in classical mechanics. According to the Hamiltonian formulation

---

[6]See Wallace and Timpson (2010). For a more recent assessment of this problem, see also Rivat (2016).

[7]See, for instance, Goldstein and Zanghì (2013, p. 97, n. 2).

[8]In Bohmian mechanics, one naturally works with the wave function in the representation of positions and from this follows the prominence of configuration space. However, in other quantum theories, the quantum state is defined as a ray of Hilbert space and its interpretation as a field would be even more problematic.

[9]For an excellent and extended discussion of this point, see Rivat (2016, Section 5).

of classical mechanics, the motion of a system of N particles is given by Hamilton's equations:

$$\frac{dq_k}{dt} = \frac{\partial \mathrm{H}_{\mathrm{class}}}{\partial p_k} \quad \frac{dp_k}{dt} = -\frac{\partial \mathrm{H}_{\mathrm{class}}}{\partial q_k} \tag{9.1}$$

where $q = (q_1, \ldots, q_N) \in \mathbb{R}^{3N}$ are the coordinates for the generalized positions and $p = (p_1, \ldots, p_N) \in \mathbb{R}^{3N}$ the generalized momenta and $\mathrm{H}_{\mathrm{class}}$ is the classical Hamiltonian defined in the 6N-dimensional *phase space* of the system. DGZ note these two equations can be expressed in a compact form by:

$$\frac{d\xi}{dt} = Der\,(\mathrm{H}_{\mathrm{class}}) \tag{9.2}$$

where ξ represent the classical state variables and '*Der*' stands for a suitable derivative operator. In turn, one can rewrite the GE of Bohmian mechanics in the following form,

$$\frac{dQ}{dt} = Der'\,(\log(\Psi)) \tag{9.3}$$

where $\log(\Psi)$ is the logarithm of the wave function and, again, 'Der'' stands for a derivative operator.

There is an obvious formal resemblance between the Eqs. (9.2) and (9.3) that may suggest that the wave function (or, perhaps, its logarithm) plays in Bohmian mechanics a role that is analogous to that played by the Hamiltonian in classical mechanics. And, indeed, further scrutiny indicates that this is so. If one looks at Hamilton's Eq. (9.1), one can see that the role of the Hamiltonian is to provide the equations of motion: once the specific form of the Hamiltonian of a system is computed, plugging it into Eq. (9.1) one gets the equation of motion whose integration delivers the trajectories of the particles. In more abstract terms, one can say that the Hamiltonian defines a vector field in phase space and that this field induces a flow; the trajectories representing the possible evolution of the system are the integral curves that are tangent to the Hamiltonian field at each point. But something similar can be said of the wave function in Bohmian mechanics: once it is known, plugging it into GE, it defines a velocity vector field in configuration space and the particles' trajectories are obtained through an integration. Thus, given that the classical Hamiltonian is not regarded as representing a physical object but as encoding the law that sanctions how the physical objects evolve in time, DGZ invite us to think of the wave function in analogous terms, that is, as being nomological in nature.

The analogy with the Hamiltonian can be carried out further. We have just pointed out that the classical Hamiltonian is defined in a *phase space* that is of even greater dimension than configuration space. Yet, since the classical Hamiltonian is regarded not as a part of the physical state but as the generator of the evolution of the state, nobody complains about its dimensionality. In a similar vein, if the wave function is nomological, the fact that, mathematically, it is a field defined in the $3N$-dimensional configuration space should raise no worries. In addition to this, given that the Hamiltonian has the status of a law, one easily understands that it determines the particles' positions without there being a back reaction of the particles on the Hamiltonian. Therefore, if one advocates a similar law-like status for the wave function, the fact that it acts upon the Bohmian particles but that there is no back reaction of the particles on it should neither be considered worrisome.

The analogy with the classical Hamiltonian is fruitful in order to illuminate another important consequence of the nomological interpretation of the wave function that is often ignored in the literature but that DGZ make explicit. Notice that the Hamiltonian not only is a parameter in the laws of motion Eq. (9.1) but also codifies further nomological structure of the theory. This can be easily seen by taking into consideration how Hamiltonians are computed. Typically, one starts by considering the laws of interaction that hold among the particles, for instance, whether particles attract/repel each other according to the Universal Law of Gravitation or to Coulomb's Law. Knowing the form of these interactions, one works out the potential function and computes the Hamiltonian. Hence, the Hamiltonian can be understood as encoding the above-mentioned laws of interaction. Different theories with different laws will then correspond to different Hamiltonians.

We want to stress that an analogous consideration holds of the wave function if it is interpreted as being nomological in nature. The wave function is then regarded as a parameter in the law of motion such that, if this parameter changes, the law itself changes. Since, for reasons that will be detailed in the next section, only the universal wave function should be treated as law-like, it follows that *considering a different wave function of the universe amounts to considering a different law*. We take that this is what Goldstein and Zanghì mean with the following remark:

> "If the wave is nomological, specifying the wave function amounts to specifying the theory." (Goldstein and Zanghì 2013, p. 102)

Now, consider that Ψ is the universal wave function of the universe. The fact that this is nomological then entails that a possible world with universal wave function $\Psi' \neq \Psi$ has different laws. If physical laws constrain what is physically possible, we would have to say that a possible world with universal wave function $\Psi' \neq \Psi$ is not physically possible. We will return to this point later, since we take it that this is a rather revisionary consequence of the nomological interpretation.

9.3 The Nomological Interpretation: Negative Analogies and DGZ's Way Out

So far, so good. We have this amazing idea about the nomological nature of the wave function based on the positive analogies between the former and the classical Hamiltonian. Unfortunately, the considerations above seem to exhaust these positive analogies and, in fact, many of our most firm intuitions about laws clearly run against a nomological interpretation of the wave function. In this regard, the three most obvious difficulties faced by such an interpretation are the following:

1. The wave function typically has a *non-trivial temporal evolution* that fits poorly with it being part of a law—something that we usually assume to be immutable.
2. As a solution of the Schrödinger equation, the wave function is *contingent* upon a choice of initial conditions—a very unusual feature for a nomological object. Moreover, the contingency of the wave function seems to be clearly entrenched with physicists' regular experience in the laboratory, since they are used to preparing systems in a given quantum state, for instance, by means of a selective pre-measurement. This controllable character of the wave function is at odds with it being part of a law, since we would not say that we have a similar control over physical laws.
3. The wave function is a solution of the Schrödinger equation, which is typically regarded as a *fundamental* equation; but lawlike-parameters in fundamental laws (here, GE) are not typically, in our experience, *solutions* of *other* fundamental laws.

We can refer to these apparent difficulties for the nomological interpretation as the **problem of time-dependent laws** (1. above), the **problem of contingency** (2.) and the **problem of a hierarchy of laws** (3.).

These three problems have been recognized and addressed in the literature; we will comment on DGZ's responses to them in a moment. Yet there are other sources of disanalogy between the (Bohmian) wave function and the (classical) Hamiltonian that have gone largely unnoticed.

In classical mechanics, we have an intuition about what the Hamiltonian is and why it "encodes" the law. The Hamiltonian is, in general, the total energy function, the sum of the kinetic energy and the potential energy terms. In turn, the potential energy is closely connected with the notion of force and this latter with the notion of law. Given a system, if we know the forces at stake (how the different constituents interact), we know what its Hamiltonian is. The same does *not* apply to the wave function in quantum/Bohmian mechanics: if we know the forces that operate in a system (e.g., the Coulomb force binding the electron to the nucleus in a hydrogen atom), we still do not know what its wave function is like (indeed, there are infinite wave functions compatible with that system.)

A second thing to notice is that both in (orthodox) quantum mechanics and in Bohmian mechanics we have the Hamiltonian (understood here as an operator) actually appearing in the Schrödinger equation. The (quantum) Hamiltonian determines the temporal evolution of the wave function. This is in complete analogy with what happens in classical mechanics, since, there, the (classical) Hamiltonian determines the evolution of the classical state (the classical state representing what there is, i.e., the physical systems). Now, if we want to take the analogy further, since in classical mechanics the thing whose evolution is determined by the Hamiltonian is the state that represents the physical system, we should conclude that, in quantum mechanics, the thing whose evolution is determined by the Hamiltonian –the wave function– also represents the physical system. But this would obviously go against the nomological interpretation of the wave function. Thus, we arrive at the somehow paradoxical result that, motivated by an analogy between the wave function and the Hamiltonian we conclude in favor of the nomological interpretation of the wave function which, in turn, runs against establishing a close analogy between the quantum and the classical Hamiltonian and the roles they play in their respective theories.

9.3.1 The DGZ Response to 1. – 3.: Conditional Wave Functions to the Rescue

These difficulties can be mitigated somewhat (perhaps a lot), if we notice that at least the intuitions about the contingency and time-dependence of the wave function arising from usual experience in the laboratory concern not the *universal* wave function but the wave function attributed to specific subsystems of interest. From the standpoint of Bohmian mechanics, this latter is the so-called *conditional* or *effective* wave function of the system. Now, it is only the wave function of the universe, but not the conditional wave function of smaller subsystems, that deserves to be interpreted nomologically. In this regard, DGZ's argumentative strategy consists in showing that even assuming a static, uniquely determined universal wave function, one can nevertheless obtain conditional wave functions with the desired phenomenology, behaving according to their own Schrödinger dynamics.

To see how this can be, let us consider a subsystem of the universe, A, made up of M particles with generic configurations represented by $x \equiv (x_1, \ldots, x_M)$. Let $y \equiv (x_1, \ldots, x_L)$ be the variables for the generic configurations of the rest of the particles of the universe. A's *conditional* wave function at time t, ψ_t^{A}, is defined as follows:

$$\psi_t^{\mathrm{A}}(x) \equiv \Psi_t\,(x, Y(t)) \tag{9.4}$$

where Ψ_t is the universal wave function at t and $Y(t)$ is the actual configuration at t of the particles not included in A. Given this definition, it is easy to see that even if the universal wave function is static, the conditional wave function ψ_t^{A} can still be time-dependent since typically the configurations $Y(t)$ are so. The idea here is that, according to the GE, a static but non-spatially constant wave function of the universe still generates a current for the particles and this latter endows ψ_t^{A} with a non-trivial time dependence.

It is worth noticing, in addition, that from the very definition of a conditional wave function, it follows straightforwardly that the temporal evolution of A's particles is given in terms of A's conditional wave function in the usual Bohmian way:

$$\frac{dX_k}{dt} = \frac{\hbar}{\mathrm{m}_k}\,\mathrm{Im}\left(\frac{\vec{\nabla}_{x_k}\psi_t^{\mathrm{A}}\,(X_1, \ldots, X_M)}{\psi_t^{\mathrm{A}}\,(X_1, \ldots, X_M)}\right) \tag{9.5}$$

It is not generally the case that $\psi_t^{\rm A}$ itself obeys Schrödinger dynamics with A's own Hamiltonian. It can be shown, however, that it will be so if the conditions for $\psi_t^{\rm A}$ to be the *effective* wave function of system A are met.[10] To demonstrate this one has to assume, crucially, that the universal wave function is also a solution of the SE. Now, DGZ discuss a particular example in which a time-dependent conditional wave function that obeys its own Schrödinger equation emerges from a static wave function of the universe that does *not* obey, strictly speaking, the Schrödinger dynamics. In addition, the authors provide a complicated argument to the conclusion that this situation should be expected to occur with more generality.[11] With this, DGZ attempt to show that it is not necessary to assume that SE holds at the universal level in order to have subsystems of the universe that evolve according to the Schrödinger dynamics. The SE is therefore not regarded as fundamental but rather as an emergent, "phenomenological" dynamics.

The way in which systems' wave functions are typically controlled in the laboratory is explained from a Bohmian point of view not in terms of the contingency of the universal wave function but in terms of a contingent choice of the initial configuration of the Bohmian particles. This explanation is pretty intuitive, since systems are typically known to be in a specific quantum state because they are picked from one or another output channel after a selective pre-measurement. As each of these channels corresponds to a macroscopically different configuration, it is the physical contingency of the initial configuration (and not that of the wave function itself) that ultimately accounts for the possibility of preparing systems in a given quantum state.

As already mentioned, the upshot of DGZ's line of argumentation is showing that it is possible to preserve our intuitions, and to have subsystems with the desired Bohmian dynamics, even *if* the wave function of the universe is a static object with the required properties to be interpreted as law-like in nature. Now, DGZ attempt to motivate that this is actually the case by invoking the so-called Wheeler-De Witt equation,[12] the fundamental equation for the wave function of the universe in canonical quantum cosmology, which can be schematically represented as,

$$\mathcal{H}\Psi_{\rm WdW} = 0 \tag{9.6}$$

and whose solutions are static. (In the expression above, $\Psi_{\rm WdW}$ is the wave function; $\mathcal{H}$ represents the Hamiltonian constraint in quantized quantum gravity and it involves no explicit time-dependence; this is *not* the Hamiltonian used in ordinary non relativistic quantum mechanics nor is the $\Psi_{\rm WdW}$ appearing in Eq. (9.6) the Ψ appearing in non-relativistic Bohmian mechanics; $\Psi_{\rm WdW}$, for instance, is a function of spacetime curvature). DGZ consider that the universal wave function

[10]See Dürr et al. (1992, 860ss) for the definition of the effective wave function of a system and an argument to the conclusion that effective wave functions obey the Schrödinger dynamics.

[11]See Dürr et al. (1997, Section 13).

[12]See DeWitt (1967).

must be obtained as a solution (ideally, *the* solution) of Eq. (9.6), or a more general equation, interpreted as a sort of generalized Laplace equation that allows us to obtain the central parameter Ψ of what is regarded as the *only* genuine law of motion of Bohmian mechanics, namely, the GE.

DGZ's attempt to work out a nomological interpretation of the wave function surely constitutes a very stimulating research program that, if pursued further, might greatly contribute to our understanding of the wave function. However, when resorting to the Wheeler-de-Witt equation to motivate their claim that the universal wave function is constant in time (and, perhaps, unique given some further cosmological constraints), they are abandoning non-relativistic Bohmian mechanics. According to non-relativistic Bohmian mechanics, the universal wave function of the universe is a solution of the Schrödinger equation including the Hamiltonian of the universe. We can think of this Hamiltonian as including—at least—the kinetic energy term and the term for the potential energy due to the gravitational attraction among the Bohmian particles. As solution of this equation, the (non-relativistic) universal wave function will typically have a non-trivial temporal dependence and—if our aim is to assess the prospects of the nomological view qua interpretation of non-relativistic Bohmian mechanics—we have to see whether the time-dependence of the universal wave function is compatible with a nomological interpretation thereof.

9.4 Problems. 1. – 3.: Other Responses in the Literature

9.4.1 Problem 1.: Time Dependent Laws

If the wave function depends on time and is nomological in nature, how should we understand it? Here we have Belot's take on this:

> "We can think of a solution to the Schrödinger equation as determining a tenseless law-proposition that is temporally indexed in the sense that the sort of motion it decrees for a given configuration of Bohmian particles depends on the time at which that configuration obtains." (Belot 2012, p. 75)

Now, given that the wave function only decrees a specific motion for the Bohmian particles once inserted into the GE, clearly, one should think of Ψ, or a proposition expressing what Ψ is, not as being a law *per se,* but rather as specifying a parameter that plays an essential role in the fundamental dynamical law, GE. The temporal indexing of the laws, thus, arises in the GE. A simpler, toy example of the basic idea can be had if we imagine that the gravitational constant G in Newton's law of gravity had, instead of being a constant, been some simple function of time, e.g., $G(t) = G_0 + (G_0/8\pi) \sin(t)$. Then Newton's law would have had an explicitly time-dependent parameter in it. $\Psi(t)$ may be thought of as a time-dependent parameter in GE, analogous *in this respect* to $G(t)$.[13]

Several authors have discussed time dependency of laws, giving different diagnoses. Belot, for instance, admits that several (philosophical) accounts of law are compatible with the idea of a time-dependent nomological structure; yet he regards as impossible to digest that this possibility is forced upon as by any of our central physical theories.

> "[The problem of time dependent laws] seems to me to provide the more daunting obstacle. Most philosophical accounts of laws of nature allow that temporally-indexed laws are possible. But one is used to regarding temporal-indexing as a remote possibility—certainly not one forced upon us by any of our central physical theories. And this does much, I think, to undermine the salience of the analogy between the role of the wavefunction Ψ in the Bohmian law of motion and the role of the classical Hamiltonian H in Hamilton's equations (note H is time-independent in paradigmatic decent physical theories)." (Belot 2012, pp. 75–76).

But once it is accepted that time-dependent laws are possible, what is the force of saying that we are accustomed to regarding this possibility as "remote"? As a mere appeal to existing intuitions, this does not seem like a very strong argument.

Other authors have a more drastic diagnosis of the time-dependence issue. For instance, Suárez (2015) defends, simply, that a time-dependent law that constraints the temporal evolution of a physical object is logically inconsistent. This is how he puts it:

> "It is extremely hard to see how the law can determine—as it must for a law—the temporal evolution of the objects in its domain if the law itself is subject to constant temporal evolution. For what would it mean for the law at time *t* to prescribe a certain future state at time *t'* of some object in its domain when the law itself may be a completely different one by the time *t'*, and therefore establish a completely different prescription at that time? How can such a law be said to have any modal force? [...] If the law genuinely determines the state of the particles at any given time with nomological force, it must not itself vary in time on pain of potentially failing to determine uniquely such states, and thereby possibly incurring a contradiction." (Suárez 2015, p. 3215)

Now, we do not see any logical contradiction here if the content of the time-dependent law of temporal evolution is adequately assessed. Consider, again, the case of a system of non-relativistic Bohmian particles evolving over time. The wave function is a tenseless, time-indexed law-proposition that constrains, at each time, how the configuration of the particles obtaining at that time evolves into the next instant. At time *t*, it decrees how the configuration $Q(t)$ evolves into $Q(t + \Delta t)$.

[13]Despite both $G(t)$ and $\Psi(t)$ being time-dependent parameters in a law, there are important disanalogies between these two cases. First, the wave function may well be much more complicated than $G(t)$. Second, $G(t)$ is a spatially constant parameter that only determines the strength of the gravitational force but not its form. In the case of the wave function, however, it has a non-trivial spatial dependence and the specific form of the law of motion of the Bohmian particles cannot be grasped without knowing $\Psi(t)$. As a consequence, while it does not make sense to claim that $G(t)$ *defines* the law of gravitation, it is more plausible to consider—together with DGZ—that $\Psi(t)$ *defines* the law of motion of the Bohmian particles.

Then, at time $t' \equiv t + \Delta t$, the law (indexed at that time t'), sanctions how the state Q(t') evolves into Q($t' + \Delta t$), and so on. We do not see any contradiction in this process. Indeed, if we are lucky enough, we can even know, given the configuration Q(t), what this configuration will be at a distant time t'', taking into account how the law changes in the period [t, t'']. We can do so just by solving the right differential equation that of course will include an explicit temporal parameter in the right-hand side.

If we are convinced that a time-dependent law is not an inconsistent notion, whether or not we are ready to admit such laws in our ontology may ultimately depend on which account of laws we find most plausible. There are certainly accounts of laws that allow for time-dependent laws. This will clearly be the case for the Humean view, which can admit any kind of true statement as a law, as long as it is an axiom of the best system balancing strength simplicity, and probabilistic fit. But what about other accounts? Let us mention four others briefly. (i) On a strong necessitarian view (e.g., Bird (2007))—a view according to which the fundamental laws of nature are metaphysically or even logically necessary—it may seem that the laws could not change over time, because what is metaphysically or logically necessary is so forever, timelessly. But the conclusion does not actually follow, at least not without some further premises. Until we are told more about the metaphysical necessity of the laws, we can't rule out that it may be *metaphysically necessary that a certain time-variable mathematical formula governs the relationship between certain physical things or properties.* (ii) On a weak necessitarian view, such as the accounts of Armstrong (1983) and Dretske (1977), similar remarks apply. But since the "necessitation relations" posited as underlying the laws are avowedly contingent, i.e., can be different in different metaphysically possible worlds, it is hard to see what could rule out an explicitly time-dependent necessitation relation existing in some possible worlds. (iii) Marc Lange (2009) urges us to see the laws as sets of generalizations with maximal counterfactual stability under counterfactual antecedents that do not contradict any member of the set. As far as we can see, nothing in his account rules out a time dependent generalization being part of the set. (iv) Tim Maudlin (2007) defends a primitivist view of laws, insisting that the laws of nature cannot be given a reductive definition in terms of anything else. Nothing about this stance rules out the idea of a time-dependent law and of taking our world's specific universal wave function as a law-like parameter defining the law of motion of the Bohmian particles.

9.4.2 Problem 2: Contingency

If it is only the universal wave function that is ascribed nomological status, the reply to the contingency problem offered by DGZ seems mostly satisfactory. The wave functions that we can choose and (to some extent) control are merely effective wave functions and given that we can control (to some extent) the positions of Bohmian particles, it is unsurprising that we can choose or control the effective wave functions guiding their short-term motions under certain conditions.

That said, it remains the case that, according to non-relativistic Bohmian mechanics, the specific form of the universal wave function Ψ appears to be quite contingent, with many (probably infinitely many) possibilities compatible with the Hamiltonian for our universe. On a non-nomological understanding of the universal wave function, one can think of these possibilities as analogous to the many possible initial configurations of the particles; in fact, one may think of all the possible wave functions, $\{\Psi(t)\}$, as simply the full solutions of the Schrödinger equation given the possible initial states, $\{\Psi(t_0)\}$, at the start of the universe. By contrast, once we ascribe the wave function nomological status, we drastically cut down the overall space of physical possibilities of our theory. Automatically, we make it the case that a world with the same number and types of Bohmian particles as our world, but with a different wave function is physically impossible. This amounts to a significant revision to our ordinary notions of what sorts of worlds should count as "physically possible according to quantum mechanics."

The natural mitigating reply to this inconvenience is to introduce hierarchies of physical possibility. The innermost sphere of physical possibilities would be the set of worlds sharing our universal wave function and number of particles, but having different particle configurations; the next, larger sphere of possibilities would have worlds with all the possible wave functions for worlds with the same number and type of Bohmian particles as our world; a still larger sphere of possible Bohmian worlds would be composed of all possible spheres of the first two types, for all possible numbers and types of Bohmian particles. The set of worlds we normally think of as "physically possible according to Bohmian QM" would then correspond to this largest sphere; and one should distinguish the nested spheres inside it declaring that, strictly speaking, only the innermost sphere of worlds with our wave function are physically possible. While one of us (A. Solé) considers that this move amounts to an important revision of our intuitions about what is possible according to (non-relativistic) Bohmian mechanics, the other (C. Hoefer) regards it as a harmless non-objectionable revision to our terminological conventions. With this reflection, we have already begun addressing the third problem for the nomological view.

9.4.3 Problem 3: A Hierarchy of Laws?

The idea of a hierarchy of laws—at least, one with two levels—is not unfamiliar in the philosophy of lawhood. Certain statements, which tend to be dubbed 'principle' rather than 'law', are viewed by physicists as stating higher-level constraints that any acceptable physical theory at the level of dynamics must satisfy. Some prominent examples are: the principle of Lorentz covariance (which should be satisfied at least locally by relativistic theories of matter and dynamics); principles of conservation of energy, energy-momentum or stress-energy (depending on theoretical context);

and the principle of general covariance (holding that reasonable physical theories must be formulable in coordinate-independent mathematical language). Consistent physical theories can be written down that violate any or all of these principles. But they seem to be satisfied (or satisfiable) by correct dynamical theories in our world, and physicists adopt them as constraints to be respected in the search for new and improved dynamical theories, whether quantum or non-quantum. In this sense they seem to possess—or to be *treated as though they possessed*—a higher level of modal force than that of the physical necessity possessed by the full set of true laws of nature. Sometimes philosophers speak metaphorically of these principles as "governing" the laws underneath them, although this seems to be a still-more-metaphorical use of the already-metaphorical "governing" role that is used to describe the function of laws of nature.

Lange's account of laws is especially well suited to accommodating this sort of hierarchy of laws. Physical intuition readily assents to counterfactuals such as: "Had the Lagrangian for strong-force interaction been different, it still would have been Lorentz-invariant." And we think that this sort of hierarchy of laws can be accepted by law primitivists, strong necessitarians, and weak necessitarians alike.

As Belot (2012) notes, however, the Humean best-system approach to laws does seem to be in tension with a hierarchy of laws. Lawhood is simply a matter of being an axiom in the best system for our world, period. It is not clear what sense there is to be made of singling out some axiom and claiming it to be stronger or higher-level than (some of) the others. In fact, if we look at the three examples just discussed above, it is pretty clear that they would not appear as laws in a best system for a world. They are simply true statements that characterize features of the dynamical laws ("All the laws respect energy conservation", "All the laws are expressible in generally covariant fashion"). As such, adding them as axioms in their own right to the best system would be redundant, making the system less simple with no gain in strength!

Coming back to Bohmian mechanics and the nomological view of the wave function, it is not clear that the hierarchy that is faced here is of the same kind as that discussed just above. Rather than a higher-level "law" (or principle) governing dynamical laws, we have a lawlike parameter that appears in one law (the guidance equation (GE)) being determined (or constrained) by a different law (the Schrödinger equation (SE.)). As we saw at the end of the last subsection, this relationship generates a hierarchy of increasingly more permissive senses of 'physically possible.' But it is not clear that it gives us any hierarchical relationship among the laws *per se*, namely GE and SE. It is not clearly the case that SE stands above GE, or vice-versa, just because SE constrains a parameter that appears crucially in GE.

Consider in this respect the following analogy. The Einstein field equations of General Relativity with cosmological constant are written:

$$G_{ab} + \Lambda g_{ab} = 8\pi T_{ab} \tag{9.7}$$

in units where the gravitational constant and speed of light are set to unity. Now, the cosmological constant is sometimes thought to not represent a mere free parameter, but instead to represent some sort of vacuum energy whose value should be determinable from quantum field theories. The idea does not seem to work, since the theoretical calculations from the quantum side give values many orders of magnitude too large. But imagine that it did work out. Then this would be a case of one (set of) law(s) determining a parameter that figures crucially in another law. But it would not, we feel, entail that quantum field theories are either higher or lower than General Relativity in the nomological hierarchy, though Λ itself would clearly be determined by and thus subordinate to the quantum field theories.

In the case of the nomological view of the wave function, then, we do not see a clear hierarchy of laws being presupposed, rather only the universal wave function being subordinate to the SE, just as Λ might be subordinate to quantum field theories. This seems perfectly compatible with any account of laws with which we are familiar, including the Humean best system view.

Let us mention, finally, that there is a reading of the nomological interpretation according to which the issue of the hierarchy of laws does not even arise. It may well be that this reading is the most akin to DGZ's original view. Suppose that $\Psi(t)$, a specific function of the particles' positions coordinates and time, is the actual wave function of the universe. Once inserted in the GE, we obtain the velocity of each particle as a specific function of all particles' positions and time. And this prescription for the particles' velocities can be thought as the *only* fundamental law of motion of the theory; indeed, if the wave function of the universe had been $\Psi' \neq \Psi$, then we would have had a different function for the particles' velocity, that is, we would have had a different fundamental law and, therefore, a different theory.[14] Given this reading, the fact that $\Psi(t)$ is, mathematically, a solution of some equation (i.e., the Schrödinger equation) does *not* entail that this latter equation is itself a law since the only law, as we have remarked, is a law of motion that sanctions the particles' velocities as a function of its positions (and time).

9.5 Dispositionalism About Laws and the Nomological View

In the recent literature, the nomological view of the wave function has been explored in the context of primitivism about laws, Humeanism about laws, and dispositionalism about laws. In this last section we will make a few remarks about the latter.

Dispositionalism about laws in general is a view that dethrones the laws of nature, taking them to be simply expressions or codifications of the dispositions

[14]Recall Goldstein and Zanghì's remark, already quoted, that "if the wave is nomological, specifying the wave function amounts to specifying the theory." (Goldstein and Zanghì 2013, p. 102).

possessed by existing entities and their properties. Causal powers may be involved in these dispositions; both powers to affect other existing things in certain ways, and tendencies (whether deterministic or probabilistic) to react to other things or to situations in certain ways. The view is easily understood through simple examples from classical physics. Particles or bits of matter possessing the property of *mass* have a disposition to attract other things with mass, and in turn to be attracted by other things with mass; those dispositions are captured by Newton's law of gravity. Similarly, bits of matter with a positive charge have the power to attract bits of matter with negative charge, to repel bits of matter with positive charge, and in turn to be attracted/repelled by those same bits of matter. These dispositions and powers are codified in Coulomb's law. And finally, all massive particles or bits of matter are disposed to accelerate when subject to a force, in inverse proportion to their masses and in the direction of the force vector, as is codified in Newton's 2nd Law.

In the context of Bohmian mechanics, the dispositionalist views the universal GE, with specified universal wave function Ψ, as codifying the dispositions of the world's particles to move with certain velocities. There are two ways of understanding the dispositions of Bohmian systems that have arisen in the literature:

Dispositionalism$_1$: The universal wave function and the GE jointly codify the velocity dispositions of the *totality* of the Bohmian particles in the world depending on their global configuration. This disposition is holistic, possessed by the totality as a whole, not based on or reducible to the dispositions of individual particles. It manifests itself *spontaneously* at every moment, i.e., no "trigger" is required for the manifestation. This view has been defended, for instance, by Esfeld et al. (2014).

Dispositionalism$_2$: The universal wave function and the GE jointly codify the individual dispositions of each and every Bohmian particle—dispositions to move with certain velocities depending on the positions of all the other particles in the universe. These dispositions are "triggered" by the other particles' occupying certain positions relative to the given particle. This view has been defended, for instance, by Suárez (2015).

Both varieties of dispositionalism are ways of taking the wave function to be something real, namely, a dispositional property of the whole universe in the case of *Dispositionalism$_1$* or a collection of dispositions, each attributed to a single Bohmian particle, in the case of *Dispositionalism$_2$*. We consider, however, that either of them has important drawbacks, as we argue in what follows.

First, *Dispositionalism$_1$* posits a kind of disposition that requires no trigger (and cannot possibly have one, since the bearer of the disposition exhausts physical reality). This in itself is not unheard-of, because spontaneous stochastic dispositions have been contemplated since Lucretius, whose atoms notoriously "swerved" spontaneously from time to time. And more recently, the tendency to radioactive decay has sometimes been described as a disposition whose manifestation requires no trigger. But in both these cases one is dealing with a stochastic event, and it seems natural to think that if nature contains such events, i.e., if any existing thing is disposed to behave in a truly stochastic manner, it is understandable that there need be no trigger. It seems practically built into the notion of a genuinely stochastic event. The untriggered dispositions of *Dispositionalism$_1$*, however, are perfectly deterministic, in the following strong sense: given a specification of (i) the positions of all the particles (in the same coordinates used for the expression of $\Psi(t)$) and (ii) the time t, the instantaneous velocities (manifestations) are fully determined.[15] This may or may not be seen as an awkward feature of the view. (*Dispositionalism*$_2$ does involve *non-local* triggers for each manifestation of a disposition and is subject to other concerns to be discussed below.)

Other disanalogies between Bohmian dispositions and ordinary physical dispositions may be more worrisome. Ordinary physical dispositions are typically seen as defeasible or not perfectly reliable.[16] The power of aspirin to cure headaches sometimes fails, struck matches sometimes do not light, and a positive charge will fail to attract a negative charge if the latter is fully shielded from electromagnetic fields. But the Bohmian dispositions of *Dispositionalism*$_1$ and *Dispositionalism*$_2$ are not defeasible or subject to random failure. Secondly, ordinary physical dispositions are possessable by individual objects or things, small sub-parts of the universe, independently of what other objects may exist. But on *either Dispositionalism*$_1$ or *Dispositionalism*$_2$, the velocity dispositions that the particles possess (whether as a whole or individually) are, in general, *completely different* in a world in which just one more particle exists. Finally, ordinary physical dispositions tend to be fairly easy to express in ordinary language, and if they are codifiable using mathematical formulas, those formulas tend to be fairly simple as well. But the Bohmian velocity dispositions (again, on either *Dispositionalism*$_1$ or *Dispositionalism*$_2$) cannot be described in ordinary language at all and can only be mathematically expressed by a complex scalar field defined on a space with more than 10^{80} dimensions.

This point brings us to our final concern with the dispositionalist reading of Bohmian mechanics. Bohmian dispositions cannot be described or expressed without giving the wave function and GE. By contrast, Bohmian mechanics can be fully expressed and understood without using the language of dispositions. Our concern, then, is that dispositionalism here may amount to a mere verbal gloss added to the nomological interpretation (of a non-Humean variety) of the wave function, rather than a significantly different ontology.

---

[15]This is not quite the same determinism as the determinism one normally ascribes to Bohmian mechanics. The latter can be expressed in brief like this: Given the positions of all the particles at some time t_0, and the universal wave function $\Psi(t_0)$, the full history of the universe is mathematically determined. The determinism of the dispositions described here should be expressed instead as: Given all the positions at a moment t_0, and the universal wave function $\Psi(t_0)$, the velocities of all the particles at t_0 (i.e., the manifestation of the global disposition) are determined.

[16]An exception to this is classical gravity, which is universal (both on the active and passive side) and impossible to shield or thwart. Precisely this universality is what suggested to Einstein that it might not be a force at all, which led to the geometrization of gravity in General Relativity. In General Relativity gravity is still universal and impossible to shield, if we understand it as the disposition of all massive/energetic substances to curve spacetime in the fashion prescribed by Einstein's equations.

References

Albert, D. (1996). Elementary quantum metaphysics. In J. T. Cushing, A. Fine, & S. Goldstein (Eds.), *Bohmian mechanics and quantum theory: An appraisal* (pp. 277–284). Dordrecht: Kluwer Academic Publishing.

Armstrong, D. (1983). *What is a law of nature?* Cambridge: Cambridge University Press.

Belot, G. (2012). Quantum states for primitive ontologists. *European Journal for Philosophy of Science, 2*(1), 67–83.

Bird, A. (2007). *Nature's metaphysics: Laws and properties*. Oxford: Oxford University Press.

Bohm, D. (1952). A suggested interpretation of the quantum theory in terms of 'hidden' variables I and II. *Physical Review, 85*, 166–193.

Bohm, D., & Hiley, B. J. (1993). *The undivided universe: An ontological interpretation of quantum theory*. London: Routledge & Kegan Paul.

DeWitt, B. S. (1967). Quantum theory of gravity. I. The canonical theory. *Physical Review, 160*, 1113–1148.

Dretske, F. I. (1977). Laws of nature. *Philosophy of Science, 44*(2), 248–268.

Dürr, D., Goldstein, S., & Zanghi, N. (1992). Quantum equilibrium and the origin of the origin of absolute uncertainty. *Journal of Statistical Physics, 67*, 843–907.

Dürr, D., Goldstein, S., & Zanghi, N. (1997). Bohmian mechanics and the meaning of the wave function. In R. S. Cohen, M. Horne, & J. Stachel (Eds.), *Experimental metaphysics: Quantum mechanical studies for Abner Shimony* (pp. 25–38). Berlin: Springer.

Earman, J. (1989). *World enough and space-time: Absolute vs. relational theories of space and time*. Cambridge, MA/London: MIT Press.

Esfeld, M., Lazarovici, D., Hubert, M., & Dürr, D. (2014). The ontology of Bohmian mechanics. *British Journal for the Philosophy of Science, 65*, 773–796.

Goldstein, S., & Zanghi, N. (2013). Reality and the role of the wave function in quantum theory. In A. Ney & D. Albert (Eds.), *The wavefunction: Essays on the metaphysics of quantum mechanics* (pp. 91–109). Oxford: Oxford University Press.

Lange, M. (2009). *Laws and lawmakers: Science, metaphysics, and the laws of nature*. Oxford: Oxford University Press.

Maudlin, T. (2007). *The metaphysics within physics*. New York: Oxford University Press.

Rivat, S. (2016). *On the metaphysics of quantum mechanics: Why the wave function is not a field*. Unpublished manuscript available at: http://philosophy.columbia.edu/files/philosophy/content/Sebastien_Rivat_On_the_metaphysics_of_quantum_mechanics_-_why_the_wave_function_is_not_a_field.pdf

Suárez, M. (2015). Bohmian dispositions. *Synthese, 192*, 3203–3228.

Wallace, D., & Timpson, C. (2010). Quantum mechanics on spacetime I: Spacetime state realism. *British Journal for the Philosophy of Science, 61*(4), 697–727.

Part III
Local Scientific Realism

Chapter 10
Scientific Realism Meets Metaphysics of Quantum Mechanics

Juha Saatsi

Abstract I examine the epistemological debate on scientific realism in the context of quantum mechanics (QM), focusing on the empirical underdetermination of different formulations (and interpretations) of QM. This underdetermination is unsurprising in the light of the realism debate, since much of the interpretational, metaphysical work on QM transcends those epistemic commitments of realism that cohere well with the history of science. I sketch a way of demarcating empirically idle metaphysics of QM from the empirically well-confirmed aspects of the theory in a way that withholds realist commitment to what quantum state $|\Psi>$ represents. I argue that such commitment is not required for fulfilling the ultimate realist motivation: accounting for the empirical success of QM in a way that is in tune with a broader understanding of how theoretical science progresses and latches onto reality.

10.1 Introduction

Epistemological scientific (anti-)realism has hitherto made little contact with philosophy of quantum physics. The latter mostly revolves around metaphysical controversies, recent developments of which raise a serious epistemic demarcation problem for the scientific realist: the realist needs to outline a principled way to demarcate empirically well-confirmed aspects of quantum physics from the quantum metaphysics that is a hotbed of controversy, disagreement, and (seemingly) radical speculation. Here I explore the nature of this demarcation problem and propose a way for the realist to approach it.

The scientific realism debate in general philosophy of science has a core epistemological dimension. According to realists we are justified in optimism regarding sciences' ability to represent the reality beyond observable phenomena.

J. Saatsi (✉)
School of Philosophy, University of Leeds, Leeds, UK
e-mail: J.T.Saatsi@leeds.ac.uk

A. Cordero (ed.), *Philosophers Look at Quantum Mechanics*,
Synthese Library 406, https://doi.org/10.1007/978-3-030-15659-6_10

There is much variation in *how* optimistic we should arguably be in this regard, but all realists are optimistic compared to anti-realists – empiricists and instrumentalists – who trust scientific theories mainly regarding observable matters. The principal motivation for realism comes from the empirical success of science. What primarily drives the realists, in connection with theoretical sciences in particular, is the impressive predictive and explanatory success of our best theories. Arguably realism best accounts for this success, while antirealists' complete lack of optimism about theoretical progress and the reach of theory-based knowledge fit badly with the systematic success of science in yielding impressive predictions and good explanations.

Given realists' emphasis on empirical success, one should expect this epistemological debate to *rage* in connection with quantum physics. For, on the one hand, quantum physics is one of the most successful areas of science of all time, and as such it elicits the realist intuition from empirical success as forcefully as anything in modern science: if there is anything we *should want* to be realists about, it is quantum physics.[1] On the other hand, there are well-known challenges in pinning down what quantum physics purportedly says about the unobservable reality, making it exceptionally challenging to say what realism 'about the quantum' actually amounts to. In the light of this obvious tension, one would indeed expect quantum physics to be *the* battle ground for scientific realism. Digging into the large literature on scientific realism reveals very limited discussion focused on quantum mechanics (QM), however.[2]

While surprisingly little has been said about the implications of QM to the core epistemic issues in the realism debate, a *huge* deal has been written about the metaphysical implications of quantum physics, both in relation to quantum field theory and non-relativistic quantum mechanics. From the early days of quantum physics there has been extensive investigation into metaphysical issues that naturally arise from a realist outlook: what could the world literally described by quantum theories be like? Although this metaphysical question is naturally associated with realism, it should not be identified with the epistemological issues that occupy much of the realism debate in general philosophy of science. The latter issues – the topic of this paper – largely concerns the level of optimism we are justified in having towards quantum physics as a representation of reality: In what sense (if any) are we justified in regarding QM as partially or approximately true, or (more generally) as latching onto unobservable reality? What kind of epistemic optimism about QM best coheres with the historicist anti-realist evidence regarding the pessimistic track-record of science in figuring out, with empirically highly successful theories, the fundamental nature of light, heat, gravity, and so on? What kind of epistemic optimism best

---

[1]The realist intuition about quantum mechanics is driven by countless novel predictions and explanatory achievements with respect to various distinct phenomena regarding atomic spectra, the periodic structure of elements, and the band structure of the semiconductors, to name a few.

[2]There are some notable exceptions, of course, such as Cordero (2001), Cushing (1994), Barrett (2003), Belousek (2005), and van Fraassen (1991).

coheres with the kind of underdetermination exhibited by the foundations of QM, or the possibility of there being yet further foundational variants of QM hitherto unconceived by theorists?

The last of these questions has become particularly pressing of late, as metaphysical explorations of QM have taken an increasingly radical turn. A casual survey of the blooming metaphysical literature on QM raises pressing questions about the epistemological status of the competing claims regarding the nature of the quantum state involved in various competing accounts. This is made all the more pressing by the striking lack of consensus amongst the experts: the current state of the art exhibits an unprecedented and radical underdetermination of the different world-views associated with a scientific theory that enjoys extraordinarily solid and varied empirical evidence.[3] Many philosophers are rightly alarmed by this underdetermination, because it appears to make it extraordinarily difficult to say what realism about QM amounts to.

A detailed examination of this interaction between the epistemology and the metaphysics of QM is long overdue. Here I will contribute to this task by delineating an epistemic attitude towards QM that coheres well not only with the current state of affairs regarding quantum metaphysics, but also with the anti-realist arguments from the history of science. I will outline a sense in which a realist can regard QM as more than a mere instrument for prediction, allowing for quantum theoretic understanding of various empirical phenomena. I will sketch a realist account of the empirical success of QM that demarcates empirically confirmed aspects of QM from quantum metaphysics, withholds commitment to what quantum state $|\Psi>$ represents in the world, and avoids the brunt of the underdetermination problem.

10.2 The Epistemic Demarcation Problem

Most scientific realists are naturally wary of the deeper reaches of metaphysics when it comes to delineating their epistemic commitments. While realists do not want to renounce metaphysics altogether in the way e.g. constructive empiricists do – think of van Fraassen (1980), for example – they generally acknowledge the pressure of the anti-realist arguments from either the history of science, or underdetermination, or both. Realists have toiled hard to render their epistemic commitments compatible with the features of past and present science emphasised by the anti-realists. In the light of these features it would be an obvious folly for the realist to commit to anything like the *literal* truth of any piece of current physics. More generally, realists should want to be less committal towards the more deeply metaphysical claims about the nature of reality, given the evidence that firmly points to the unreliability of theoretical reasoning regarding such claims (Laudan 1981; Stanford

[3] This lack of consensus is equally true amongst philosophers of physics and physicists themselves. For one snapshot, see Schlosshauer et al. (2013).

2006). For this reason, prominent realist positions (further discussed in Sect. 10.3) tend to radically reduce their epistemic commitments from a face-value reading of theoretical science, and only bank on what our theories say about the 'structure' of reality, for example, or about the core causal features of the unobservable world as opposed to peripheral metaphysical embellishments. (See e.g. Chakravartty 2007; Frigg and Votsis 2011; French 2014.)

Recent philosophy of QM stands in stark contrast to this broad anti-metaphysical trend in the epistemology of scientific realism. Over the past couple of decades much of the philosophical work inspired by QM has gained an increasingly deep metaphysical flavour. Various radical ideas about the fundamental nature of reality have emerged as philosophers have attempted to spell out what the world described by the different variants of QM could be like if they are taken to truly represent the unobservable world behind the appearances. Many of these ideas are not only radical, but also rather indirectly connected to the actual scientific practice of using quantum theory to predict, manipulate, and explain things. For this reason, I call them 'deeply' metaphysical.

Consider, for example, the debate about wavefunction realism. This debate is about the nature of the quantum wavefunction |Ψ> construed as a field-like feature of the world– as a literal reading of QM might suggest – was sparked by the recognition that the central posit of quantum mechanics, the wavefunction |Ψ>, can be naturally interpreted as representing a field, but only if one takes seriously a very high-dimensional 'configuration' space, quite different from the familiar (or 3+1 spacetime) that we are directly acquainted with (Ney and Albert 2013). This line of thought immediately calls for a deep metaphysical account of quantum reality, since any interpretation of QM involving a realist commitment to |Ψ> thus construed cries out for a story of how the familiar 3-space (whether as a real-but-not-fundamental space, or merely as a matter of appearances) 'emergences' from, or relates to, the very different kind of space occupied by the wavefunction (see essays in Ney and Albert 2013).

Wavefunction realism is partly motivated by a fairly literal realist reading of the theory's formalism, but it is by no means forced upon the realist. A much-discussed alternative is to demote the wavefunction to a different ontological category altogether, construing it rather as representing dynamical-cum-nomological features of a *primitive* ontology that occupies the familiar 3-dimensional space. Relegating the ontological status of |Ψ> to a law-like feature of reality avoids the need to tell a deep metaphysical story of how what we see around us relates to (or emerges from) the 'fundamental wavefunction', but in its stead it requires a commitment to some kind of 'primitive stuff' (Maudlin 2007; Allori 2013). As to the nature of such 'stuff' occupying spacetime, a broad array of alternatives has been entertained by its advocates, ranging from relatively sparse momentary flashes in spacetime, to an esoteric mass density field, to individual particles that are entirely featureless in term of their intrinsic properties, and so on (see e.g. Esfeld 2014). There is a clear sense in which such a primitive ontology is deeply metaphysical in that it is, indeed, posited as an ontological primitive, as opposed to being something that needs to be written into the theory in the interest of improved predictions or explanations of empirical phenomena. Furthermore, regarding the wavefunction as a purely dynamical-cum-nomological feature of the world is itself an interpretational move that is far from obvious from the perspective of scientific practice. Such a move can be philosophically motivated in various ways, of course, but these largely hinge on deep metaphysical issues surrounding laws of nature.[4]

There is no shortage of exciting alternatives in the quantum metaphysical marketplace. A further option is to regard the wavefunction as a representation of quantum superpositions and take the notion of quantum superposition itself at face-value, as a primitive and fundamental feature of reality. This is what the (in)famous Everettian many-worlds interpretation does. This line of thought is frequently defended by its advocates as being metaphysically light weight, introducing no further metaphysical posits or assumptions to what is to be found already in quantum physics (both in QFT and QM) pure and simple (Saunders et al. 2010; Wallace 2012a, b; Vaidman 2014). To an extent this seems right: there is no need to posit a primitive ontology, or to adopt a particular stance regarding the metaphysics of laws of nature, for instance. On the other hand, the metaphysical picture of reality painted by the many-worlds interpretation relies on a way of making sense of how effectively stable classical branches (or 'worlds') 'emerge from the quantum multiverse. The Everettian understanding of classical worlds' as quasi-independent, stable patterns of an unimaginably richly structured fundamental quantum state of the universe relies not only on important features of quantum theory itself – environment-induced decoherence, in particular – but also on a deeply metaphysical account of how we can relate our (mostly 'classical') experiences to the quantum formalism that describes the fundamental quantum multiverse.

This broad-brush run-through of the metaphysical aspects of the most central theoretical posit of QM, the wavefunction, in the most prominent 'realist interpretations' of QM highlights a couple of things relevant to scientific realism. Firstly, as I will further discuss below, interpretations of QM, when spelt out in detail required for their defence, become deeply metaphysical due to indispensably involving ideas about quantum reality, and its relationship to observable features of the world, that are far removed from the actual scientific use of quantum theory to predict and explain empirical phenomena. Secondly, assuming that the different interpretations are underwritten by variants of QM that are all sufficiently empirically adequate, the realist faces a radical underdetermination of the metaphysical alternatives.

A natural knee-jerk realist response to the deeply metaphysical claims associated with QM is an *incredulous stare*. Should we seriously regard ourselves as having discovered, by carefully reflecting on an empirically extremely well confirmed scientific theory, that tables and chairs are stable, effectively non-interacting parts of an incredibly complex quantum multiverse à la Everett.? Or that they are in some sense reducible to a fundamental wavefunction that 'lives' in an extremely high-

[4] See, e.g., Esfeld et al. (2017), Esfeld (2014), and Bhogal and Perry (2015), on the role played by Humean metaphysics of laws as a backgrop to Bohmian QM.

dimensional configuration space? Or that material objects are galaxies of relatively rare flashes associated with a sui generis dynamical collapse law that doesn't give rise to any new predictions? Such claims are all well and good as part of a metaphysical endeavour and as exploratory science, and for all we know one of them might depict the world more or less correctly. But in the light of the well-motivated anti-metaphysical trend in the epistemology of scientific realism, a realist should be very wary of any such claim as an empirically well-confirmed part of current science, falling under the realist's epistemic commitments. A realist operating with appropriate epistemic caution should rather regard such claims as belonging to some different, more speculative epistemic plane.[5]

The response of incredulous stare is partly an indication of the high epistemic stakes of the radical revisions that interpretations of QM call for with respect to our everyday image of reality. We have, of course, become quite accustomed to the idea that the features of fundamental reality revealed to us by modern physics are unfathomably unlike our 'everyday reality'. But it is still reasonable to require that the evidence in support of any proposed metaphysical image of empirical reality should be commensurate with how revisionary that image is. The realists' worry about deeply metaphysical stories about quantum reality is that they are just that: *just-so* stories, devised so as to make sense of quantum mechanics literally construed, but without all the qualities that render scientific theories well confirmed by empirical evidence. This worry is bolstered by the fact that there is serious competition for any particular metaphysical image of quantum reality, making it harder to justify the adoption of any specific alternative as being firmly a part of the scientific realist's commitments.[6]

Assuming the realist is rightly worried about any particular interpretation of QM, what epistemic attitude should she have towards it then? Should she give up realism about QM altogether in the light of the historical track record of theorists' unreliability in pinning down the metaphysics of empirically successful theories? This would, of course, give the realist game away entirely in connection with one of the empirically most successful areas of science, making it thereby also harder to maintain the realist motivation–which, recall, just turns on empirical success of science– in relation to other areas of science that deal with fundamental features of reality. Or can the realist appeal to some notion of 'approximate truth' or 'selective' realism that does not take QM at anything like its face value, but nevertheless maintains that the theory 'latches onto' reality in ways that account for its empirical success? There is significant pressure for the realist to find a way of doing this, but it is not easy, as it requires a principled criterion to demarcate justified epistemic commitments from what the realist should be inclined to view as 'metaphysical

[5]Note that none of this speaks against the rationality, meaningfulness, or purposefulness of these metaphysical ideas. The point is purely epistemological.

[6]Peter Lewis (2016: 182) aptly summarises the state of play at the end of his book length review of quantum metaphysics: "Very little can be concluded unconditionally on the basis of quantum mechanics . . . The best we can say is that not everything in our received classical worldview can be right."

hubris', as far as the empirical evidence for the theory is concerned. Delineating such a criterion is thus the prime task for scientific realists in relation to QM.

In attesting to some such demarcation criterion scientific realists resist the kind of confirmational holism that metaphysicians often appeal to (cf. Saatsi 2017b). Even if relying on confirmational holism is a way of doing metaphysics and justifying it as a rational endeavour, scientific realists should maintain that there is a more fine-grained demarcation to be done in relation to the epistemic reach of empirical evidence. The indispensability of such a demarcation can be further motivated by considering, by way of an analogy, sensible realist attitudes towards metaphysics of other scientific theories, e.g. classical mechanics or biology. Various philosophers, in the spirit of naturalistic metaphysics, have drawn deeply metaphysical conclusions from classical mechanics, for instance. According to Quinean naturalists, scientific realists should say that numbers exist, given their indispensable theoretical and explanatory role in e.g. classical mechanics (Colyvan 2015; cf. also Saatsi 2017a). According to Lewisian genuine modal realists, the modal features of classical physics can support very substantive theses in modal metaphysics (Lewis 1986; see also Timothy 2016). According to others, classical physics provides evidence for the reality of dispositions (e.g. Bigelow and Pargetter 1990). According to the Humeans, the laws of classical physics are best-system regularities (e.g. Cohen and Callender 2009). Given that classical mechanics is a hugely successful theoretical framework empirically, a scientific realist attitude towards it is very well motivated. But does the empirical success of classical mechanics suggest that we should extend scientific realist commitments to the kinds of things that metaphysicians naturally associate with this theory's ontology: e.g. numbers, dispositions, particular metaphysics of laws? Friends of confirmational holism may think so (see e.g. Ellis 2009), but most philosophers engaged in the scientific realism debate rightly worry that there is a slide to speculative metaphysics here: notwithstanding their 'naturalistic', science-driven credentials, metaphysical claims about abstracta, the ontology of laws of nature, and modality, for example, transcend the empirical evidence in a way that outstrips the kind of empirical justification that realists rely on.

Resisting confirmational holism in this way requires more than a mere assertion, of course. I have said more to this effect elsewhere, e.g. with respect to mathematical Platonism and scientific realism (Saatsi 2007, 2017b). Here I just want to stress that many scientific realists do not want to slide into committing themselves to the various posits and explanations that the best metaphysical analyses may associate with that theory. In a similar vein, many scientific realists about biological theories of the evolution of proteins, say, or speciation processes, do not want to be saddled with having to pick a metaphysical account of species, or, proteins as natural kinds. This is largely due to the fact these metaphysical analyses are simply *too indirectly connected to the empirical successes* of the relevant theories that motivate realism in the first place. I think we should follow this intuition regarding quantum metaphysics as well: the realist should not feel pressed to choose between the competing metaphysical packages, because those metaphysical accounts are too indirectly connected to the empirical successes that motivate realism about QM in the first place. I will base my realist analysis on this intuition after critically reviewing, in the next section, some realist 'recipes' that one might try to appropriate to QM.

10.3 Realist 'Recipes' to Rescue?

In response to the anti-realist challenges scientific realists have come up with various ways of demarcating the belief-worthy contents of science from what seems, in the light of the history of science in particular, rather more speculative and less trustworthy. These demarcation principles are typically given in the abstract, recipe-like, so as to be applicable to different scientific theories, more or less independently of their specific subject matter or content. Familiar monikers include 'structural realism', 'entity realism', and 'semi-realism', each of which stands for a particular recipe for extracting from a given scientific theory its belief-worthy content, so as to allow the realist to be agnostic in a principled way about the rest of the theory, which can function as a mere heuristic crutch, or as a vehicle of a pleasing (but not necessarily truth-tracking) sense of intelligibility. One might think that the right way to approach the demarcation problem in the context of QM is also a matter of first identifying and then applying the right realist recipe.

I seriously doubt this is the best way for the realist to proceed, partly due to my misgivings about the spirit of (what I have called) recipe-realism in general (Saatsi 2015b). Instead of aiming to provide an abstract recipe for extracting realist commitments from any given theory, it is better, I believe, to attend to the nature and subject matter of the theory in question and ask how that theory's empirical successes are best accounted for in a realist spirit. There is no reason whatsoever to expect the answer to not vary from one theory (or area of science) to another in substantial ways that are not well captured by any abstract recipe (without such a recipe becoming rather contentless and disjunctive, at least). Rather, we should be open to the possibility that science itself, as well as realist commitments towards it, vary in such a way that the realist is better off by providing various more local *exemplars* of the sense in which the realist wants to commit herself to a given theory latching onto unobservable reality, without reducing that sense to any general definition of 'partial', 'approximate', or 'structural' truth (Saatsi 2016). Let's now briefly consider some prominent realist recipes in relation QM more specifically.

Structural realism, as first proposed by John Worrall (1989) in connection with Fresnel's ether theory of light, relies on a distinction between a theory's structural content (or what it says about the structure of the world), on the one hand, and its non-structural content (or what it says about the nature of the world), on the other hand. Structural realism aims to capitalise on structural commonalities between different theories in order to provide a sense in which false theories – theories we struggle to view as 'approximately true' at the level of ontology – can nevertheless be taken to latch onto unobservable reality. Its advocates have suggested that structural commonalities between classical physics and quantum mechanics also fit this image, even if not as neatly as Worrall's main example does.[7] Ladyman and Ross (2007, p. 94), for example, offer simple examples of "continuity in the mathematical structure of successive scientific theories", even across "the most radical cases of theory change in science, namely the transition from classical mechanics to Special Relativity, and the transition from classical mechanics to quantum mechanics."

The transition from classical mechanics to theories of relativity is a rich area of study, which has been discussed in the realism context in detail by Barrett (2008) and Saatsi (2016). The subtle correspondence between Newtonian gravity and Einstein's general theory of relativity is where the action really is, given that the general theory is more fundamental than the special theory, and given the particularly stark ontological disparity between general relativity and Newtonian gravity. Spelling out how the latter 'approximates' the structure of general theory of relativity arguably requires ideological resources specific to this area of physics, and properly accounting for the empirical successes of the classical theory, with its radically mistaken face-value ontology of gravitational forces acting at-a-distance, involves much beyond the notion that there is 'partial continuity of mathematical structure' between the two theories. A realist's account of what makes Newtonian gravity empirically successful can ultimately have little in common with the realist's account of what makes Fresnel's ether theory empirically successful. In particular, I (for one) do not see any useful abstract characterisation of structure that furnishes a unified explanatory sense in which Newtonian gravity and Fresnel's ether theories can both be regarded as 'getting the structure right'.

How about QM then? Here the structural realists point to various well-known results that capture one or another aspect of the quantum-classical correspondence. For example, Ladyman and Ross (2007) mention Ehrenfest's theorems, and Bohr's 'correspondence principle' which requires that quantum mechanical models ought to mathematically reduce to their classical equivalents in the limit of large numbers of particles or when Planck's constant is taken to zero.[8] French (2014) additionally points to the two theories' symmetry features, such as the relationship between Poisson brackets (classical) and Moyal brackets (quantum), which is naturally captured in group-theoretic terms. All these important relationships between the classical and the quantum – and there's plenty more, cf. Landsman (2007) – no doubt have a role to play in our best scientific understanding of the quantum-classical correspondence, as well as in a realist account of her epistemic commitments

[7]I do not endorse a structuralist reading of the Fresnel-Maxwell theory shift either (Saatsi 2015a, b).

[8]Ehrenfest's theorem shows how quantum mechanical expectation values of momentum and position operators obey an equation that structurally corresponds to Newton's equations.

towards QM. But the account itself is again not reducible to the existence of such 'structural' correspondences. Rather, the account crucially involves sui generis dynamical features of QM, falling under the heading of decoherence, in particular, as I will discuss below (§10.4). Again, as we will see, the ideological resources required for a realist account of how classical physics relates to quantum physics are specific to quantum dynamics, and they involve much beyond the notion that there is a partial continuity of mathematical structures between the two theories.[9]

Let's now move to the other side of the realist spectrum, as it were, where Hacking (1982, 1983) and Cartwright (1983), amongst others, have defended a very different kind of realist recipe for delineating realist commitments. The central idea of entity realism is that realists should be committed to those (and only those) aspects of electrons, for example, that are required to account for scientists' ability to build finely-tuned 'electron spraying' instruments, such as the electron 'guns' that produce beams of polarised electrons, widely used in atomic and condensed matter physics. As Hacking's famous slogan has it, "if you can spray them, they are real." (1983: 23) Electrons are of course exactly the kind of thing that QM is used to study and understand, but Hacking regards as entirely unnecessary such high-level quantum theoretical grasp of electrons. Allegedly one simply need not appeal to a high-level theory to successfully build and operate an electron gun; all that is needed is knowledge of lower-level phenomenological causal regularities regarding electron behaviour.

There are well-known difficulties in spelling out what the entity realist is actually committed to in terms of our epistemic access to the unobservable entities that are 'sprayed' or manipulated to some empirical effect. Consider the entity realist's commitment to electrons, for instance. The idea is to capitalise on various kinds

---

[9]French (2014) furthermore takes the continuity and *enrichment* of the theories' symmetry features to signal the need to shift from (merely) epistemic structural realism (ESR) to ontological structural realism (OSR):

> But if ESR is going to [incorporate the kinds of structures that matter in QM, such as the structures encoding permutation symmetry], then it will have to take on the metaphysical consequences of this symmetry and those, I argue, lead us to abandon the notion of object, hidden or otherwise. In other words, if structural realism is to broaden its grasp and seize the kinds of structures that modern physics actually presents to us, then it is going to have to shift from ESR to OSR. (p. 19)

As far as the scientific realism debate in general philosophy of science is concerned, this shift is in tension with the epistemological motivations that led to the idea of structural realism in the first place. The degree of epistemic humility that Worrall recommended by placing the realist's commitment to mere structure (as opposed to 'nature') is quite drastic from the point of view of 'standard' realism. If we take this degree of humility to be epistemically well motivated in the first place, and if we think that the distinction between structure and nature can be sensibly drawn, then we should see it as indicating scientists' unreliability in theorising about the nature of light and the nature of all other things (ultimately) quantum mechanical. But this level of scepticism about scientists' reliability to theorise about the fundamental nature of the world would also, it seems, speak against the philosopher's reliability to figure out whether the structural features of our best theories correspond to a structuralist ontology or otherwise (see also Saatsi 2007).

of instruments that use electrons effectively as a tool to some well-controlled effect. Spintronics provides a great example of modern instrumentation of this kind, relying on scientists' ability to manipulate electrons in intricate ways with electric and magnetic fields on the basis of their electric charge and a quintessentially quantum mechanical feature of spin. But what does the entity realist's existential commitment to spin-1/2 electrons amount to? An essential part of Hacking's realist brief is his advocacy of the causal theory of reference (as developed by Putnam 1975) to underwrite the truth of the existential claim 'electrons exist'. With the causal theory of reference, the realist commitment to the referent of 'electron' does not presuppose descriptive accuracy of our current theory of electrons and spin: knowing of the existence of electrons can come apart from knowing (much) about what electrons are like.

Unsurprisingly, many commentators (e.g. Musgrave 1996) have found this difficult to stomach: what sense does the entity realist's existential claim make in the absence of corresponding commitment to our best theory of what these entities are actually like? In a broadly similar spirit, Stanford (2015) has argued that given how very thin the referential-cum-existential commitment is, antirealists can effectively agree that atoms and electrons probably exist, since all that really matters for the antirealists is whether or not we actually have some substantial knowledge of what electrons are like! I think this line of criticism undermines reference-focused realism committed to the existence of entities called 'electrons'. However, as will become clear shortly, I prefer to think of realist commitment (at least in relation to fundamental physics, such as QM and spacetime theories) in a way that does not boil down to claims regarding existence.

Entity realism is furthermore problematic, since it is not clear how the entity realist recipe *accounts* for the empirical success of QM at large. This difficulty is accentuated in the context of the metaphysics of QM. Faced with the radical divergence in the characterisation of spin, charge, and mass in Bohmian versus Everettian variants of QM, for example, the entity realist is all the more pressed to spell out her commitment to electrons. For the Bohmian it is not the case that in spintronics electrons are manipulated on the basis of their intrinsic property spin, for instance. Rather, Bohmians can regard spin entirely as a feature of the quantum wavefunction (or whatever $|\Psi>$ represents) – it is not a property instantiated by the particles, which only have positions (Brown et al. 1996; Norsen 2014). So, according to this variant of QM the 'entities' being manipulated in spintronics, for example, are not electrons, but $|\Psi>$. By contrast, the Everettian regards spin as a property of the entities which instantiate it. In this way the metaphysical underdetermination leads to radical uncertainty as to what exactly is causally 'sprayed' or manipulated, effectively deflating the realist commitment.

This is a challenge also for a more sophisticated, latter-day entity realism known as semirealism, which shifts the focus from entities to the *core causal properties* in effort to say something more substantial about the objects of realist commitment. Semirealism, as developed by Chakravartty (1998, 2007) and Egg (2012, 2016) in particular, is committed to knowledge of "causal properties that one has managed to detect" (Chakravartty 2007: 47). Semirealists contrast such 'detection' properties with 'auxiliary' properties, which are "any other putative properties attributed to particulars by theories" (ibid.), regarding which we are meant to be agnostic.

> Detection properties are connected via causal processes to our instruments and other means of detection. One generally describes these processes in terms of mathematical equations that are or can be interpreted as describing the relations of properties. [One] can thus identify detection properties as those that are required to give a minimal interpretation of these sorts of equations. (Chakravartty 2007: 48)

But what kind of 'minimal interpretation' in terms of causal detection properties can we give, for example, of the equations that predict the behaviour of a Stern-Gerlach detector, or quantum cyclotron, or a solid-state physics device in spintronics? On the face of it, it looks like our handle on spin is merely formal and mathematical, at the 'minimal' level shared by the different metaphysical interpretations of QM (cf. also Morrison 2007). Viewing spin as a 'detection property' *of* the entities involved already presupposes a layer of metaphysics unsupported by the empirical success at stake, and it is not even clear how well the causal ideology of semirealism fits the understanding of spin in e.g. Bohmian mechanics (Brown et al. 1996). One begins to worry that in order for the semirealist's epistemic commitments to be consistent with the varied landscape of quantum metaphysics, these commitments have to be *so minimal* that they do little to account for the empirical success of quantum physics.

***

The prominent realist 'recipes' reviewed in this section have been developed largely independently of the specifics of quantum theory and the metaphysical issues that challenge realism about QM in particular.[10] I will next argue that instead of any of these popular recipes we should approach the epistemic demarcation problem by asking how to best account for the empirical success of QM given its relationship to classical physics.

10.4 Accounting for the Empirical Success of QM

The underdetermination problem challenges the realist, as discussed in Sect. 10.2: arguably the realist needs to pick a specific (realist) interpretation of QM in order to express her epistemic commitments regarding the relationship between QM and reality, but she has no empirical grounds for doing so.[11] The challenge appears

[10]Ontic structural realism is a clear exception to this, but for the reasons given (cf. footnote 10) I will here focus on structural realism merely as a form of epistemic humility.

[11]Musgrave (1992) argues that a realist can appeal to general metaphysical criteria to eliminate all but one competing interpretation. In the light of the anti-metaphysical trend in the contemporary

rather pressing, for how could one claim to be a realist about a theory without being able to say what the world is like according to it? Antirealism beckons, unless one can respond to the epistemic demarcation problem in a way that does not require choosing any specific interpretation.

I think the realist can hope to provide a satisfactory response to the epistemic demarcation problem while maintaining a kind of *quietism* about the interpretational issue. The resulting position is somewhat minimal in its epistemic commitments, yet sufficiently realist in its spirit. Also, it is no more minimal than the epistemic commitments that would have been appropriate for the Newtonians or the ether theorists, for example, regarding their theories (Saatsi 2015a, 2016). The key is to identify the point at which theorising about quantum phenomena slides into deep metaphysics that goes beyond empirically justified realist commitments. Although it is difficult to pin down the exact point at which this happens, we can reflect on the general principles that determine the answer. I view the following, in particular, as hallmarks of deeply metaphysical aspects of scientific theories: (A) the inability to give rise to any predictions, and (B) the inability to support bona fide scientific explanations.

The theoretical framework of QM is hugely successful, of course, in terms of both its predictive capacity and its explanatory power with respect to various phenomena, and at minimum the realist is committed to claiming that these varied empirical successes are due to QM latching onto unobservable reality. But the realist can regard the extant attempts to spell out what $|\Psi\rangle$ represents as deeply metaphysical – thus lying outside her epistemic commitments – *because they neither generate new testable predictions nor support explanations that are bona fide scientific*. Therefore, as they stand, the realist can deem the different interpretations of $|\Psi\rangle$ as an exercise in metaphysics or exploratory science that transcends her epistemic commitments.

The realist's epistemic commitments are thus determined by what she thinks accounts for the theory's undeniable empirical successes. For sure, the realist is unable to provide a fully-fledged account of these successes in the absence of a *complete* grasp on the relationship between quantum and classical physics, which would involve both a complete quantum theoretical understanding of this inter-theoretic relationship and the role of decoherence therein. (A *complete* account of the theory's empirical success would of course involve also the correct metaphysics of $|\Psi\rangle$.) But this does not mean that the realist cannot say anything about what accounts for QM's empirical success, since she can rely on the broad outlines of an emerging scientific understanding of the relationship between quantum reality and classical appearances, and she can analyse the way in which the modal features of scientific explanations supported by QM are independent of the specific interpretations of $|\Psi\rangle$.

realism debate such general metaphysical criteria for theory-choice are difficult to motivate as a reliable source of justification, however.

Let me now elaborate on this sketch, beginning with (A). We can begin with the truism that the work on quantum metaphysics and the measurement problem by and large does not generate any new predictions. The aim of this work is rather to *make sense* of QM and to spell out what the world could be like according to this or that empirically adequate variant of the theory. This is all well and good as a foundational and metaphysical endeavour, but there is an obvious sense in which interpretational and foundational work is not responsible for QM's immense predictive and practical successes that motivate realism in the first place.[12] To the extent that the empirical successes of quantum physics can be regarded as independent from such metaphysical-cum-foundational work, the realist is justified in bracketing the fruits of that labour (as they stand) outside of her epistemic commitments.

There is a long tradition in the realism debate at large, as well as in the philosophy of QM more specifically, to think otherwise. This has been partly motivated by shortage of coherent realist interpretations of QM and lack of understanding of the quantum-classical correspondence, and partly by presuppositions about what realism about QM should amount to. In particular, it has been common presupposition that realists should be able to tell us what the nature of reality (quantum or otherwise) *is like*; that they should be able to specify what the key theoretical terms (e.g. $|\Psi>$, 'entanglement', etc.) *refer to*; that they should be able to tell what the world must be like to underwrite the theory's *approximate truth*. However, more recent developments on the epistemic side of the realism debate have driven realists – myself, at least – to forgo these kinds of commitments in reaction to the challenges from the history of science and elsewhere. The thought is that one should delineate one's realist commitments towards current science in a way that is applicable to, for example, Newtonian gravity in the day of Newton, and to Fresnel's theory of light in his day, in advance of the subsequent scientific developments that we can now (with the benefit of hindsight) employ to account for those past theories' empirical successes from our current vantage point. (See Stanford 2006, 2015; Saatsi 2015a). Such historical applicability of the realist perspective is forced upon us, lest one is to argue for some kind of exceptionalism about the epistemic standing of current fundamental physics. And arguably in the light of the history of science it is simply indefensible to maintain the traditional realist hope that our best theories reveal the nature of reality. (The structural realist intuition has been an important step in this direction, but as already indicated in Sect. 10.3, I don't think it's the best way of spelling out the realist commitments).

Let's now move on to consider (B), regarding the inability of deep metaphysics to support bona fide scientific explanations. Drawing a distinction between scientific and metaphysical explanations is subtle business, but the core idea here is quite simple: scientific explanations turn on counterfactual information that by scientific lights is regarded as justified by empirical evidence. By contrast, the different

[12]This is of course not to say that such work cannot *become* responsible for such successes, but this potential has no bearing on our current epistemic commitments.

interpretations of QM furnish metaphysical explanations in terms of the nature of |Ψ> and its relationship to observable matters, such that the explanatory information in question does not boil down to counterfactual information that is empirically justified by the lights of science. Drawing the distinction in these terms is motivated by recent accounts of scientific explanation, which explicitly capitalise on counterfactual information of this sort: arguably many (if not all) scientific explanations, causal and non-causal alike, involve counterfactual information that links the values of an explanans variable to the state of the explanandum so as to answer change-relating what-if-things-had-been-different questions (e.g. Woodward 2003a, b; French and Saatsi 2018; Jansson and Saatsi 2018). And arguably the different metaphysical accounts of |Ψ> do not provide further explanatory information of this sort, since they do not involve further explanans variables, such that some empirical explanandum could be regarded as depending on those variables in an empirically well-grounded way.

Metaphysical explanations supported by interpretations of QM can be distinguished from scientific explanations in epistemological terms, even if they have the same basic structure as scientific explanations. For example, Schaffer (2017) argues that metaphysical and scientific explanations share the same tripartite structure of 'source', 'principle', and 'result', where the connecting explanatory principle can be e.g. causation (in science) or grounding (in metaphysics), and the explanatory connection can be represented by structural equation models that capture how variation in the source is explanatorily connected to variation in the result.[13] Applying this unifying analysis to QM, Schaffer argues that it allows us to make sense of the wavefunction realists' metaphysical explanation of how objects and facts about 3-space are grounded in the fundamental wavefunction. This is Schaffer's response to the worry that Maudlin (2010) amongst others have voiced about the impossibility of comprehending how the fundamental wavefunction ontology can give rise to regularities in 3-space. My present point is that even if we can make sense of the nature of the quantum metaphysical explanation in these broadly modal terms, the explanation need not be regarded as involving the kind of explanatory connection for which we have good empirical evidence, and hence the realist should still deem it deeply metaphysical.

Which explanatory successes of QM should the realist aim to account for? I think the answer to this question is determined by scientists' own assessment of the various explanations that QM furnishes: the realist can take the scientific community as a (hopefully) reliable judge as to which quantum mechanical

[13] As Schaffer (2017, p. 2) explains:

> With causal explanation, there is the structure of cause (such as the rock striking the window), law (laws of nature), and effect (such as the shattering of the window). Metaphysical explanation has a parallel structure, involving ground (the more fundamental source), principle (metaphysical principles of grounding), and grounded (the less fundamental result). One finds a similar structure with logical explanation, involving premise, inference rule, and conclusion.

explanations should be regarded as undeniable successes. Healey (2015: 2) rightly observes that "the continuing failure to agree on any specific realist interpretation or reformulation [of QM] contrasts strikingly with the widespread acceptance in the scientific community of the enormous explanatory power of contemporary quantum theory", before discussing in detail accepted quantum theoretic explanations of single-particle interference phenomena, the stability of matter, and interference of Bose-Einstein condensates. It is natural for the realist to rely on scientists' own assessment of these kinds of explanatory successes, since in the present dialectic it is the explanatory success of science (as opposed to metaphysics of science) that the realist aims to account for, and realists typically furthermore argue that scientists' own assessments of explanations are a reliable guide to theoretical progress.

It is notable that in providing quantum theoretic explanations of various phenomena scientists by and large do not feel the need to appeal to any particular explication of the nature of the quantum state. Also, more specific explanations of e.g. interference phenomena that indispensably turn on specific interpretational choices can be ruled out, since they do not properly count as successes by virtue of not possessing sufficiently wide-spread scientific agreement qua actual explanations. I furthermore conjecture that the explanations that physicists largely agree upon are associated with reasonably precise and empirically well-founded counterfactual information, amenable to a counterfactual account of explanation and explanatory understanding (cf. Healey 2015). The realist can thus account for these explanatory successes in terms of QM getting the appropriate explanatory counterfactuals right, since this is what really matters for providing the explanatory information, and this can be achieved even when the theory we are operating with is only in some sense a limited 'approximation' to a better theory we don't yet have (and may never have). Whatever the theory says about the world beyond those counterfactuals is supererogatory with respect to accounting for its explanatory success. In a similar way a minimal realist can capture the explanatory successes of Newtonian mechanics and gravity, for example. The posit of gravitational force, acting at a distance, or Newtonian absolute simultaneity, are not involved in accounting for the explanatory successes of Newtonian gravity. By the same token, these genuine explanatory successes are not undermined by the fact that there are various features of the world that the theory simply got wrong (Woodward 2003b; Bokulich 2016). It is in this same spirit that the realist can regard the metaphysical accounts of the quantum state as simply supererogatory in accounting for the explanatory successes of QM.

To summarise, the appropriate realist response to the underdetermination challenge is to insist that the underdetermination takes place at the level of deep metaphysics going beyond realist commitments. The different variants of QM, in as far as they are empirically adequate, all latch onto reality in ways that account for their empirical success. Getting a more complete handle on this account is something that will gradually take place alongside future scientific advancements (Saatsi 2016). The underdetermination problem is thus neutralised by a natural, substantial reduction in realist commitments, which is furthermore incentivised (for reasons given in Sect. 10.2) independently of the underdetermination challenge:

even without the predicament of underdetermination the issue of separating the empirical wheat from the metaphysical chaff looms large. For the sake of the argument and to illustrate, consider a possible counterfactual history where theorists only ever come up with the de Broglie-Bohm variant of QM and are unable to conceive of any serious alternatives to it. In the light of the anti-realist challenges from the history of science, the realist would face the epistemic demarcation problem even in the absence of any actual alternative underdetermined by the empirical evidence. For example, the realist should not want to commit herself to an interpretation of |Ψ> as a peculiar law of nature, say, even if the de Broglie-Bohm variant of QM seemed like the only game in town.

10.5 Is This Realism at All?

One may feel that the epistemological stance sketched above is insufficiently realist. At least a couple of potential objections immediately crop up. First, what can we be realists about, if we bracket the different interpretations of |Ψ> as 'deep metaphysics' that transcend realist commitments? Secondly, what about the notorious measurement problem: how do we respond to it if we cannot help ourselves to the resources afforded by a fully-fledged interpretation? Isn't solving the measurement problem a sine qua non for realism about QM?

Let's address the latter question first. There is, of course, a long-standing tradition to think that a realist must give an account of what the quantum state represents in the world in order to deal with the measurement problem. This line of thought goes as follows. The standard ('textbook') QM, which incorporates the collapse postulate, is not amenable to a realist attitude towards the dynamics of the theory, given the irreducible role played by the notion of measurement as yielding determinate observable measurement outcomes. The upshot, then, is that the orthodox QM, unvarnished with a 'realist interpretation', is best regarded as a mere instrument or recipe for making predictions. Avoiding such blatant anti-realism about QM – as the realist desires – thus requires articulating and defending a variant of QM that does not involve the problematic collapse postulate inconsistent with the unitary quantum dynamics. That is, it requires articulating and defending a variant of QM amenable to a realist interpretation.

This standard story is too black-and-white, however, from the perspective of the kinds of fairly minimal and unambitious epistemological stances that many (e.g. structural) realists have adopted towards fundamental physics in general. Doing without the collapse postulate, and defending a particular realist interpretation of QM, are not one and the same thing. There are degrees of epistemological commitment that fall between adhering to the 'orthodox' QM with the collapse postulate, on the one hand, and committing to one or another variant that does without it, on the other. Since the collapse postulate drops out of the picture in all (current) variants of QM seriously entertained by the realists, and since it arguably plays less of a role in the physicists' actual (more interpretation-independent) use of QM than the above line of thought suggests, it is natural to consider what can be said of the relationship between classical and quantum physics independently of any 'realist interpretation'.[14]

It is particularly noteworthy that the unitary quantum dynamics by itself gives rise to environment-induced decoherence that is at the heart of many physicists' own understanding of the relationship between quantum and classical physics, in a way that is independent of any particular variant of QM (Schlosshauer 2007; Wallace 2012a, b). Decoherence does not 'solve' the measurement problem in and of itself, of course, because it does not answer the metaphysical question of what $|\Psi>$ represents. An answer to the metaphysical question is a 'necessary coda' (as Rosaler 2016 puts it) to any decoherence-based account of how (approximately) classical dynamical and kinematical structures are compatible with a fundamentally quantum reality. But, as Rosaler (2016) forcefully argues, one can say a good deal about classical-quantum correspondence even without the interpretational coda. More specifically, Rosaler argues for the potential for combining technical, foundational understanding of (i) decoherence, (ii) Ehrenfest's Theorem for open quantum systems, and (iii) a decoherence-compatible mechanism for collapse, in providing a local interpretation-neutral reduction between particular models' of quantum and classical theories. Such a foundational programme points to the kind of interpretation-independent account of the empirical success of quantum mechanics that a realist like myself is committed to being there to be fully worked out as a part of future science. Although metaphysical issues concerning effective 'wave function collapse' and the ontology underpinning a scientifically kosher reductionist account is an ineliminable part of a fully-fledged account of quantum-classical correspondence, Rosaler shows how these concerns can be effectively decoupled from "the bulk of technical analysis necessary to recover localised, approximately Newtonian trajectories from quantum theory" (p. 54). Correspondingly, in defending a realist attitude towards QM one does not need to solve the measurement problem – to provide the interpretational coda – since the interpretation-neutral part of the analysis is enough to support the realist belief that the theory's empirical success is due to, and can be accounted for, in terms of it latching onto the unobservable reality in appropriate ways. As Rosaler (2016: 59) puts it:

> [O]ne can go quite far in providing a quantum-mechanical account of classical behavior without taking on the speculative commitments associated with some particular interpretation of quantum theory. Of course, we must also keep in mind that at most one of these interpretation-specific accounts can be correct as a description of the collapse mechanism that nature itself employs.

What does it take to account for a theory's empirical success exactly? This is an important question that requires further analysis. I will limit myself here to noting a couple of complicating issues. For one, the realist's optimism about a

[14]Wallace (2019) argues against philosophers' commonplace idea that collapse (or 'projection') postulate is central to 'orthodox' or 'standard' QM that physicists employ in practical applications.

theory should be compatible with the possibility that we can more fully account for its empirical success only with the benefit of hindsight furnished by a currently unavailable successor theory that advances on our present science. (Else, the history of science contains powerful cases against realism). But even in the absence of such future science the realist can commit to optimism about there existing such an account, and one that we can hopefully give in due course. That is, the realist can express confidence in the fact the theory relates to reality in objective ways that are responsible to its success. This kind of attitude towards QM is clearly different from instrumentalism or empiricism; hence I associate it with the realist tradition. Secondly, there is a difference between a realist account of empirical success, which can be given in scientifically kosher terms that do not transcend the reach of available empirical evidence, and a complete account of empirical success, which can only be given from a (scientifically chimerical) omniscient point of view, involving also deep metaphysics of reality.

One may be inclined to associate more lofty ambitions with 'scientific realism', of course. For example, one may think that a scientific realist attitude towards a theory must entail *knowledge claims* about what kinds of things are real; what there is; what our theoretical terms refer to (see e.g. Stanford 2015). Admittedly, by those lights the optimistic epistemic stance I have sketched does not qualify as realism, given that this stance indeed does not defend realism about the quantum wavefunction, or spin, or quantum particles, in anything like the way that standard 'convergent realism' does regarding a theory's central posits (cf. Laudan 1981). If one is strongly inclined to stipulate that 'scientific realism' must entail such commitment, a new label is needed for the kind of optimism that I have argued for. ('Theory-progressivism' perhaps?) As I see it, this optimism should be directed towards a more abstract sense in which we are justified in regarding QM as latching onto unobservable reality in ways that drive the theory's empirical success, both predictive and explanatory. This latching is a matter of the theory's central kinematic and dynamical aspects representing the world's kinematical and dynamical structures sufficiently faithfully, in appropriate respects, along the lines studied by e.g. Rosaler (2016) and Landsman (2007).

Acknowledgments Support from Arts and Humanities Research Council (as part of Scientific Realism and the Quantum project) is gratefully acknowledged. I received helpful feedback on presentations in Leeds, San Sebastian, NYU Abu Dhabi. Special thanks to Fabio Ceravolo, Kevin Coffey, Steven French, Simon Newey, Kohei Morita for helpful discussions.

References

Allori, V. (2013). Primitive ontology and the structure of fundamental physical theories. In D. Z. Albert & A. Ney (Eds.), *The wave function: Essays in the metaphysics of quantum mechanics* (pp. 58–75). Oxford: Oxford University Press.

Barrett, J. A. (2003). Are our best physical theories (probably and/or approximately) true. *Philosophy of Science, 70*(5), 1206–1218.

Barrett, J. A. (2008). Approximate truth and descriptive nesting. *Erkenntnis, 68*(2), 213–224.
Belousek, D. W. (2005). Underdetermination, realism, and theory appraisal: An epistemological reflection on quantum mechanics. *Foundations of Physics, 35*(4), 669–695.
Bhogal, H., & Perry, Z. (2015). What the humean should say about entanglement: What the humean should say about entanglement. *Noûs*, 1–21.
Bigelow, J., & Pargetter, R. (1990). *Science and necessity*. Cambridge: Cambridge University Press.
Bokulich, A. (2016). Fiction as a vehicle for truth: Moving beyond the ontic conception. *The Monist, 99*(3), 260–279.
Brown, H., Elby, A., & Weingard, R. (1996). Cause and effect in the pilotwave interpretation of quantum mechanics. In J. T. Cushing et al. (Eds.), *Bohmian mechanics and quantum theory: An appraisal* (pp. 309–319). Dordrecht: Kluwer.
Cartwright, N. (1983). *How the laws of physics lie*. Oxford: Oxford University Press.
Chakravartty, A. (1998). Semirealism. *Studies in History and Philosophy of Science Part A, 29*, 391–408.
Chakravartty, A. (2007). *A metaphysics for scientific realism: Knowing the unobservable*. Cambridge: Cambridge University Press.
Cohen, J., & Callender, C. (2009). A better best system account of lawhood. *Philosophical Studies, 145*(1), 1–34.
Colyvan, M. (2015). *Indispensability arguments in the philosophy of mathematics*. Stanford Encyclopedia of Philosophy.
Cordero, A. (2001). Realism and underdetermination: Some clues from the practices-up. *Philosophy of Science, 68*, S301–S312.
Cushing, J. T. (1994). *Quantum mechanics: Historical contingency and the Copenhagen hegemony*. Chicago: University of Chicago Press.
Egg, M. (2012). Causal warrant for realism about particle physics. *Journal of General Philosophy of Science, 43*(2), 259–280.
Egg, M. (2016). Expanding our grasp: Causal knowledge and the problem of unconceived alternatives. *British Journal for the Philosophy of Science, 67*(1), 115–141.
Ellis, B. (2009). *Metaphysics of scientific realism*. London: Acumen.
Esfeld, M. (2014). The primitive ontology of quantum physics: Guidelines for an assessment of the proposals. *Studies in History and Philosophy of Modern Physics, 47*, 99–106.
Esfeld, M., Lazarovici, D., Lam, V., & Hubert, M. (2017). The physics and metaphysics of primitive stuff. *The British Journal for the Philosophy of Science, 68*, 133–161.
French, S. (2014). *The structure of the world: Metaphysics and representation*. Oxford: Oxford University Press.
French, S., & Saatsi, J. (2018). Symmetries and explanatory dependencies in physics. In A. Reutlinger & J. Saatsi (Eds.), *Explanation beyond causation: Philosophical perspectives on non-causal explanations*. Oxford: Oxford University Press.
Frigg, R., & Votsis, I. (2011). Everything you always wanted to know about structural realism but were afraid to ask. *European Journal of Philosophy of Science, 1*(2), 227–276.
Hacking, I. (1982). Experimentation and scientific realism. *Philosophical Topics, 13*, 71–87.
Hacking, I. (1983). *Representing and intervening: Introductory topics in the philosophy of natural science*. Cambridge: Cambridge University Press.
Healey, R. (2015). How quantum theory helps us explain. *The British Journal for the Philosophy of Science, 66*, 1–43.
Jansson, L., & Saatsi, J. (2018). Explanatory abstractions. *British Journal for the Philosophy of Science*. Advance Access. axy026.
Ladyman, J., & Ross, D. (2007). *Everything must go: Metaphysics naturalised*. Oxford: Oxford University Press.
Landsman, N. P. (2007). Between classical and quantum. In J. Butterfield & J. Earman (Eds.), *Handbook of the philosophy of science: Philosophy of physics* (pp. 417–554). Amsterdam: Elsevier.
Laudan, L. (1981). A confutation of convergent realism. *Philosophy of Science, 48*(1), 19–49.

Lewis, D. (1986). *On the plurality of worlds*. Oxford: Basil Blackwell.
Lewis, P. (2016). *Quantum ontology: A guide to the metaphysics of quantum mechanics*. Oxford: Oxford University Press.
Maudlin, T. W. (2007). Completeness, supervenience and ontology. *Journal of Physics A: Mathematical and Theoretical, 40*(12), 3151.
Maudlin, T. (2010). Can the world be only wavefunction? In S. Saunders, J. Barrett, A. Kent, & D. Wallace (Eds.), *Many worlds? Everett, quantum theory, & reality* (pp. 121–143). Oxford: Oxford University Press.
Morrison, M. (2007). Spin: All is not what it seems. *Studies in History and Philosophy of Modern Physics, 38*(3), 529–557.
Musgrave, A. (1992). Realism about what. *Philosophy of Science, 59*(4), 691–697.
Musgrave, A. (1996). Realism, truth and objectivity. In R. Cohen, R. Hilpinen, & Q. Renzong (Eds.), *Realism and anti-realism in the philosophy of science* (pp. 19–44). Dordrecht: Kluwer.
Ney, A., & Albert, D. Z. (2013). *The wave function: Essays on the metaphysics of quantum mechanics*. Oxford: Oxford University Press.
Norsen, T. (2014). The pilot-wave perspective on spin. *American Journal of Physics, 82*, 337–348.
Putnam, H. (1975). *Philosophical papers: Volume 2, mind, language and reality* (Vol. Volume 2). Cambridge: Cambridge University Press.
Rosaler, J. (2016). Interpretation neutrality in the classical domain of quantum theory. *Studies in History and Philosophy of Modern Physics, 53*, 54–72.
Saatsi, J. (2007). Living in harmony: Nominalism and the explanationist argument for realism. *International Studies in the Philosophy of Science, 21*(1), 19–33.
Saatsi, J. (2015a). Historical inductions, old and new. *Synthese*. First Online: doi:https://doi.org/10.1007/s11229-015-0855-5.
Saatsi, J. (2015b). Replacing recipe realism. *Synthese, 194*(9), 3233–3244.
Saatsi, J. (2016). What is theoretical progress of science. *Synthese*, First Online.
Saatsi, J. (2017a). Dynamical systems theory and explanatory indispensability. *Philosophy of Science, 84*(5), 892–904.
Saatsi, J. (2017b). Explanation and explanationism in science and metaphysics. In M. Slater & Z. Yudell (Eds.), *Metaphysics and the philosophy of science: New essays* (pp. 162–191). Oxford: Oxford University Press.
Saunders, S., Barrett, J., Kent, A., & Wallace, D. (Eds.). (2010). *Many worlds? Everett, quantum theory, & reality*. Oxford: Oxford University Press.
Schaffer, J. (2017). The ground between the gaps. *Philosophers' Imprint, 17*(11), 1–26.
Schlosshauer, M. (2007). *Decoherence and the quantum-to-classical transition*. Berlin: Springer.
Schlosshauer, M., Kofler, J., & Zeilinger, A. (2013). A snapshot of foundational attitudes toward quantum mechanics. *Studies in History and Philosophy of Modern Physics, 44*(3), 222–230.
Stanford, P. K. (2006). *Exceeding Our Grasp: Science, History, and the Problem of Unconceived Alternatives*. Oxford: Oxford University Press.
Stanford, P. K. (2015). "Atoms exist" is probably true, and other facts that should not comfort scientific realists. *Journal of Philosophy, 112*(8), 397–416.
Timothy, W. (2016). Modal science. *Canadian Journal of Philosophy, 46*(4–5), 453–492.
Vaidman, L. (2014). Many-worlds interpretation of quantum mechanics. *Stanford Encyclopedia of Philosophy*.
van Fraassen, B. C. (1980). *The Scientific Image*. New York: Oxford University Press.
van Fraassen, B. C. (1991). *Quantum Mechanics: An Empiricist View*. Oxford: Oxford University Press.
Wallace, D. (2012a). Decoherence and its role in the modern measurement problem. *Philosophical Transactions of the Royal Society A, 370*(1975), 4576–4593.
Wallace, D. (2012b). *The Emergent Multiverse: Quantum Theory according to the Everett Interpretation*. Oxford: Oxford University Press.
Wallace, D. (2019). *What is orthodox quantum mechanics?* This volume.

Woodward, J. (2003a). Experimentation, causal inference, and instrumental realism. In H. Radder (Ed.), *The Philosophy of Scientific Experimentation* (pp. 87–118). Pittsburgh: University of Pittsburgh Press.
Woodward, J. (2003b). *Making things happen: A causal theory of explanation*. Oxford: Oxford University Press.
Worrall, J. (1989). Structural realism: The best of both worlds? *Dialectica, 43*, 99–124.

Chapter 11
Structural Realism and the Standard Model

Steven French

Abstract The Standard Model of elementary particle physics is one of the best theories that we currently have and thereby invites realist engagement. Adopting a realist stance towards it involves careful consideration of the nature of the symmetries that it incorporates. Here I begin with such a consideration and argue that it leads us to a form of structural realism that, following Cassirer might be called 'Parmenidean'. I conclude with some thoughts on how this meshes with 'local' forms of realism.

11.1 Introduction

The claim that our 'best' theories 'latch onto' the world lies at the heart of scientific realism. Standardly 'best' here is rendered as not only empirically adequate and explanatorily powerful but also as capable of providing novel predictions (that are then confirmed). And 'latching onto' has been standardly understood in terms of the relevant linguistic terms in the theory referring to certain entities in the world and the theory itself rendered as true or 'approximately' so (however that is then cashed out). Alternatively, if one has qualms about this insertion of the philosophy of language into the philosophy of science one might prefer to talk of the theory 'faithfully' representing the world, drawing on recent work on scientific representation. However we decide to cash out these notions, it surely cannot be denied that the so-called 'Standard Model' of elementary particle physics is currently the best theory in this area that we currently have and one that many take to 'latch onto' the world in the above respects. Some might rest content with asserting that the relevant terms of the model refer, or that the model as a whole faithfully represents the relevant systems and that, as a result, it can be regarded as approximately or partially or quasi-true. Others may wish to press on and articulate

S. French (✉)
School of Philosophy, Religion and History of Science, University of Leeds, Leeds, UK
e-mail: s.r.d.french@leeds.ac.uk

A. Cordero (ed.), *Philosophers Look at Quantum Mechanics*,
Synthese Library 406, https://doi.org/10.1007/978-3-030-15659-6_11

a conception of how the world is, according to the Standard Model. It is the latter articulation that I shall be concerned with here.

11.2 The Standard Model

The Standard Model famously, encompasses the electromagnetic, the weak nuclear and the strong nuclear interactions and classifies all known elementary particles. Crucially, it embraces certain kinds of symmetries, including Permutation Invariance, that is associated with the so-called indistinguishability of quantum particles and is represented by the permutation group (see French and Krause 2006), the global Poincaré symmetry that all relativistic quantum field theories incorporate plus the local SU(3) x SU(2) x U(1) gauge symmetry that effectively characterises the model and covers the above three fundamental interactions. The first of these symmetries divides up the relevant state space into non-combining sectors, each corresponding to a certain fundamental kind of particle, the two most well-known being fermions, which obey Fermi-Dirac statistics and bosons, obeying Bose-Einstein statistics. Thus, the most fundamental kinds into which elementary particles can be divided, effectively 'drop out' of the imposition of this symmetry. Furthermore, as Wigner famously showed, the second symmetry – that of Minkowski spacetime—yields a classification of all elementary particles in terms of their mass and spin. Hence these fundamental properties can also be said to 'drop out' of this particular symmetry.

Finally, gauge symmetry refers to the way in which the Lagrangian of a system – which basically captures the system's dynamics – remains invariant under a group of transformations, where the 'gauge' aspect denotes certain redundant degrees of freedom of that Lagrangian. The generator of this group of transformations represents a field and when such a field is quantised, we get certain gauge bosons that 'carry' the interaction. Thus, in the case of electrodynamics, the relevant gauge symmetry group associated with the property of charge is U(1) and the requirement of gauge invariance yields the photon. Thus, particles like the photon also 'drop out' of the imposition of this symmetry.

How this 'dropping out' of kinds and properties and particles is to be captured in terms of some philosophical framework for explanation remains to be discussed (see French and Saatsi 2018) but certainly one must acknowledge that it is here we see some of the explanatory force of the Standard Model. As for novel predictions, consider the famous case of the ω^-particle: it was observed that the nine spin 3/2 baryons then known 'fitted' into the 10-dimensional representation of the SU(3) group (see Bangu 2012, p. 81). Given that the tenth node was formally similar to the other nine, there was good reason to expect that it too could be interpreted as representing another spin 3/2 baryon. The subsequent discovery of this particle thereby helped to establish the Standard Model as our current 'best' theory of elementary particle physics.

11.3 The Challenge for the Realist

Given the status of the Standard Model and given the role of symmetry principles within it, as sketched above, it is incumbent upon the realist to say something about this role. Lamentably, very little has been presented on this in the realist context. With some notable exceptions, most realists still seem content to hash over old debates about the ether or caloric, rarely if ever extending their epistemic considerations to quantum physics, much less the standard model (here I'm deliberately not including the extensive literature regarding different 'realist' *interpretations* of quantum mechanics). And where the realist dips her toes into the deeper waters of naturalized metaphysics, there tends to be a focus on laws and their supposed governance, with little if anything said about symmetry principles and the role they play. Given their prominence in twentieth century fundamental physics, this is an astonishing oversight.

Perhaps it is thought that, given the extensive literature on laws, there is no need to say anything particular about symmetries, since accounts of the former can simply be extended to the latter. Unfortunately, however, neither of the more prominent analyses of laws appear capable of accommodating symmetries. Consider the dispositionalist account: Bird famously spelled out how, from the 'stimulus and manifestation' characterization that is central to the dispositionalist project, we are able to recover, apparently the relevant laws (Bird 2007). But, at the very least, it remains utterly unclear how this can be extended to symmetry principles such as the above (see Psillos 2006; Lange 2012; and for further discussion in this specific context see Cei and French 2014, French forthcoming). Indeed, dispositionalism seems to get the order of dependence the wrong way round and seems incapable of accommodating the way in which kinds, properties and gauge bosons 'drop out' of the relevant principles, as indicated above.

The alternative, 'Humean' account fares no better. Here the underlying metaphysics is that of a 'mosaic' of 'perfectly natural' properties instantiated at space-time points. This mosaic exhibits certain regularities and those that meet certain criteria (traditionally articulated in terms of simplicity and informativeness) are represented in our 'best' system and thereby deemed to be laws (see Cohen and Callender 2009). However, little, if anything, has been said on how the Humean might accommodate the above symmetry principles. An obvious move would be to understand them, metaphysically, to be 'meta-regularities' that in a sense span the ordinary 'lawlike' regularities of the mosaic. But again, this is to begin, metaphysically, with the properties and to build up the regularities (laws) and meta regularities (symmetries) from those—talk of the former 'dropping out' of the latter is then going to have to be dismissed as question begging. That's not an unreasonable move, perhaps, but it does reveal how the Humean is not going to be able to take what physics seems to tell us literally and is going to have to engage in some revisionary manouevres.

Neither of these concerns are insurmountable but it seems to me that if the realist is going to stake her stance to our 'best' theories and if she is going to say something about such metaphysical matters, then she should pay close attention to what those theories appear to be telling us. And in the case of the Standard Model, what this is telling us is that symmetries play a fundamental role in our understanding of what the world is like. Fortunately, there is a realist position that can accommodate that! As well known as it now is, let me approach it *via* a somewhat different route than the usual.

11.4 Symmetries and Laws

Let us consider the relationship between symmetry principles and laws in a little more detail, beginning with the connection between such principles and conservation laws. The standard view is that (given certain conditions) these two come as a package: each space-time symmetry entails a conservation law (within the Lagrangian framework) and each conservation law entails a space-time symmetry (given a dynamical law such as Hamilton's principle). Thus, according to Brown and Holland (2004) the two should simply be regarded as correlated, with neither to be regarded as conceptually or explanatorily more fundamental than the other. Indeed, Noether's famous first theorem, which establishes the connection between global symmetries and conservation laws, can also be proved in reverse (Brown and Holland, op. cit., pp. 1137–1138). Her second theorem focuses on local symmetries and considers the different status of the conservation laws when the global symmetry group is a subgroup of some local symmetry group of the theory in question (see Brading and Brown, 2003; Brading and Castellani 2008).

However, Lange (2007, 2009) has argued that symmetry principles have explanatory priority over conservation laws, insisting that all that such proofs show is that Noether's Theorem is irrelevant when it comes to accounting for the explanation of conservation laws by symmetry principles, as suggested by the long history of giving such explanations *prior* to the establishment of Noether's result. Now, of course, if x and y are correlated, there may be all sorts of reasons why one would begin with x and use it to obtain y, rather than the other way round: heuristic reasons spring to mind most obviously (and of course, symmetry principles have famously been deployed in such a heuristic capacity; see Post 1971), but it may be that epistemically x is more accessible than y (although that may not be the case here) or for broadly 'ideological' reasons to do with what one takes to be relatively more fundamental. And Noether's results could be seen as a corrective to these previous, historical moves, effectively revealing that although it was earlier assumed that symmetry principles had priority, this is in fact incorrect—they are 'on a par' with conservation laws.

Let us look at this in more detail. Both conservation laws and symmetry principles can be regarded as either 'by products' of laws or requirements imposed upon them. Taking a conservation law to be a 'by product' means that the law in question is a logical consequence of the relevant dynamical law (such as Newton's Second Law), together with the relevant force laws, plus a closure requirement to the effect that there are no other forces apart from those described in the aforementioned force laws (Lange op. cit., pp. 466–467). As such a conservation law is still a law, possessed of a necessity that distinguishes it (or rather its statement) from accidental generalisations, but its holding is simply a result of there happening to be no forces that fail to conserve the relevant quantity. By contrast, if the conservation law is understood as a requirement, then it is no coincidence that the relevant quantity is conserved, since the given conservation law explains why only those interactions in which the quantity is conserved are permitted. This distinction can be cashed out in modal terms: 'if a given conservation law is a requirement that the force laws must satisfy, then the conservation law would still have held even if the universe had been populated by different forces.' (ibid., p. 467)

Thus we obtain four obvious combinations:

CL (by) + Symm (by)
CL (by) + Symm (req)
CL (req) + Symm (by)
CL (req) + Symm (req)

However, Lange insists, if symmetries are regarded as by products that would render them too 'weak' to explain conservation laws as requirements and hence CL (req) + Symm (by) must be ruled out. Furthermore, taking CL (by) + Symm (by) implies that neither should be taken as more fundamental than the other. However, if a symmetry principle or conservation law is understood as a requirement, it has to be taken as dictating what laws and kinds of forces there could be. In that case, 'if "the real physics" includes such a symmetry principle or conservation law, then (contrary to Brown and Holland) not all of the real physics is in laws like the fundamental dynamical law and the force laws.' (ibid., p. 470) Now, symmetry principles and conservation laws that are requirements add the obtaining of certain counterfactual conditionals to the first order laws, just as a law's lawhood goes beyond its truth in asserting that the law would still have held under certain counterfactual circumstances. It is in these terms that Lange argues that symmetry principles as requirements have explanatory priority over conservation laws as requirements (pp. 473–474).

So, on Lange's account, laws can be distinguished from 'accidental' generalizations in virtue of possessing 'counterfactual stability'. The idea is that lawlike generalisation remain true under logically independent counterfactual circumstances that are accidental. If we call those propositions that do not contain the phrase 'it is a law that' or any modal operator, 'sub-nomic' propositions, then the set of all such propositions can be defined as stable if the members of the set remain true under every sub nomic supposition consistent with the set. A generalisation is then regarded as lawful if and only if it belongs to the largest non-maximal stable set of true propositions. Putting it simply, the necessity that distinguishes laws from accidental generalizations '... involves a kind of maximal persistence under counterfactual suppositions.' (Lange 2007, p. 472).

Now consider a non-nomically stable set from which certain force laws have been excluded: the non-nomic stability of the set requires that the conservation laws in that set would still have held even if the force laws had been different. This counterfactual holds if the conservation laws are seen as requirements but fails if they are merely by-products of what force laws there happen to be. As a requirement on these laws, the conservation laws possess a stronger variety of natural necessity. Now the difference between symmetry principles and conservation laws is that whereas the regularity expressed by a conservation law does not itself mention laws, and so a statement of that regularity is a non-nomic claim, that expressed by a symmetry principle precisely concerns laws; it is a meta-regularity. To explicate this, Lange takes the relationship between laws and the non-nomic facts they 'govern' and reproduces it at the meta-level. Here he obtains a nomic analogue to non-nomic stability in terms of which he expresses the demand that some symmetry principles (at least) would still have held had the fundamental dynamical laws been different, or had the force laws been different or had there been additional forces besides those there actually are. Crucially, the conservation laws do not join these symmetries in forming a nomically stable set; thus, had the relevant fundamental dynamical law (e.g. F = ma) been different, the symmetries would still have held but the relevant conservation laws need not have (ibid. p. 475). Hence, he writes, 'Symmetry principles as requirements possess a stronger variety of natural necessity than conservation laws as requirements do, empowering the symmetry principles to explain the conservation laws and preventing the reverse' (p. 474).

What about Noether's theorem? How can this be reconciled with the supposed priority of symmetries over conservation laws? Brading and Brown take the three theorems in total, as '... mathematical tools that enable us to explore and extract the structural properties of our theories that are associated with symmetries.' (2003, p. 90). As such, the (first) theorem can be understood as expressing a fundamental relationship such that neither symmetries nor conservation laws can be taken to be modally, and hence, in Lange's terms, explanatorily prior to the other. But what, then, of Lange's argument above? The crucial step is the requirement that symmetries *qua* meta-laws belong to a nomically stable set that excludes the relevant dynamical laws, force laws etc. It is in such terms that we can understand the claim that the symmetries *would* have held had the other laws been different. Now as he notes, such counterlegal claims might well be dismissed as inaccessible to empirical investigation. His response is to insist, first, that their counterlegality does not make them any more remote to such investigation than other counterfactuals and second, that scientists take the relevant evidence as confirming that not only actual, but as yet undiscovered laws obey a given symmetry principle or conservation law, but also that the kinds of forces that would have existed under various counterfactual suppositions do so (op. cit. p. 478).

Taking the second point first, note that what is acknowledged is that scientists take the evidence (whatever that is) as confirming obedience to a given symmetry principle or conservation law, where these are taken to be on a par. More importantly, however, Lange argues that '[f]acts about what would have been are confirmed right along with facts about what is.' (op. cit. p. 479), giving the following example: the fact that all emeralds are green confirms not only that actual but undiscovered emeralds in Brazil, say, are green but also that had there been an emerald in my pocket it would also have been green. However, it is not clear that such an example is entirely apposite in this context. Compare this with the examples from physics that Lange gives, which involve conservation of energy and the reluctance of physicists to entertain its violation (*ibid.*, p. 468). Thus consider the Bohr-Kramers-Slater theory of the emission and absorption of radiation by atoms, which implied that energy and momentum were only conserved statistically overall and not necessarily in each interaction; or Pauli's account of β-decay which retained conservation of energy – that appeared to be experimentally violated—by introducing a new particle, subsequently discovered and dubbed the neutrino.

In both these latter cases, and whatever degree of reluctance physicists may have expressed at the time, it was *experimental* evidence that was ultimately crucial in determining whether to retain the conservation law. In the case of emeralds, we have a pretty good fix on what makes an emerald an emerald and our degree of confidence in the 'fact' that all emeralds are green is so strong that we might well be inclined to agree that it not only confirms the counterfactual claim that had I an emerald in my pocket (in which case my wife would be very glad to see me!) it would have been green, but that it confirms this claim to the same degree as the claim that an undiscovered emerald in Bahia, Brazil is also green. However, do we have such a strong fix on the kinds of forces, say, that would have existed under various counterfactual suppositions? Consider: prior to 1956 we might have entertained with some degree of confidence the claim that all forces obey parity symmetry. This itself should give us pause in taking such a claim to 'confirm' further claims involving even more recondite counterfactuals than those entertained by Yang et al.

How then do we account for physicists' attitudes in these cases? The answer is that they do not take the relevant claims themselves as evidence, but rather, as Post first noted, that they take the relevant symmetries and conservation laws as heuristic principles which may serve to construct new theories, both within the same domain or in new domains (Post 1971). If such new theories are empirically successful then we may take *that* as further evidence for the universality of the relevant symmetry, as in the extension of gauge invariance to the strong nuclear force. We may then further speculate as to the structural similarities between the relevant domains that this common symmetry reveals. However, as the case of parity violation indicates, this is a fallible procedure, hence the idea that claims regarding symmetries can stand as evidence for counterfactual claims seems problematic.

Now Lange himself acknowledges the case of parity violation (op. cit., p. 470) but the surprise that scientists supposedly felt over this result (although actually it had been suspected for some years before) is taken to help legitimate the distinction between symmetries as by-product and as requirement. Understanding them in the latter sense is of course perfectly compatible with the view of symmetries as heuristic resources. Indeed, I would suggest that it is this feature of the practice of physics that motivates taking them as requirements, where this is understood as defeasible as already noted. What these cases do not do is help ground the relevant counterlegal on which Lange's assertion of their explanatory priority depends; or at the very least, it does not help ground it sufficiently strongly as to overcome the inherent symmetry of Noether's theorem. Thus we may remain unconvinced that symmetries should be taken as explanatorily *prior* to conservation laws.

What about the very notion of a symmetry principle, or conservation law, acting as a 'requirement'? For Lange this notion is explicated in terms of the nomic stability of the relevant set of claims but that, as we have just seen, involves the acceptance of certain counterlegal claims which appear to be problematic. Notice, however, that in order to get a grip on this notion, Lange does just what the dispositionalist does with regard to the necessity of laws, namely he introduces a kind of modal 'gap', such that he can effectively hold the symmetries fixed, and then entertain the (meta?) possibility of the laws being different. Here then we see a similar presuppositional move being made as in the case of the relationship between laws and objects. There we allow for the possibility of a metaphysical 'gap' between objects and laws, such that we can then articulate the issue of the necessity of the latter in terms of prospective variations across possible worlds consisting only of the former. The dispositionalist closes that gap by conceiving of the objects in dispositional terms and then showing how the laws flow from or supervene on those dispositions; since this holds in all worlds in which there are such objects, the necessity of the laws is thereby accounted for.

In the case of conservation laws, the distinction between being a by-product and being a requirement is similarly cashed out in terms of a prospective metaphysical gap between the forces and the given law: if, given a different set of forces than is realised in the actual world, we would obtain different conservation laws, then the latter are mere by-products; and if the converse, then they are requirements. Concern about the above counterlegals might lead one to suggest that opening up such a gap is problematic, or, more strongly, that on the basis of the view of physics practice sketched above, it should not be opened up in the first place.

11.5 The Parmenidean Structuralist

Indeed, as far as the structuralist is concerned, there is no gap to begin with, as once you fix the laws (and symmetries), as part of your fundamental base, you get the (putative) objects (i.e. the elementary particles). Can we make any sense of the idea of symmetries as a 'requirement' without such a gap? We might begin by thinking of such symmetries (and conservation laws, since we're using Noether's theorem in the way Brading and Brown suggest) as aspects or features of the structure of the world, along with (first order) laws. Thinking of laws and symmetries in this way allows us to distinguish them to the extent that we can now think of their inter-relationship. But even in those cases where the relationship is such that the relevant symmetry is associated with the conservation of a quantity the inter-relationship between whose instances is described by the relevant law, the fact that the symmetry can be taken to express a regularity at the (meta-)level of the laws themselves does not in itself imply any kind of priority, explanatory or otherwise.

How then might we characterise this relationship? Here we can turn to an earlier attempt to philosophically reflect on the implications of newly emerging physics, namely Cassirer's *Determinism and Indeterminism in Modern Physics* (1936). Of course, Cassirer himself was not a realist, and certainly not in the sense we understand that term today, and his book was written long before the Standard Model came into being. Nevertheless, by virtue of considering the implications of the recently developed quantum mechanics from a perspective that had already encompassed General Relativity, for example, Cassirer's work offers a framework that can be adapted here (see French 2014). Summarising, this consists in a form of 'Parmenidean' structuralism in which symmetries, laws and measurement outcomes mutually support and condition one another in a kind of 'reciprocal interweaving and bonding' (Cassirer 1936, p. 35). Thus the structure of the world is not to be conceived of in terms of a kind of pyramid, with symmetries constraining laws, which in turn determine and govern measurement outcomes; rather, it is a kind of 'well-rounded sphere' in which these three features can be conceptually distinguished but which should not be regarded as modally independent.

From this perspective, the above discussion, regarding counterlegals and symmetries acting as 'requirements', reveals a presumption of just the kind of spatial metaphor that Cassirer urged we should reject, with the symmetry principles at the top, the laws in the middle and the results of measurement at the bottom. This would suggest that one or other layer could be removed, as it were, without affecting the others, in just the kind of counterlegal move that Lange envisages and that produces the 'gap' between laws and symmetries. However, from Cassirer's perspective, this would be untenable since the truth of all such statements at whatever level is due to their interconnectedness. In these terms, there can be no such modal gap; at least not while preserving the structure of the world.

Adopting (and adapting) this framework in the context of the Standard Model (French 2014) yields a form of structural realism according to which 'the structure of the world' can be characterised in terms of this 'well-rounded' and 'inter-locking' arrangement. In a sense, then, this blurs the distinction between symmetries as 'by-products' and as 'requirements'. If the latter is understood in terms of some further modal strength that symmetries are supposed to have, then this is ungrounded in the practice of physics. On the other hand, if the symmetries can be said to constrain the laws, then they only do so by representing the interconnections between the latter but in such a way that, as Cassirer indicated, we should not conceptually imagine the symmetries as existing distinct from the laws; in that sense, they are like by-products.

11.6 Tune In and Drop Out

How, then, does this framework accommodate, or take seriously, the above talk of kinds, properties and particles 'dropping out' of the symmetries? Consider again the way that Bose-Einstein and Fermi-Dirac statistics 'drop out' of Permutation

Invariance: mathematically the relationship in focus here is that between the permutation group and two of its representations, namely the symmetric and anti-symmetric respectively (there are, again, others, corresponding to so-called parastatistics). What metaphysics can be hung on this formal relationship in order to explicate the manner in which such kinds 'drop out' of the symmetry? Here we can adopt the 'toolbox' approach (French and McKenzie 2012), according to which current metaphysics can be seen as a kind of 'toolbox' of devices, moves and manouevres that we may appropriate. What tools are available?

One such is supervenience: *x* supervenes upon *y* just in case there can be no difference in *x* without a difference in *y*. Applying that here, the particle kinds could be said to supervene on Permutation Invariance, if and only if there could be no difference with regard to the former without a difference in the latter. One way to capture this sense of 'no difference on one without a difference in the other' would be for the former to be instantiated in every possible world in which we have the latter. However, not all of the possible representations and associated kinds allowed by the permutation group are instantiated in a given world. Consider this, the actual world: as we have noted, only the symmetric and anti-symmetric representations, corresponding to bosons and fermions, are instantiated. Any of the infinite number of other representations corresponding to paraparticles of different orders, are not. As is well-known, there was a time when it was thought that quarks obeyed parastatistics but this model came to be replaced, with a new property, known as 'colour', introduced to account for the apparently anomalous quark statistics (French 1995). Indeed, insofar as this move led to the development of quantum chromodynamics, an element of the Standard Model is grounded in this 'ruling out' of parastatistics. Such a move suggests that the distinction between describing quarks in terms of colour or as paraparticles of order three is merely one of conventon (French ibid.), a suggestion that has been formally 'firmed up' in the context of quantum field theory (Baker et al. 2015). But of course, that doesn't restore supervenience since there are still the infinitely many other paraparticle kinds that are not instantiated (corresponding mathematically to so much 'surplus structure'; see Bueno and French 2018). Hence the particle kinds do not supervene on the symmetries and the relationship between the two cannot be adequately captured in these terms (Wolff 2011; McKenzie 2014, p. 1097).

As an alternative, consider the notion of dependence: *x* depends upon *y* just in case *x* exists only if *y* exists (see Lowe 2005). Now, clearly both the bosonic and fermionic representations are dependent on the permutation group, since the (irreducible) representations in general are given, *mathematically*, by the group theoretic structure. However, it is not enough to simply reiterate the mathematical dependence in this context. Indeed, it has been argued that for the relevant symmetry, represented group-theoretically, to be regarded as *physical*, it must yield determinate, measurable kind properties, via the relevant representations. Hence, it has been claimed, reference to the latter cannot be avoided if Permutation Invariance is to be taken to be an element of *physical* reality and, in that sense, particle kinds (and, more generally, properties) and symmetries (conceived of as physical) must be taken to be on a par ontologically (McKenzie op. cit., p. 1101). In that case, the former cannot be physically dependent on the latter.

However, there is a third option: the determinable-determinate relationship (see French 2014, Ch. 11). Thus consider colour, in general, and red, more specifically: the property of being coloured can be regarded as the determinable of which the property of being red is the determinate; and of course, one can continue, with 'red' as the determinable of which 'crimson' is the determinate. The core features of this relationship can be summed up as follows:

1. if a determinate concept (e.g. red) can be predicated of something, then at least one determinable concept (e.g. coloured) must also be predicable of that thing;
2. if a determinable concept (e.g. coloured) can be predicated of something, then there must be some determinate concept (e.g. red) that is also predicable of that thing;
3. two determinates (e.g. green and red) of the same determinable (coloured) cannot characterize something at the same time (Johnson 1921).

The relationship between Permutation Invariance and particle kinds satisfies these features. Thus, for example, with Permutation Invariance understood to be the relevant determinable, the bosonic kind, as mathematically represented by the appropriate irreducible representation, is then one of that determinable's determinates, just as 'scarlet' is a determinate of the determinable 'red'. And crucially, bosonic and fermionic kinds, seen as two determinates of the symmetry determinable, are mutually exclusive and cannot characterize the same particle at the same time, or indeed at different times – this is due to the effect of the group which, as noted above, divides the state space up into non-combining sectors and the fact that the Hamiltonian commutes with the particle permutation operator means that once in such a sector, a particle can't switch to the other via any interaction (bosons always remain bosons and fermions always remain fermions). And the inclusion of such determinates within the structure allows them to function as 'existential witnesses' (Wilson 2012), so that this structure is the structure of *the* (i.e. this) world.

11.7 Measurement Outcomes

The third feature of Cassirer's 'well-rounded' structure concerns statements of measurement outcomes. How are these accommodated within the above picture? It's all well and good to say that kinds are determinate features of Permutation Invariance, as a determinable, and similarly, spin is a determinate feature that drops out of the Poincaré group. But what about the outcome of a definite spin measurement? Here we bump up against the infamous measurement problem! Can we give a specifically structuralist solution to this?

There are obvious, not to say glaring, obstacles, not the least of which, to put it crudely, is how one motivates the shift from a superposition to a definite value, 'in structuralist terms'. As it stands, the picture sketched above (and in French 2014) is entirely too general; it takes us from determinable symmetries and laws to determinate kinds and properties but not to definite values of such properties (Esfeld sees this as a source of criticism of this framework; Esfeld 2015). One option would be to adapt one of the 'standard' solutions to the measurement problem and give that solution a structuralist gloss. Thus, one might offer a patterns based structuralist rendering of the Everett interpretation (Wallace 2012; but see Ladyman and Ross 2007 pp. 179ff); or one could take a structuralist stance on the Bohm interpretation (French 2001), with the Bohmian 'particles' understood as objects in only the thinnest sense with all their properties encoded in the relevant structure (Esfeld et al. 2014). One could even take a form of the venerable 'consciousness' solution, with 'subject' and 'object' emerging as two 'poles' in a general phenomenological structure (French 2002).

Here I want to sketch an alternative line of approach, which has the virtue of keeping the focus on the actual practices of physics. Consider the impact of decoherence: interaction with the environment leads to a dampening of interference terms in the description of the superposition, a process that proceeds extremely quickly in the case of the kinds of macroscopic objects involved in measurements. As Cordero notes, this effect blocks attempts to experimentally decide between the various 'solutions' to the measurement problem or, as he puts it, decoherence phenomena '...burden experimental access to the world with a kind of effectively irreducible "experimental astigmatism" ' (Cordero 2001). Thus we are faced with an experimental obstacle that is contingent upon some 'unfortunate' facts about the actual world (ibid., p. 307). Here we might recall Cassirer's admonition to take the 'conditions of accessibility' as 'conditions of the objects of experience' and trim our metaphysical sails accordingly. Nevertheless, as Cordero goes on to remark, this does not impact on the overall picture presented above:

> Aspects as profound as those regarding the group-theoretic symmetries are untouched by the debate at the stochastic level. That is, above a certain descriptive depth all the models yielded by the [various solutions to the measurement problem] converge both structurally and semantically in terms of effective partial isomorphisms that reach deeply into the respective theoretical fabrics, and do so to a very high degree of approximation (ibid., p. 308).

Consequently, the different solutions to the measurement problem all share a 'thick body' of modeling and relevant prior knowledge in terms of which, for example, the kind of experimental support that has accrued to the Standard Model (via the Large Hadron Collider) can be understood.

And although that experimental support is often expressed via talk of 'particles' of various kinds, this in and of itself does not require us to step outside the structuralist framework.

11.8 To Eliminate or Not, That Is the Question

Structural realism has been characterised in terms of a shift in focus away from objects to structures (Ladyman 2016). One can understand that shift either in terms of downplaying the ontological status of objects or eliminating them entirely.

Nothing in what has been said above conflicts with either view. So, one could opt for the so-called 'moderate' form of structural realism and retain a 'thin' notion of object, perhaps understood in Quincan terms as simply that which is represented by the appropriate variable in the regimented form of the model, with all the properties of those objects encoded in the relevant structure (see French and Ladyman 2011). Alternatively one could reject the need even for this minimal conception and insist that 'objects' as a metaphysical device, can and should be eliminated (French 2014).

Note that this does not imply anti-realism about 'particles' such as electrons, pions, quarks, whatever. It just means that these need to be reconceived not as objects – whether understood as bits of substance in which properties inhere or as bundles of said properties – but as features or aspects of the structure of the world. Again, recall that the nature of these particles, whether they are bosons or fermions, whether they have spin etc., is entirely given in structural terms. All that is left that might be taken to be the core 'object' is just a placeholder, kept, I believe, through some misguided metaphysical prejudice!

Nevertheless, it has been pointed out, the very mathematics that is deployed to capture these symmetries assumes the very notion of 'object' that the 'radical' structural realist is attempting to eliminate. Consider group theory, understood in quite abstract terms: we have a set of elements, a, b, g ... on which certain transformations – permutations say – are defined; thus we obtain axb = g, say. The elements and the transformations seem entirely distinct, conceptually speaking. However, the language of mathematics, constructed originally to deal with the putative objects we apparently find all around us, may be misleading! Eddington, for example, insisted that just because we can write down the symbols separately, this does not mean that which they designate can be so separated. Instead of decomposing the above as a, b, g, and x, he suggested, we should see it in terms of ax and xb so that the entity and the transformation are understood as forming an indivisible whole (French 2003). This is not quite the same as the 'moderate' form of structural realism indicated above. That suggests that we can conceptually distinguish the object, thin as it is, from the structure. Eddington, on the other hand, proposed that the entity and the transformation be considered as merely different aspects of the structure as a whole, just as the symmetries and the measurement outcomes are different features of Cassirer's 'well-rounded' structure of the world.

11.9 Returning to Realism

There is a great deal more that can be said about the metaphysics of this picture and how we might accommodate this elimination of objects (again see French 2014). However, let us return to the issues with which we began this essay. The realist urges us to take seriously our 'best' theories of the world. One of the best we have right now is the Standard Model. Taking that seriously means taking the role of symmetries seriously and the way in which kinds and properties 'drop out' of that framework. I claim that 'ontic' structural realism, which holds that the world *is* structure, does just that. Perhaps other forms of realism can do likewise, but the onus is on their advocates to come up with the goods.

Furthermore, nothing I have said here suggests that these features must be preserved through theory change. Although a recent form of structural realism—generally known as 'epistemic' structural realism (Worrall 1989)—was developed in response to the problems posed for realism by theory change and although I personally believe that one can mount a form of continuity of structure argument, the above picture is limited in scope insofar as it is grounded in one particular model. In that respect, what I am proposing here is in the spirit of Cassirer and Eddington's efforts (granted neither was a realist of course): they did not defend their structuralist conceptions on the basis of some commonality with earlier theories; rather they presented them as a way of making philosophical sense of what they both saw as the most important theoretical development of their age, namely that of quantum mechanics.

In this sense, the above meshes with recent moves that eschew global 'recipes' that demand one should be a realist about some element that features across a range of temporally related theories, such as certain entities, or descriptions, or structures, in favour of 'local' accounts that leave one free to say 'given theory T one should be a realist about such and such features', where said features may be different when it comes to theory T' (Saatsi 2016). Here I am suggesting that to be a realist about the Standard Model one should be a realist about the symmetries and laws that it embodies and hence one should be a structural realist. Of course this is to step away from the motivation of responding to theory change but as I've said elsewhere (French 2006) that motivation and that of accommodating our best recent physics may well come apart and, as just noted, that former motivation has not always been the driving force in the history of structuralism. Indeed, that the two do come apart may not appear so surprising when one reflects on the shifting prominence of symmetry principles as we turn from classical to quantum physics!

References

Baker, D. J., Halvorson, H., & Swanson, N. (2015). The conventionality of parastatistics. *British Journal for the Philosophy of Science, 66*, 929–976.

Bangu, S. (2012). *The applicability of mathematics in science: Indispensability and ontology*. Houndmills: Palgrave Macmillan.

Bird, A. (2007). *Nature's metaphysics: Laws and properties*. Oxford: Oxford University Press.

Brading, K., & Brown, H. R. (2003). Symmetries and Noether's theorems. In K. Brading & E. Castellani (Eds.), *Symmetries in physics: Philosophical reflections* (pp. 89–109). Cambridge University Press: Cambridge.

Brading, K., & Castellani, E. (2008) *Symmetry and symmetry breaking*. The Stanford Encyclopedia of Philosophy, Edward N. Zalta (ed.). http://plato.stanford.edu/archives/fall2008/entries/symmetry-breaking/.

Brown, H., & Holland, P. (2004). Dynamical versus variational symmetries: Understanding Noether's first theorem. *Molecular Physics, 102*, 1133–1139.

Bueno, O., & French, S. (2018). *Applying mathematics: Immersion, inference and interpretation*. Oxford: Oxford University Press.
Cassirer, E. (1936/1956). *Determinismus und Indeterminismus in der Modernen Physik*. Göteborgs Högskolas Årsskrift 42. Translated as *Determinism and Indeterminism in Modern Physics*. New Haven: Yale University Press, 1956.
Cei, A., & French, S. (2014). Getting away from governance: Laws, symmetries and objects. *Méthode – Analytic Perspectives*, 3. https://doi.org/10.13135/2281-0498%2F4
Cohen, J., & Callender, C. (2009). A better best system account of lawhood. *Philosophical Studies, 145*, 1–34.
Cordero, A. (2001). Realism and underdetermination: Some clues from the practices-up. *Philosophy of Science, S68*(2001), 301–312.
Esfeld, M. (2015). Review of *The structure of the world. Mind, 124*, 334–338.
Esfeld, M., Lazarovici, D., Hubert, M., & Dürr, D. (2014). The ontology of Bohmian mechanics. *British Journal for the Philosophy of Science, 65*, 773–796.
French, S. (1995). The esperable uberty of quantum chromodynamics. *Studies in History and Philosophy of Modern Physics, 26*, 87–105.
French, S. (2001). Quantum physics and the ontology of structures. In *Physis: Proceedings of the IVth International Ontology Congress (A tribute to John Bell)*, San Sebastian, pp. 175–181.
French, S. (2002). A phenomenological approach to the measurement problem: Husserl and the Foundations of Quantum Mechanics. *Studies in History and Philosophy of Modern Physics, 22*, 467–491.
French, S. (2003). Scribbling on the blank sheet: Eddington's structuralist conception of objects. *Studies in History and Philosophy of Modern Physics, 34*, 227–259.
French, S. (2006). Structure as a weapon of the realist. *Proceedings of the Aristotelian Society, 106*, 167–185.
French, S. (2014). *The structure of the world*. Oxford: Oxford University Press.
French, S. (Forthcoming). Doing away with dispositions: Powers in the context of modern physics. In A. S. Meincke (Ed.), *Dispositionalism: Perspectives from metaphysics and the philosophy of science*. Springer Synthese Library, Springer.
French, S., & Krause, D. (2006). *Identity in physics*. Oxford: Oxford University Press.
French, S., & McKenzie, K. (2012). Thinking outside the (tool)box: Towards a more productive engagement between metaphysics and philosophy of physics' with K. McKenzie. *The European Journal of Analytic Philosophy, 8*, 42–59.
French, S., & Ladyman, J. (2011). In defence of ontic structural realism. In A. Bokulich & P. Bokulich (Eds.), *Scientific structuralism* (pp. 25–42). Dordrecth: Springer.
French, S., & Saatsi, J. (2018). Symmetries and explanatory dependencies. In J. Saatsi & A. Reutlinger (Eds.), *Explanation beyond causation*. Oxford: Oxford University Press.
Johnson, W. E. (1921, 1922, 1924). *Logic* (Vol. 1: 1921; vol. 2: 1922; vol. 3 1924). Cambridge: Cambridge University Press for the hypertext version see http://www.ditext.com/johnson/toc.html
Ladyman, J. (2016). In E. N. Zalta (Ed.), *Structural realism. The Stanford Encyclopedia of philosophy* (Winter 2016 ed.). URL = https://plato.stanford.edu/archives/win2016/entries/structuralrealism/.
Ladyman, J., & Ross, D. (2007). *Every thing must go: Metaphysics naturalized*. Oxford: Oxford University Press.
Lange, M. (2007). Laws and meta-laws of nature: Conservation laws and symmetries. *Studies in History and Philosophy of Modern Physics, 38*(457–481), 457–481.
Lange, M. (2009). *Laws and lawmakers: Science, metaphysics, and the laws of nature*. Oxford: Oxford University Press.
Lange, M. (2012). There sweep great general principles which all the laws seem to follow. In K. Bennett & D. Zimmerman (Eds.), *Oxford studies in metaphysics* (Vol. 7, pp. 154–185). Oxford. Oxford University Press.
Lowe, E. J. (2005). *Ontological dependence*. The Stanford Encyclopedia of Philosophy. Edward N. Zalta (ed.). http://plato.stanford.edu/archives/spr2010/entries/dependence-ontological/.

McKenzie, K. (2014). On the fundamentality of symmetries. *Philosophy of Science, 81*, 1090–1102.

Post, H. R. (1971) (1971 [1993]). Correspondence, invariance and Heuristics. *Studies in History and Philosophy of Science*, *2*, 213–255; reprinted In S. French & H. Kamminga (Eds.), *Correspondence, invariance and Heuristics: Essays in honour of heinz post*. (Boston Studies in the Philosophy of Science 148), Dordrecht: Kluwer Academic Press (1993), 1–44.

Psillos, S. (2006). What do powers do when they are not manifested? *Philosophy and Phenomenological Research, 72*, 135–156.

Saatsi, J. (2016). Replacing recipe realism. *Synthese,* First Online. https://doi.org/10.1007/ s11229-015-0962-3.

Wallace, D. (2012). *The emergent multiverse*. Oxford: Oxford University Press.

Wilson, J. (2012). *Fundamental determinables*. Philosophers Imprint 12. http://www.philosophersimprint.org/012004/.

Wolff, J. (2011). Do objects depend on structures? *British Journal for the Philosophy of Science.*, *63*, 607–625. https://doi.org/10.1093/bjps/axr041.

Worrall, J. (1989). Structural realism: The best of both worlds? *Dialectica* 43: 99–124; reprinted In D. Papineau (Ed.), *The philosophy of science*. Oxford: Oxford University Press, pp. 139–165.

Part IV
Individuals, Individuation, and QM

Chapter 12
The Problem of Individualism from Greek Thought to Quantum Physics

Peter Mittlestatedt

Abstract Individuals in the strict sense do not exist in quantum physics. This paper argues that unsharp observables, almost repeatable and weakly disturbing measurements allow for the definition of unsharp individuals which is sufficient for all practical purposes. Many quantum physical experiments and the obvious existence of individuals in the classical world can be explained in this way. On the other hand, if quantum mechanics is considered as universally valid then there is no classical world in the strict sense. The paper includes a Divertimento on an analogy between the motion of individual quantum systems and the motion of angels according to the treatment of Thomas Aquinas in his *Summa Theologica*.

12.1 Introduction

The question, in which way individuals can be determined within a class of objects and by which means they can be distinguished, has been discussed since the Ancient Greek philosophy. We will treat this problem here from the philosophy of Aristotle up to the present philosophy of science, in particular to the philosophy of quantum mechanics. It turns out that in spite of the important conceptual differences there are common ways to solve these problems. The constitution of objects by their properties proves to be a guiding principle from Aristotle to quantum physics.

Here, we will treat this topic in several steps roughly following the historical development. First, we consider briefly the investigations of Aristotle about individual objects and their behaviour, as treated in the books *Metaphysics* (**Z**, **H**) and *Physics* (**Z**). The Aristotelian ontology is concerned with the identity and distinguishability of natural objects including human beings, e.g., Socrates and

Peter Mittlestatedt (deceased, 2014) was Professor of Physics at the University of Cologne, Germany.

P. Mittlestatedt (✉)
Department of Physics, University of Cologne, Cologne, Germany

A. Cordero (ed.), *Philosophers Look at Quantum Mechanics*,
Synthese Library 406, https://doi.org/10.1007/978-3-030-15659-6_12

Callias (Sect. 12.2).—Secondly we consider the interpretation of the Aristotelian ontology by Thomas Aquinas and its extension with respect to the Christian tradition. Accordingly, Thomas Aquinas investigates the individuality of entities which are beyond the scope of the Aristotelian ontology, e.g., the souls of human beings after their death and angels (Sect. 12.3).—Our next section is concerned with the problem of individuality in the philosophy of Leibniz which partly goes back to the arguments of Thomas Aquinas and Duns Scotus. It is an interesting topic to compare the principle of the *"Identity of Indiscernibles"* with the completely different approach by Kant (Sect. 12.4).

The Kantian arguments are very near to the way of reasoning in classical mechanics. Indeed, the constitution of objects by permanent properties and the determination of individuals by their position can be made explicit within the framework of classical mechanics (Sect. 12.5).

The limits of this way of reasoning become apparent if we finally discuss the problems of individuals in quantum mechanics. Here, the situation is much more complicated and the traditional attempts to characterize individual objects fail (Sect. 12.6). In quantum mechanics we are confronted with all problems for determining individual objects, which were discussed in the philosophical tradition. The various principles of individuation discussed in the literature turn out to be insufficient. We mention here the individualisation by *matter*, by *complete concepts,* by *history,* by *position,* etc. These principles can only partly be applied to quantum objects. We close with a discussion of the intricate problem whether a joint but incomplete application of these methods to quantum objects is possible, and whether there are at all individual quantum objects.

12.2 Ancient Greek Philosophy: Aristotle

In the Aristotelian philosophy the ontology of individual entities is treated in volumes **Z** and **H** of the book *Metaphysics.* For a given individual object Aristotle distinguishes its *form (eidos)* and its *matter (hyle).* Generally, we assume that a given object preserves some properties which characterize this object as such, whereas other properties may or may not pertain to the object without thereby invalidating its persistent identity. A certain stone may change its temperature or change its position in space, without thereby losing the properties which determine this particular stone. Hence, we can distinguish two kinds of properties, the essential properties which determine the object as such, and the accidental properties which are varying in time. However, it is not meant here that there is first a well-defined object and in addition essential and accidental properties which may pertain to it. Instead, according to the Aristotelian philosophy the object is constituted by those properties which characterize and determine the object, irrespective of the varying accidental properties. Hence, it seems to be correct to identify these essential constituents with the *form* of the object.

It is a controversial question whether the *form* in the Aristotelian ontology refers to a certain kind of objects and hence to a class of things with the same essential properties, or whether the form refers to an individual object. In the former case an additional principle of individuation is needed for characterizing an individual object. In the philosophical tradition several principles of individuation were formulated and we will discuss them in the following sections (Thomas Aquinas, Locke, Kant, etc.). An individualistic interpretation of the Aristotelian concept of *form* was adopted by philosophers of the Stoa, by Duns Scotus, and by Leibniz. For a quite recent and most interesting individualistic interpretation of the Aristotelian *form* we refer to the work of *Frede* and *Patzig.*[1]

Individuality of the *form* means distinguishability of the given object from other objects of the same kind and reidentifiability at a later time - which means again distinguishability from other objects. It is obvious that the distinction of individual forms can also be performed by comparing the matter in which the forms are realized and by distinguishing properties of the material objects. We recognize a certain person by very few characteristic properties and not by comparing its *form* with the *form* of other persons. Hence, in many cases there is no need for directly comparing two *forms* since their distinction by the respective material realisations is completely sufficient. We will come back to this intricate problem in our discussion of Leibniz' *Principle of the Identity of Indiscernibles.*

In addition to the characterization of individual objects by their *form* and *matter* we mention here another interesting feature of material bodies, the continuous motion of bodies in space. In the Aristotelian ontology this problem is not directly related to the question of individuality and reidentifiability. Since, however, other authors connected the problems of motion and individuality in various ways, it is of interest for the following investigations to mention here how the motion of material bodies is treated by Aristotle. In volume Z of his *Physics,* Aristotle shows that the continuity of a motion implies that the moving body is itself continuous. Moreover, if an object is continuous, then it is unrestrictedly divisible. It is obvious that this argument applies to material bodies which are extended in space. However, it follows also from this way of reasoning that an indivisible entity, whatever it may be, cannot be subject to a continuous motion (*Physics,* Z 10). An indivisible attribute can be moved continuously only if it pertains accidentally to an object, which as such is continuous and in continuous motion.- These problems will become relevant for Modern Physics as well as for Thomas Aquinas.

The ontology and the conceptual framework of Aristotle were developed with respect to natural objects, *i.e.,* for stones, trees, animals, and human beings. These objects are given in our experience as individuals by some persistent and constituting properties. In addition, these objects possess accidental and variable properties like temperature, position, etc. The transition from this conceptual framework to artificial objects like buildings, boats, or pieces of art is difficult, since the individuality of these objects is not quite obvious. E.g., the Cathedral

[1] Frede and Patzig (1988).

of Cologne is in permanent reconstruction and it is clear that after a few hundred years the material of this church is completely replaced by new material- It is still more complicated to apply the Aristotelian concepts to very primitive objects like indivisible atoms or elementary particles. Hence, we should be aware of these limits of the concepts mentioned and we should be prepared to be confronted with serious difficulties if the concepts of form and matter are applied to things beyond the scope of the Aristotelian ontology.

12.3 Medieval Scholasticism: Thomas Aquinas

The Aristotelian ontology was interpreted by Thomas Aquinas in such a manner that the *form (eidos)* is general and characterizes a class of objects, whereas an individual object of this class is determined by its *matter (hyle)*. For this reason *matter* is considered by Thomas Aquinas as the principle of individuation. An individual substance is given by its *matter, i.e.,* by the material realisation of the respective form. An individual person is a material realisation of the *form*, "man".

Thomas Aquinas extended the application of this conceptual framework with respect to the Christian theology In *De Ente et Essentia* where he investigates the following two problems. First, he considers two men, e.g., Socrates and Callias, who are determined as individuals by their bodies *(matter)*. If *matter* is the principle of individuation, how can Socrates and Callias be distinguished after their death? According to Thomas Aquinas two souls are distinguished by their different histories in which they were received in their bodies. This means that two immaterial souls can be distinguished by their material past which is considered here as a real property of these souls. The second generalization is concerned with angels which are immaterial beings without any material history. Hence, angels cannot be distinguished by their past. However, since angels are individuals Thomas Aquinas concludes that their form is sufficient for their individualization. This means that there are exactly as many angels as kinds of angels. Every angel is unique and uniquely defined by its form or its essential properties. Hence, there are no problems to give proper names to angels.

Another interesting topic which was investigated by Thomas Aquinas extending the Aristotelian ontology is the motion of angels. In *Physics* Z Aristotle had shown that an indivisible entity cannot be subject to a continuous motion. This result becomes relevant if one tries to describe the motion of angels within the context of the Aristotelian concepts.[2] This problem is treated extensively by Thomas Aquinas in *Questio 53* of his *Summa Theologica.* According to Thomas Aquinas angels are spiritual, simple, and indivisible beings. They are immaterial and do not belong to the visible reality which we know from our daily physical experience. (Correspondingly, the angels were created prior to the creation of the physical universe in the *hexaemeron.*[3] Since angels are indivisible, they cannot be subject

[2]Cf. Wieland (1973).

[3]Cf., e.g., Johannes Philoponos. De opificio miindi, 1.10 ff., Herder, Freiburg, (1977).

to a continuous motion. Hence, the question arises how the motion of an angel looks like if it comes in contact with our visible world. From a conceptual point of view this means that we must investigate the question whether there are motions of indivisible and immaterial entities which are compatible with the Aristotelian principle of continuity. According to Aristotle, a material body which is spatially extended has always a position (*topos)* since it is contained in space. In contrast to this well-known doctrine the angel is not genuinely an entity in space and time. However, according to the *Bible*, he has the ability to assume a certain position in space at a certain time and then the angel appears in our visible world.

From these assumptions we obtain some more detailed information about the motion of angels. In *Questio 53* Thomas Aquinas investigates three problems[4]:

1. Whether an angel can be moved locally (*Utrum angeluspossit moveri localiter*)
2. Whether an angel passes through intermediate space (*Utrum angelus transeatper medium*)
3. Whether the movement of an angel is instantaneous (*Utrum motus angeli sit in instanti*)

Since the angel is not subject to the *"topos"* doctrine, he is not permanently localized. This means that at some time instant t he assumes a position x and at a later instant $t' > t$ of time he assumes another position x′. Consequently, his orbit consists of discrete space-time events and cannot be represented by a continuous trajectory.

> *Sed angelus non est in loco ut commensnratus et contentus, sed magis ut continens. Unde non opertet quod motus angeli in loco commensuratur loco, ... sed potest esse motus ejus continuus et non continuus. (Questio 53.1).*[5]

In the Newtonian or the Minkowskian space-time this kind of motion gives the impression that the angel "jumps" from one event (x, t) to another event (x′, t') without thereby travelling on a continuous trajectory which connects these events.

> *Si astern motus angeli non sit continuos, possibile est quod pertranseat de aliquo extreme in aliud, non per- transito medio. (Questio 53.2).*[6]

Since it is only said here that the angel assumes different positions x, x′, x′′, ...at successive instants $t < t' < t''$ of time it is completely open whether two successive events (x, t), (x′, t') are timelike, null or spacelike. This opens the possibility that the angel's motion is superluminal or even instantaneous. Problems with Einstein

[4]English translation by Fathers of the English Dominican Province. Westininstei (Maryland), 1981.

[5]But an angel is not in a place as commensurate and contained, but rather as containing it. Neither then is the local movement of an angel commensurate with place: ...; in fact it may be either continuous or not.

[6]But if an angel's movement is not continuous, it is possible for him to pass from one extreme to another without going through the middle.

causality should not appear; since the discrete events (x, t) and (x', t') are without causal connection, which must be continuous.

Et sit angelus in uno instanti (t) potest esse in ono loco (x), et in alio instanti (t') in alio loco (x', x' $\neq$ x), nullo tempore intermedio existe (t – t' = + 0). (Questio 53.3).[7]

Finally we mention the following point. The angel is able to assume a position at a certain instant of time. However, this is possible only with respect to one point in space. For this reason the angel cannot be simultaneously at different places, *i.e.*, he is not omnipresent.[8]

...ita est in uno loco, qu od not in alio. (Questio 53.2).[9] *...moats angeli in loco nihil aliud sit quam diver si contactus locorum successive et nun simul. (Questio 53.7).*[10]

12.4 Leibniz, Locke, and Kant

(i) *Leibniz and the identity of indiscernibles*

A second principle of individuation was formulated by Duns Scotus. He assumed that also material substances can be individualized conceptually by their essential properties. Starting from very general concepts one arrives at the individual object by successive distinction of concepts in the sense of a decision tree. This means that in contrast to the interpretation of Aristotle by Thomas Aquinas' the principle of individuation is not matter but the form or the concept of the material entity. This idea has been adopted later by Leibniz. With respect to the principle of individuation in Thomas Aquinas' interpretation of Aristotle he mentions that what Thomas Aquinas assures about the angels must be considered as valid for all substances.[11] The essential properties which are contained in the "complete concept" determine *uniquely* the individual system. Leibniz emphasizes that the position property is not a label for distinguishing two otherwise identical substances, since this distinction would work only for impenetrable objects. Two distinct substances can always be distinguished by their essential *internal* properties whereas the *external* properties like the position can be deduced from their complete concepts. On the basis of these assumptions it is obvious that two substances which cannot be distinguished by any property are identical. This is the famous *Principle of the Identity of Indiscernibles* which appears here as an immediate consequence.

---

[7]So an angel can be in one place (x) in one instant (t), and in another place (x', $x' \neq x$) in the next instant (t') without any time intervening ($t - t' = +0$). (*Formulas are added by the author).*

[8]In order to avoid that the angel is omnipresent, in *Questio 53.3 "nullo tempore"* must be understood such that $t - t' = +0$, *i. e.*, the time difference is arbitrary small but positive.

[9]. . . he is in one place, in such a manner that he is not in another.

[10]The movement of an angel in a place is nothing else than the various contacts of various places successively but not at once.

[11]G.W. Leibniz, Discourse de Metaphysique, 9. (GP IV, p. 433).

Nevertheless, for practical purposes Leibniz seems to accept also a distinction of objects by their observable and external properties. Since the internal properties determine the observable external properties including their temporal development, two different external properties indicate two distinct substances. For this reason, the identity of indiscernibles can be illustrated by observable properties of physical entities.[12] The *most famous* Leibnizian example for the identity of indiscernibles illustrates the impossibility of finding two "perfectly similar" leaves in the garden of Herrenhausen... *"une grande Princesse, qui est d'un esprit sublime, dit un jour en sepromenant dans son jar din, qu'elle ne croyoit pas, qu'il y avoit deux feuilles parfaitement semblables. Un gentilhomme d'esprit, qui estoit de la promenade, crut qu'il seroit facile d'en trouver; mais quoiqu'il en cherchât beaucoup, il fut coinvaincu par ses yeux, qu'on pouvoit toujoursy remarquer de la difference"*[13] It is obvious that Leibniz is confronted here with the same problem that appears in the Aristotelian ontology, if *forms* are interpreted individualistically.

(ii) *From Locke to Kant: Individuation by position*

Another principle of individuation goes back to John Locke who considered the position of an object as a sufficient characterization of an individual system. Clearly, for the application of this principle one must presuppose the impenetrability of the material bodies, which is of course a contingent property. However, Locke's principle is free from metaphysical concepts like the "complete concept" and can be immediately applied to empirical objects. Since the known objects are in fact impenetrable, the individuation by localization can be used for all practical purposes. Locke's principle was accepted also by Kant, who pointed out that the "*difference of locations... makes the plurality and distinction of objects, as appearances*; *not only possible but also necessary*".[14] Kant mentions this point only briefly in connection with his critique of the philosophy of Leibniz and in particular of the mentioned *Principle of the Identity of Indiscernibles.* Within the framework of his transcendental philosophy Kant has not derived systematically a new principle of individuation. For a full understanding of Kant's argument in favour of Locke and against Leibniz we must consider the constitution of objects in Kant's philosophy in more detail.

(iii) *The Constitution of Objects in Kant's Philosophy and in Classical Physics*

It is an often discussed question of traditional philosophy whether in addition to the observations of qualities there exist some entities, things, or objects which possess the qualities mentioned as their properties. In his *Treatise of Human Nature* David Hume emphasized that we never observe objects but only qualities and that it is nothing but imagination if we consider the observed qualities as properties of an object. Hence any scientific cognition begins with the observation of qualities

[12]More details about this point can be found in E. Castellani and P. Mittelstaedt (1998).

[13]G. W. Leibniz, Nouveaux Essais sur l'entendement humain, Chap. XXVII, § 3. (GP V, p. 214).

[14]*I. Kant,* Critique of Pure Reason, *A 272.*

and it seems to be merely a question of interpretation whether in addition to the observed phenomena a fictitious object, "an unknown something", is used for their description. At first glance, there is no reason to expect that general laws like the conservation of substance or some causality law hold for the observations.

However, in contrast to Hume, Kant emphasized that "objects of experience" are not arbitrary imaginations but entities which were *constituted* from the observational data by means of some conceptual prescriptions, the categories of substance and causality. Hence the interpretation of the observed data as properties of an object can be justified, if an object was constituted as carrier of properties by means of the categories mentioned. Kant formulated necessary conditions which must be fulfilled by the observational data, if these data are considered as properties of an "object of experience." Accordingly, if we have objective cognition of the reality, *i.e.,* if our observations refer to an element of the exterior reality and not to the observing subject, then the observations in space and time must have been ordered and interpreted according to the categories of substance and causality. Hence these categories are necessary preconditions of objects of experience which fulfill the *a priori* laws of substance and causality. They are, however, only preconditions of *kinds* of objects with the same "essential" properties but not of *individual* systems.

In everyday experience and in the domain of classical physics, to the formal preconditions of experience, the categories of substance and causality, material preconditions can be added which correspond to the material possibilities to measure and to observe properties. These material preconditions of experience specify the formal possibilities for the constitution of objects. In the present case there are no obvious restrictions for measuring all possible predicates P_i jointly on a system which is thus subject to the principle of complete determination: "*Every thing as regards its possibility is likewise subject to the principle of complete determination according to which if all possible predicates are taken together with their contradictory opposites, then one of each pair of contradictory opposites must belong to if* ".[15]

A system of this kind or a "thing" possesses each possible "accidental" property P either positive (P) or negative ($-$P). In this case the causality law leads to a strict and complete determination of all properties. In particular, it follows that "things" or objects possess always a well-defined position in space, *i.e.,* they are permanently localized. If in addition impenetrability is assumed, then the permanent localization can be used for a determination of individual objects by their trajectories in space and time—Kant mentioned this way to determine individuals only very briefly in connection with his critique of Leibniz' principle of the identity of indiscernibles.[16]

[15] I. Kant, Critique of Pure Reason, B 600.

[16] Kant, Critique of Pure Reason, A 272.

12.5 The Constitution of Objects in Classical Mechanics

(i) *Objectivity and Invariance*

The Kantian way of reasoning can be made explicit within the framework of classical mechanics. The goal of physics and in particular of classical mechanics is the cognition of the external reality and not of the observing subject. Accordingly, observations or measuring results should refer to the external reality and not to the observer's subjective impressions. This requirement of objectivity implies that the cognition of the external reality must be independent in some sense of the observer's preconditions. The subjective, observer dependent component of a measuring result is given by the observer's space-time coordinates. Hence the requirement of objectivity can only be fulfilled if the laws of the external reality have some *invariance* properties. If an observer changes his space-time coordinates, then the observations should be changed in such a way that they refer to the same but equivalently changed object. In this way the objectivity of the measuring results can be achieved.[17]

The fundamental laws of classical mechanics are invariant against the transformations of the ten parameter Galilean group G_{10}. If the observer is "moved" in accordance with a Galilean transformation, the translations in space, say, then the observations which refer to the external object will transform "covariantly" with respect to this transformation. Since also the observers, represented by measurement instruments, are physical objects, they will be subject to the same invariance laws. This implies a symmetry between active and passive transformations: The transformation of the measurement results does not depend on whether the observer is moved according to a Galilean transformation or whether the object is moved according to the inverse transformation.

(ii) *Covariance and Observables*

The symmetry between active and passive transformations allows for clarification of the concept of an "observable". Intuitively an observable may be understood as a measurable quantity or a property of an object system *S*, which belongs to the external reality and which is clearly distinguished from the measuring apparatus. "Properties" correspond to yes-no propositions P_i or to the most simple observables with values 0 and 1. The set $\{P_i\}$ of elementary, propositions can be extended by introducing the logical operations $\cup$, $\cap$, $\sim$, and the relation $\leq$. In this way one arrives at the propositional system of classical mechanics which is given by a Boolean lattice L_C.

One can then define an "observable" in a more formal sense as a relation between numbers on the reading scale of the measurement apparatus and properties of the object system. Hence, an observable may be considered as a mapping Φ from the Borel sets *B* on the real line $\Re$ onto the Boolean lattice L_C of propositions. An

[17] Weyl (1966).

observable is connected with the group G_{10} of Galilean transformations in a twofold way. Firstly, the properties of the system are changed by an *active* transformation, when the transformation group acts on the system and its propositional lattice. Secondly, the observer's coordinate system is changed by a *passive* transformation, when the transformation group acts on the measurement device, *i.e.,* on the Borel sets of the reading scale.

Within this conceptual framework the symmetry between active and passive transformations leads to the following important *covariance postulate* (C), which must be fulfilled by an observable: The actively transformed properties of the system, *i.e.* the transformed propositions, coincide with the propositions which are obtained by passively transforming the observer's coordinate system and hence the reading scale of the apparatus. The covariance postulate (C) is the abstract formulation of the invariance of classical mechanics with respect to the Galilean group of transformations. It determines those functions which may be considered as "observables" and it shows how these observables are transformed under a special transformation.

On the basis of the covariance postulate (C) and the Galilean group one can now define the fundamental observables p (momentum), q (position) and the observable t (time). In this way the basis quantities (p, q, t) of the state space can be shown to be "observables" in the sense explained, which satisfy the covariance postulate (C). Within the framework of classical mechanics all other observables can be written as functions $F(p, q, t)$ which depend on the coordinates p, q, and t. If an object of classical mechanics is understood as a carrier of properties, then it is obviously sufficient, to require that it is a carrier of the fundamental observables p, q, t.

One can now define the concept of a classical object S in the following way:

> "A *classical object S is an algebra* L_C *such that a representation of the (passive) Galilean group is defined by auto morphism of the lattice* L_C *which admit the observables p, q, t in the sense of the covariance postulate (C).*

This means that a classical object is a carrier of the properties $P \in L_C$ not only in one contingent situation K given by an observer and its system of coordinates, but also in all other situations K' which evolve from K by Galilean transformations. The classical object is a carrier of properties which transform covariantly under the transformations of the Galilean group.

One can further specify this concept by considering different classes. E.g., elementary systems are given by irreducible representations of the Galilean group. For elementary systems which correspond to mass points without geometrical structure, there are no *true* but only *projective* representations of the group G_{10}. These representations are characterised by one continuous parameter m which can be interpreted as the "mass" of the object.

(iii) *Individual systems*

The representations of the Galilean group characterize classes of objects with the same permanent properties. In order to denote an individual system, one has to find additional properties which distinguish the system S in question from all the other systems S', S'',... of the same class. Two questions arise at this point. Firstly, one has to make clear whether the triple (p, q, t) is a unique denotation of S, *i.e.*, whether there is only one system with these properties. Secondly, if uniqueness is guaranteed, one has to find out in which way the system S defined at time t can be reidentified at some later time $t' > t$. In order to guarantee uniqueness of S one needs an additional dynamical principle which excludes that two systems are at the same time t at the same phase point (p, q). Clearly this postulate is fulfilled if impenetrability in position space is given. This is actually the case in all known situations. However it does not follow from any dynamical principle. In order to guarantee also the reidentifiability of the system S uniquely defined at time t, at some later time value t', one needs a convenient law which connects the point $(p, q)_t$ in phase space (at time t) with the phase point $(p, q)_{t'}$ (at any other time t'). In classical mechanics a dynamical law of this kind is given by a Hamiltonian H (p, q) and the canonical equations. This means that an individual system S can be reidentified at any other time value by the (p, q)-values on its dynamical trajectory T (S), that is (p_t, q_t) in phase space. Both requirements for individual objects, the uniqueness and the reidentifiability, are usually guaranteed in classical mechanics. For this reason an individual system S can be named permanently by an arbitrary point (p_t, q_t) on its trajectory T (S).

12.6 The Constitution of Objects in Quantum Mechanics

(i) *General Remarks*

The empiricist approach first formulated by David Hume was applied to quantum mechanics by Niels Bohr within the framework of the Copenhagen interpretation. In this interpretation one considers only measurement results and their mutual relations, but without assuming that the observed predicates can be attributed to an object as its properties. However, Bohr used this "minimal interpretation" not only for philosophical reasons, but because the hypothetical assumption of objects as carriers of properties is sometimes incompatible with quantum mechanics. Indeed, the constitution of objects in quantum mechanics provides problems which are not known from Kant's philosophy and from classical mechanics.

If one tries to extend the Copenhagen interpretation by incorporating objects, then one finds that for quantum systems the laws of substance and causality are no longer generally valid. The reason for this surprising observation is that quantum systems are not "subject to the principle of complete determination". In quantum mechanics the material preconditions of experience, *i.e.* the physical laws of measurements, do not allow one to determine jointly all possible properties of a given system. In any contingent situation which is described by a state Ψ only a subset P_Ψ of properties can be measured jointly on the system S. The properties $P^i \in P_\Psi$ are mutually commensurable, which means that they can be measured in arbitrary sequence without thereby changing the results of the measurements. The measurement results of these properties (Pi or $\sim P^i$) can be related to the object system just as in classical mechanics. Hence we refer to these properties as the "objective" properties of the system in the state. However, for any state (there are also non-objective properties) $P^i \notin P_\Psi$ whose measurement provides a material change of the state Ψ of *S*.

In quantum physics, as well as in classical physics, for the constitution of objects one has to begin with the requirement of objectivity. The observed predicates should refer to an object as its properties. Again, this requirement leads to the necessary preconditions of any objective experience, the categories of substance and causality. However, in the present case the *material* preconditions of classical experience are not fulfilled, since the systems are not "completely determined". This means that a quantum object system S_Ψ can only be constituted *incompletely* by means of the restricted set of its objective properties P_Ψ.

It follows from these arguments that the causality law in quantum mechanics holds only for the set P_Ψ of objective properties which are given by the state $\Psi(t)$ at some time value *t*. The temporal development of this state is determined by the Schrodinger equation in a causal way, *i.e.*, the state determines the state $\Psi(t')$ at any later time $t' > t$. However, since the state Ψ corresponds only to the restricted set P_Ψ, of objective properties, at different time values *t*, *t*′, ... we have different sets P_Ψ, $P_{\Psi'}$, ..., of objective properties. Hence it will in general not be possible to establish a causal connection between a property $P^a(t)$ at time *t* and the same property $P^a(t')$ at a later time *t*′. Consequently, there is only a very limited causality law between the objective properties P_Ψ and $P_{\Psi'}$ at different time values.[18]

In particular, these arguments apply to the position property x. This property pertains to the system S_Ψ only, if $x(t)$ is an objective property of S_Ψ. Since this happens in general only for some discrete time values t, t', t'' ..., with objective position values $x(t)$, $x(t')$, $x(t'')$, and since the corresponding momentum values $p(t)$, $p(t')$, $p(t'')$ are objectively undetermined, any interpolation between the discrete position values is completely meaningless. Hence, individual objects cannot be determined by their space-time trajectories, even if one assumes impenetrability. On the other hand, one can determine the position of an object at any time *t* by measurement. However, since measurements of nonobjective properties disturb the system in an unpredictable way, there is no causal connection between the various position measurement results, and a continuous trajectory cannot be constructed. Hence we arrive again at the result that causally connected continuous trajectories do not exist in quantum mechanics and individual systems can thus not be determined.

There is an interesting analogy between the motion of individual quantum systems and the motion of angels according to the treatment of Thomas Aquinas in his *Summa Theologica.* Recalling the various properties and abilities of angels which we discussed above (Sect. 12.3), we obtain the following description of the path of an angel in space and time: There are several disconnected events (x_i, t_i) which correspond to the events of the angel's appearance. There is a partial ordering

---

[18] Cf. Mittlestaedt (1994), Strohmeyer (1995).

of the events in such a way that for two subsequent events (x, t) and (x′, t') we have $t \leq t'$. There is no limitation of the velocity, but the angel is not *omnipresent.* This means that the velocity must not be infinite. The trajectory of an angel is a zigzag curve since any kind of continuous interpolation is completely meaningless.

The analogy between this motion and the motion of a quantum system is obvious. Let $S(\Phi_0)$ be a quantum system which is prepared in the state Φ_0 at time $t = t_0$, such that S possesses the position property $x = x_0$. The unitary operator $U(t)$ describes the temporal development of the system's state. In general, in a state $\Phi_t = U(t)\,\Phi_0$ the position is no longer an objective property of S, *i.e.* the system does not possess a position. However, at any time $t_1 > t_0$ the position of S can be measured and the observer obtains a definite result x_1. Moreover, the two events (x_0, t_0) and (x_1, t_1) have no causal connection at all. According to a theorem by Hegerfeld et al,[19] the object S which had the position $x = x_0$ at t_0 can be detected by measurement with a nonvanishing probability at an arbitrarily distant point $x_1 >> x_0$ even if the time difference $\delta_t = t_1 - t_0$ is arbitrarily small. This means that subsequent localization events $(x_1, t_1), (x_2, t_2), \ldots,$ have four-dimensional distances which are timelike, null or spacelike and which sometime give the impression that the quantum system has moved with a superluminal velocity. However, the localization events (x_i, t_i) cannot be connected by a continuous causal trajectory, since the position of S is objectively undetermined except for few measurement events (x_i, t_i). In addition, it can be shown that the seemingly superluminal motion of the object cannot be misused for sending superluminal signals.[20]

(ii) *Objects in Quantum Mechanics*

The same way of reasoning which allows for the constitution of objects in classical mechanics can also be applied to quantum mechanics. In classical mechanics as well as in quantum mechanics we are interested in the cognition of the external reality and not in the observing subject. This leads again to the requirement of objectivity which means that the fundamental laws of physics are subject to a group of symmetry transformations. Different observers which are connected by transformations of the invariance group will then be able to describe the same object of the external reality. The invariance group is again given by the Galilean group G_{10}. The observer corresponds to a macroscopic and classical measuring apparatus, which is associated with a spacetime coordinate system. For this reason a passive Galilean transformation has a meaning, which is quite similar to the classical case. Different observers represented by measurement apparatuses are connected by transformations of the Galilean group and the measuring results will then transform "covariantly" with respect to these transformations.

Similarly as in classical mechanics also in quantum mechanics observables will be characterized by their covariance with respect to the subgroups of the Galilean group. A Galilean covariant sharp observable can then be defined as a self-adjoint operator or a projection valued measure Φ on a homogeneous space (equipped with

---

[19] Hegerfeld and Ruijsenaars (1980).

[20] Schlieder (1968).

a Borel algebra *B)* of some subgroup of G_{10}. Observables of this kind allow for sharp measurements of some properties; they are, however, subject to the well-known complementarity restrictions. The sharp properties of a quantum system S at some time value *t* which correspond to (sharp) yes-no propositions P_i are given by the subspaces of the Hilbert space of the system, or by the corresponding projection operators with eigenvalues 0 and 1. If the set $\{P_i\}$ of propositions is extended by the quantum logical operations $\cup$, $\cap$, $\sim$, and the relation $\leq$., then one arrives at the complete, atomic and orthomodular lattice L_Q of *quantum logic*. The operations introduced here are defined as intersection and span of two subspaces and as the orthocomplement.

A quantum mechanical observable Φ can then he defined as a mapping from the Borel sets *B* on the real line $\Re$ onto the propositional lattice L_Q of quantum logic, *i.e.,* as a projection valued measure. An observable is then again connected with the invariance group G_{10} in a twofold way. Firstly, the transformation group acts *actively* on the system, changing its properties. Secondly, the transformation group acts *passively* on the measuring outcomes which correspond to the Borel sets of $\Re$. The principle of covariance implies again the equivalence of *active* and *passive* transformations.[21,22,23] The difference between the covariance postulates of classical and quantum physics consists in the different propositional systems and L_C and L_Q. As in classical mechanics the general concept of an observable can again be specified by the fundamental observables of position, momentum and time.

As in the classical case, also quantum objects will be introduced as carriers of the fundamental properties which correspond to the observables *q* (position), *p* (momentum) and *t* (time). Using the covariance postulate we define a quantum object S_Q as an algebra L_Q such that a unitary representation of the (passive) Galilean group is defined in the automorphism of L_Q that admits the observables *q, p* and t in the sense of the covariance postulate. This means that a quantum object is a carrier of the properties $P \in L_Q$, but not only in one contingent situation, which is given by the apparatus and its space time coordinates, but also in all situations which can be obtained by Galilean transformations. Hence the quantum object is a carrier of properties $P \in L_Q$, which transform covariantly under Galilean transformations.

However, in spite of the similarities in the method of constitution, there are striking differences between classical objects and quantum objects which come from the different lattices L_C and L_Q, respectively. The propositional system L_C is a complete, atomic orthomodular and distributive lattice. Hence the object *S* possesses any property $P \in L_C$ either in the affirmative or in the negative sense, *i.e.,* the object *S* is "completely determined". In contrast to this well-known situation a quantum object *S* possesses at a certain time value *t* simultaneously only a limited class of commensurable properties given by elements of a Boolean sublattice of L_Q. Hence a quantum system is (at a certain time value *t*) only a carrier of a

[21] Schlieder (1968).
[22] Schlieder (1968).
[23] Piron (1976).

class of mutually commensurable properties. One can again specify this concept by considering different classes. Elementary quantum systems are given by irreducible unitary representations of the Galilean-group. For elementary objects there are only projective representations which are characterized by one continuous parameter *m* which can be interpreted as the mass of the quantum object and which characterizes a certain class of objects.

(iii) *Individual Quantum Systems*

The characterization of individual objects in quantum mechanics provides problems which are different from those discussed by Aristotle, Thomas Aquinas, Leibniz, Locke, and Kant. The reasons for the difficulties to individualize quantum objects are that—in contrast to Leibniz—the essential properties are not sufficient for the characterization of an object and that—in contrast to Locke and Kant—totality of all accidental properties which were needed for the individualization is not simultaneously available. Since, roughly speaking, only one half of the classical phase space properties pertain simultaneously to a quantum system and are thus available for the observer, the determination of quantum systems by their accidental properties will never be complete.

There are several ways to deal with this incompleteness. One could use only a set of simultaneously objective properties for partly characterizing a system. This is the way which was used by the founders of quantum mechanics in the twenties and it leads to the merely negative result that quantum systems cannot be individualized. However, one could also use the complete set of phase space properties in an unsharp sense, corresponding to unsharp properties. This latter way of reasoning which was first proposed by Heisenberg[24] in a less formal sense as early as 1930, has been applied to the problem of individuals by many authors in recent years.[25] This approach is much nearer to the experimental evidence of quantum trajectories in cloud chamber experiments and it can be applied for all practical purposes. Indeed, experimentalists have no doubts that the observed cloud chamber traces are the dynamical trajectories of elementary particles, nuclei, atoms, etc.

The individual objects which can be determined by these methods are defined only unsharply. This means that for two fuzzy individuals S_1 and S_2 there is a finite probability to confuse these objects. Even if S_1 and S_2 were clearly distinguished at a given instant of time t by their position values x_1 and x_2, after an arbitrarily short time interval $\Delta t = \zeta$ the systems can no longer be localized sharply and thus not strictly be distinguished. The position values x_1' and x_2' which could be obtained by measurement at time ζ are not uniquely correlated to the position values x_1 and x_2 obtained at time *t*. Hence the systems cannot be exactly reidentified at the time value ζ.

These arguments show that two of the procedures to determine individual systems which were discussed in the traditional philosophy cannot be applied

---

[24] Heisenberg (1930)

[25] Cf. Mittelstaedt (1984, 1995), Busch et al. (1995), Giuntini (1995), Dalla Chiara (1995).

to quantum objects. The characterization of individual quantum systems by their essential and permanent properties, which was first conceived by Duns Scotus and Leibniz, is not possible since the permanent properties define classes of objects (electrons, protons, etc.) which contain more than one object. The characterization of individual systems by their accidental properties, as it corresponds to the methods of Locke and Kant, cannot be applied since the accidental properties are not simultaneously available as sharp properties. There is, however, still the proposal to determine individuals not only by their actual properties (essential or accidental) but also by their complete historical development. The idea to distinguish two entities by their history was first applied by Thomas Aquinas to the human soul. It was not taken into account in the philosophy of Locke and Kant, since these authors considered the position of an object as a sufficient means for its individualization. It was, however, implicitly used also by Leibniz, since the "complete concept" which characterizes an individual object contains not only its present properties but also its complete past and future history.

In quantum mechanics the individualization of objects by their history is a difficult problem which has been treated only accidentally by very few authors. The pure state of a quantum system provides "maximal" information about the system and can thus not easily be extended. However, the maximality mentioned is restricted to predictions about future measurements. Aharonov and Albert[26] have shown that there are interesting situations in which the knowledge about past measurement results together with the present state provides more information about the results of measurements which could have been performed in the past. Hence, even a pure state provides non-maximal information about a system, if in addition to predictions also retrospections are considered. It is an open question whether this retrospective information can be used for distinguishing two systems which are otherwise indiscernible.

Another indication that one can know more about a system than the information contained in its pure state is given by sequential quantum logic.[27] If a system is prepared in a pure state described by the maximal (atomic) proposition W, then the acquisition of information by a sequence of measurements can formally be expressed by a sequential conjunction. A careful analysis of this proposition shows that it contains two components: First, the predictive content which corresponds to the final state of the object after the last measurement, and secondly, the information distinguishing two quantum systems S an S' whose post-measurement states have the same predictive content. It is, however, an unsolved problem whether the non-predictive part of information contained in the sequential proposition could actually be used for distinguishing two quantum systems S an S' whose post-measurement states have the same predictive content.

In classical physics individual objects can be determined by means of trajectories, provided the objects in question are impenetrable. In quantum mechanics these possibilities do not exist. Also in quantum mechanics one can define a concept of

[26]Aharanov and Albert (1984).

[27]Cf., e.g., Stachow (1985), Mittelstaedt (1983), Islam (1994).

an object as carrier of properties but this concept is too weak for the constitution of individual systems. Individuals in the strict sense do not exist in quantum physics. However, unsharp observables, almost repeatable and weakly disturbing measurements allow for the definition of unsharp individuals which is sufficient for all practical purposes. Many quantum physical experiments and the obvious existence of individuals in the classical world can be explained in this way. On the other hand, if quantum mechanics is considered as universally valid[28] then there is no classical world in the strict sense. Consequently, the deficiency of individuals in quantum physics implies that there are no individuals at all.

References

Aharonov, Y., & Albert, D. (1984). Is the usual notion of time evolution adequate for quantum mechanical systems I. *Physics Review D, 29*, 223–227.
Bunch, P., Grabowski, M., & Lahti, P. (1995). *Operational quantum physics*. Heidelberg: Springer.
Bunch, P., Lahti, P., & Mittelstaedt, P. (1996). *The quantum theory of measurement* (2nd. ed.). Heidelberg: Springer.
Castellani, E., & Mittelstaedt, P. (1998). Leibniz' principle, physics, and the language of physics. *Foundations of Physics, 30*, 1587–1604.
Dally Chiara, M. L. (1995). Unsharp quantum logics. *International Journal of Theoretical Physics, 34*, 1331–1336.
Frede, M., & Patzig, G. (1988). *Aristoteles "Metaphysik Z"* (2 Vol.) C.H. Beck, Miinchen.
Giuntini, R. (1995). Quasilinear QMV algebras. *International Journal of Theoretical Physics, 34*, 1397–1407.
Hegerfeld, G. C., & Ruijsenaars, S. N. M. (1980). Remarks on causality, localization and speeding of wave packets. *Physics Review D, 22*, 377–384.
Heisenberg, W. (1930). *Die physikalischen Prinzipien der Quantentheorie*. Leipzig: S. Hirzel.
Islam, C. J. (1994). Quantum logic and the history approach to quantum theory. *Journal of Mathematical Physics, 35*, 2157–2185.
Mittelstaedt, P. (1983). Relativistic quantum logic. *International Journal of Theoretical Physics, 22*, 293–313.
Mittelstaedt, R. (1984). Constituting, naming, and identity in quantum logic. In *Recent developments in quantum logic* (pp. 215–234). Mannheim: B.I. Wissenschaftsvedag.
Mittelstaedt, P. (1994). The constitution of objects in Kant's philosophy and in modern physics. In P. Parrini (Ed.), *Kant and contemporary epistemology* (pp. 115–129). Dordrecht: Kluwer.
Mittelstaedt, P. (1995). Constitution of objects in classical mechanics and in quantum mechanics. *International Journal of Theoretical Physics, 34*, 1615–1626.
Piron, C. (1976). *Foundations of quantum physics* (p. 93 ff, 77–90). Reading: Benjamin.
Schlieder, S. (1968). Einige Bemerkungen zu Zustandsanderungen von relativistischen quantenmechanischen Systemen durch Messung and zur Lukalitatsforderung. *Communications in Mathematical Physics, 7*, 303 331.
Stachow, E.-W. (1985). Structures of quantum language for individual systems. In *Recent developments in quantum logic* (pp. 129–145). Mannheim: B.I. Wissenschaftsverlag.
Strohmeyer, I. (1995). *Quantentheorie und Transzendentalphilosophie*. Heidelberg: Spektrum.
Weyl, H. (1966). Philosophie der Mathematik und der Naturwissenschaften, Oldenburg.
Wieland, W. (1973). Kontinuum and Engelszeit bei Thomas von Aquino. In: E. Scheibe u. G. Siiflmann (Eds.), *Einheit und Vielheit*. Gottingen: Vandenhoeck and Rupprecht.

[28]Cf. Bunch et al. (1995, 1996).

Chapter 13
Weyl, Identity, Indiscernibility, Realism

Otávio Bueno

Abstract In this paper, I reconstruct a technique originally formulated by Hermann Weyl to accommodate, in the foundations of quantum mechanics, aggregates of quantum particles despite these particles' apparent lack of identity. I defend the importance of this technique and provide a slight variant of Weyl's original formulation by avoiding altogether the use of set theory. I then offer formulations of individuals and non-individuals, inspired by considerations that Weyl made in the context of his theory of aggregates, and examine the status of non-individuals with regard to debates about realism. I conclude that there is still much to be learned from careful study of Weyl's work.

Keywords Weyl · Identity · Indiscernibility · Individual · Non-individual · Realism

13.1 Introduction

As part of his attempt to interpret the foundations of non-relativist quantum mechanics, Hermann Weyl developed a suggestive technique to accommodate aggregates of quantum particles while taking into account these particles' apparent lack of identity (see Weyl [1927/1963], pp. 237–252, and [1928/1931]). The technique is suggestive in that it attempts to make sense of the putative restrictions on the applicability of identity in the quantum domain without changing either the underlying logic or the relevant set theory.

In this paper, I reconstruct this technique and examine its significance. I offer a slight variation in the formulation of the technique by not requiring set theory at all; I also discuss associated conceptions of individuals and non-individuals, inspired by considerations made by Weyl in the context of his theory of aggregates, and discuss

O. Bueno (✉)
Department of Philosophy, University of Miami, Coral Gables, FL, USA

A. Cordero (ed.), *Philosophers Look at Quantum Mechanics*,
Synthese Library 406, https://doi.org/10.1007/978-3-030-15659-6_13

the status of non-individuals with regard to realist and anti-realist views. I conclude that there is still plenty to be learned from Weyl's work.

13.2 Identity

In the mid-1920s, when Weyl first developed the technique I describe below, the options of revising the logic or the set theory were clearly available to him. At that time, different, nonequivalent versions of set theory had already been developed. As is well known, in 1908, Ernst Zermelo provided the first axiomatization of set theory, taking sets as primitive, and providing a system that could not be finitely axiomatized (Zermelo [1908/1967]). John von Neumann, in turn, formulated an entirely different, and finitely axiomatizable, system, which takes functions as basic rather than sets (von Neumann [1925/1967]).

Given the central role that functions play in mathematical practice, which is often implemented by inferential procedures based on mappings (that is, functions) of particular structures into other structures, von Neumann's foundational approach is, in this respect, much closer to actual mathematical practice than Zermelo's theory ever was. By establishing inferential relations among structures, functions arguably are more important to that practice than collections of objects are. Such collections allow mathematicians to express, in a unified way, a variety of mathematical objects, relations, and structures, but often the resulting formulations tend to be somewhat artificial.

Consider, for instance, the different and familiar formulations of natural numbers in set theory. Zermelo ordinals specify the natural numbers as follows: $0 = \varnothing$, $1 = \{\varnothing\}$, $2 = \{\{\varnothing\}\}$, $3 = \{\{\{\varnothing\}\}\}$, and so on. In turn, von Neumann ordinals formulate them as: $0 = \varnothing$, $1 = \{\varnothing\}$, $2 = \{\varnothing, \{\varnothing\}\}$, $3 = \{\varnothing, \{\varnothing\}, \{\varnothing, \{\varnothing\}\}\}$, and so on. None of these formulations has any special connection with the objects studied in number theory, despite the fact that each formulation provides distinct set-theoretic surrogates that mimic relevant number-theoretic properties. What is gained in conceptual unification (numbers, on these set-theoretic conceptions, just are sets) is lost in naturalness. After all, whatever numbers are, they need not be sets, and, in fact, are better thought of independently of sets. Just consider that what is crucial for the characterization of number-theoretic structures are not sets, but a function: the successor function that specifies the relevant numbers. Arguably, even in the formulation of the basic objects of number theory, functions seem to be more central. (For further, more recent, developments of von Neumann theory, see Muller [2011], and Sant'Anna and Bueno [2014]. Note that part of the significance of von Neumann's innovation is lost if functions are simply characterized, as is commonly done in Zermelo's theory, in terms of sets.)

But the availability of different formulations of set theory was not the only choice available to Weyl. About a decade earlier, in the late 1910's, while developing his own formulation of analysis in *The Continuum*, Weyl had already employed non-classical techniques, with a constructivist motivation. He articulated a predicativist approach to the construction of real numbers (Weyl [1918/1987]; for additional discussion, see Feferman [1998], pp. 249–283, and Feferman [2005]). The goal was to avoid the use of impredicative definitions, that is, definitions that, in order to characterize certain objects, quantify over a totality of objects to which the objects that are being defined are supposed to belong. The concern is that such impredicative definitions involve a particular form of circularity. Quantification over a totality of objects is assumed in the characterization of these very same objects. It is unclear, however, what the quantification over such a totality amounts to without quantifying over the particular objects that belong to such a totality. But since what is to be specified in the first place are precisely such objects, the resulting definition seems to assume that the objects in question have already been defined. Clearly, this begs the question. It is not surprising that, in providing a foundational account of analysis, Weyl tried to articulate an approach that did not rely on impredicative definitions and any resulting circularity. That is the project he embarked on in *The Continuum*.

As it turns out, due to the needs that emerged from the application of mathematics to physics, which seemed to Weyl to require additional, classically formulated, mathematical theories, he eventually waived his predicativist restrictions and felt compelled to embrace a classical outlook. (Several decades later, Feferman argued that Weyl may have been able to obtain, within a predicativist setting, all the mathematics he needed; see Feferman [1998, 2005]. Weyl clearly did not realize that.) These considerations indicate that Weyl was aware of the possibility of changing the logic or the set theory as a way of accommodating foundational difficulties. In fact, a few years after the publication of *The Continuum*, he notes:

> Mathematics with Brouwer gains its highest intuitive clarity. He succeeds in developing the beginnings of analysis in a natural manner, all the time preserving the contact with intuition much more closely than had been done before. It cannot be denied, however, that in advancing to higher and more general theories the inapplicability of the simple laws of classical logic eventually results in an almost unbearable awkwardness. And the mathematician watches with pain the larger part of his towering edifice which be believed to be built of concrete blocks dissolve into mist before his eyes (Weyl [1927/1963], p. 54).

Weyl's comment on classical mathematics is very telling. Despite the epistemological and conceptual advantages of a constructivist (or, more precisely, predicativist) approach, he eventually favored a classical conception primarily for the rich resources it makes available in contexts involving the application of mathematics, all the way from quantum mechanics to biology.

But classical mathematics, at least as usually formulated in set theory, has a number of assumptions. Some of them may seem, on the surface, to be unquestionably straightforward, but upon reflection, they turn out to be problematic. This is the case with the otherwise apparently innocuous axiom of extensionality, which is part of Zermelo-Fraenkel set theory with the axiom of choice (ZFC) as well as the majority of set-theoretic systems. According to this axiom:

$$\forall x \forall y \, (\forall z \, (z \in x \leftrightarrow z \in y) \rightarrow x = y)$$

Given this axiom's content, every set has well-defined identity conditions, which are determined by their members: two sets are the same just in case they have the same members. The identity of the sets on the consequent of the conditional is determined by the identity of their members in the antecedent: the fact that the sets x and y have the same members z guarantees the identity of these sets (x and y). But this presupposes that the members themselves have well-defined identity conditions. After all, either the objects in question (understood here in a neutral way that does not assume their identity) satisfy the extensionality axiom, or they do not. If they do, then they have well-defined identity conditions, as required by extensionality. After all, the *same* objects z are members of the sets x and y, which requires their identity in the first place. If *different* objects were members of the sets x and y, these sets would *not* be the same. In other words, if the identity of a set is determined by the identity of its members, the latter's identity is demanded. However, if the objects under consideration do not satisfy extensionality and, thus, do not have well-defined identity conditions, then the identity of the resulting sets cannot be determined. It is unclear how the identity of sets whose members lack well-defined identity conditions could be specified, since the identity of the sets ultimately demands the identity of their members. In either case, the extensionality axiom's commitment to the identity of the objects under consideration is clear.

Weyl was certainly aware of the need for identity in the implementation of set theory. He was similarly aware of the challenges posed by the use of set-theoretic resources in the foundations of analysis (and mathematics more generally). In *The Continuum*, he notes: "To every primitive or derived property P there corresponds a set (P)" (Weyl [1918/1987], p. 20; italics omitted). Clearly, this amounts to a *restricted* comprehension principle, relative to primitive or derived properties. Without such a restriction, Weyl's system would be inconsistent. It is then surprising that he continues with the following claim:

> The expressions 'An object *a* has the property *P*' (or 'The relevant judgment scheme $P(x)$ containing one blank is true for $x = a$') and 'a is an element of the set (P)' have the same significance (Weyl [1918/1987], p. 20; italics omitted).

In this passage, Weyl seems to come perilously close to identify predication $P(a)$ with membership $a \in \{P\}$ by indicating that they have the same significance (that is, in light of the context, the same meaning). However, any such identification does not hold in general. As George Boolos ([1998], p. 40) notes, the identification of predication and membership does not preserve either the validity or the implication of certain second-order predications. Even though the second-order sentence '$\exists X\ \forall x\ Xx$' is valid, the corresponding set-theoretic statement, '$\exists \alpha\ \forall x\ x \in \alpha$', is not. Similarly, even though it follows from '$\forall Y\ (Yx \leftrightarrow Yz)$' by logic alone that '$x = z$', the corresponding set-theoretic expression, '$\forall \alpha\ (x \in \alpha \leftrightarrow z \in \alpha)$', does not imply '$x = z$' without the use of some set theory. Predication and membership are importantly different (see also Bueno [2010]).

The next feature in the set-theoretic background that Weyl develops in *The Continuum* deals with a version of the extensionality axiom. As he points out:

> The same set corresponds to two such properties P and P' if and only if every object (of our category) which has the property P also has the property P', and conversely (Weyl [1918/1987], p. 20; italics omitted).

Interestingly, Weyl reintroduces here an important constraint: the form of extensionality principle he advances is restricted to the objects whose properties have been previously identified as being primitive or derived (these are the objects of "our category", in Weyl's own words). Clearly, such an extensionality principle presupposes identity and thus, for the reasons discussed above, involves the commitment to the identity of the objects under consideration.

Identity is such a basic notion, and since it is arguably fundamental (see, for instance, Bueno [2014, 2015]), one may wonder why any commitment to the identity of objects should be perceived as problematic in any way at all. The source of the puzzle emerges from the impressive empirical success of non-relativist quantum mechanics and the fact that a very well-motivated interpretation of the theory, articulated by Weyl and Erwin Schrödinger, among others, insists that quantum particles lack well-defined identity conditions (for a thorough discussion and references, see French and Krause [2006]).

One of the salient features of this interpretation is that quantum statistics can be very straightforwardly obtained in light of quantum particles' lack of identity conditions. In the case of classical mechanics, given two objects, a and b, and two states, S_1 and S_2, there are four possible combinations of such objects in the relevant states:

(1) a and b in S_1 and no object in S_2;
(2) no object in S_1 and a and b in S_2;
(3) a in S_1 and b in S_2;
(4) b in S_1 and a in S_2.

Assuming that each of these combinations has the same probability, each combination then has probability ¼.

In contrast, in (non-relativist) quantum mechanics the situation is different. A symmetry principle is in place to the effect that if two quantum particles of the same kind (e.g. two electrons) are swapped, that does not change the states of the quantum system they are in. This is an expression of the fact that nothing in the quantum mechanical description of such particles allows them to be distinguished from each other. In particular, there are no individuating features that uniquely single out of any such particles. (As Weyl [1928/1931] would say: there is no alibi for an electron.) In other words, if the relevant quantum objects lack well-defined identity conditions, they cannot be distinguished from one another. They are, thus, indiscernible. Strictly speaking, they cannot even be denoted by different labels, since this might suggest that somehow there is a difference between them: a more basic and fundamental trait that goes beyond their sheer numerical difference. (There are two particles, after all.) To highlight this point, each quantum particle is denoted by *, with no individuating traits.

Given two quantum particles and two quantum states, the relevant possibilities of combination are then as follows:

(1) * and * in S_1 and no object in S_2;
(2) no object in S_1 and * and * in S_2;
(3) * in S_1 and * in S_2;
(4) * in S_1 and * in S_2.

Something quite interesting has just happened. Given the indiscernibility of the relevant quantum objects, possibilities (3) and (4) cannot be distinguished, since the objects involved in each combination are indistinguishable. As a result, rather than four combinations as in the classical mechanics case, there are only three, and the resulting probability (assuming, again, that each combination is equally probable) is 1/3 (see French and Krause [2006], which also provides additional details and discussion). A salient feature of quantum statistics emerges straightforwardly from the lack of identity conditions for quantum particles.

The same cannot be said for those interpretations of quantum mechanics that assume that quantum particles have well-defined identity conditions. They need to provide an account of how, despite the identity and distinguishability of quantum particles, the resulting probabilities are those found in quantum mechanics rather than in classical physics. Some additional story needs to be offered to generate the relevant statistics. This gives an important motivation for those views, such as Weyl's, that emphasize the lack of identity of quantum particles, even though the set-theoretic framework adopted by Weyl seems to require, due to the axiom of extensionality, the identity of the objects that are quantified over. *Prima facie*, this seems to create a tension within Weyl's view: the lack of identity of quantum particles cannot be fully expressed in the underlying set-theoretic framework, since the framework assumes the identity of *all* objects. Given set theory, one can form sets of quantum particles, and as noted above, by the extensionality axiom, such particles would need to have identity conditions. However, this conflicts with Weyl's own interpretation of quantum mechanics, which denies that quantum particles have any such identity conditions.

How can this tension be resolved?

13.3 Indiscernibility

13.3.1 Aggregates Without Sets

It is possible and, in fact, preferable to formulate Weyl's approach entirely set free. The approach that he develops to the foundations of analysis is much better implemented *without* sets. One starts, as Weyl does, with properties: precisely those that he invokes to begin with. But without forming sets from such properties, one simply considers the objects that have such properties. One is interested, in any case, in the relevant objects rather than the sets of such objects. When formulating analysis, real numbers are the objects of study rather than sets. Sets are arguably convenient devices to specify the range of the objects under consideration. But sets are not needed to formulate the objects: it is the relevant properties that do the work. That was Weyl's insight (even though he, unfortunately, ended up expressing it in terms of set theory). If a real number structure is a complete ordered field, this is a property of this structure. Resist expressing this structure in set-theoretic terms and consider directly the properties that characterize the structure. The result is a reflection on real numbers and their structure rather than on sets of such numbers.

Once this step is taken and set theory is left behind, one can consider the relations among the various properties under study without having to settle the issue whether the relevant objects have well-defined identity conditions or not. One can then consider objects that are indistinguishable from one another, whether they are complex numbers (*i* or -*i*) or electrons. Such objects form an equivalence relation (which is reflexive, symmetric, and transitive): the objects are *indistinguishable from any other objects in the relation*. But note that an equivalence relation is *not* to be thought of as a set any more than a married couple is a set. The couple is nothing more than two people related to one another by marriage; an equivalence relation is nothing more than objects related to one another by a reflexive, symmetric, and transitive relation (such as indistinguishability). One should avoid the temptation of reifying objects in terms of sets.

The resulting approach introduces aggregates as (non-set-theoretic) equivalence relations. And provided that set theory is not invoked or assumed, no assumption about the identity of the objects in question is made. This allows one to articulate Weyl's own preferred interpretation of quantum mechanics without the artificial constraints imposed by a set theory that would require quantum particles to have well-defined identity conditions.

Weyl formulates his approach to aggregates as follows:

> An aggregate of white, red, and green balls may contain several white balls. Generally speaking, in a given aggregate there may occur several individuals, or *elements*, of the same *kind* (e.g. several white balls) or, as we shall also say, the same entity (e.g. the entity white ball) may occur in several *copies*. One has to distinguish between *quale* and *quid*, between equal (= of the same kind) and identical (Weyl [1927/1963], p. 238).

An aggregate is nothing more than various objects that have some common properties. They are of the same kind in the sense that the properties they have in common specify the objects. Elements here should *not* be understood as being members of a set (kinds are not sets). Rather elements are objects that have certain properties in common, namely, the properties that specify their kind (such as, *being white balls*).

Weyl also gestures, in the passage above, at the distinction between equality and individuality. Equality involves being of the same kind, that is, having the same specifying properties. Individuality, in turn, involves having a *quid*, that is, something that makes an object the particular object it is, which requires far more than being of the same kind. Of course, this is a particular metaphysical understanding of individuality, and it is not clear that it is Weyl's. I will discuss below a less metaphysical view of the matter, which seems to be more congenial to his approach.

Weyl, however, clearly recognizes the metaphysical tradition of reflection on this issue, as he continues:

> To the question of individuation thus arising, Leibniz gave an *a priori* answer by his *principium identitatis indiscernibilium*. Physics has recently arrived at a precise and compelling empirical solution as far as the ultimate elementary particles, especially the photons and the electrons, are concerned. Closely related is the question of the conservation of identity in time; the identical 'I' of my inner experiences is the philosophically most significant instance. Our decision as to what is to be considered as equal or different influences the counting of 'different' cases [...] (Weyl [1927/1963], p. 238).

Three important points are made in this passage. (a) In light of a metaphysical understanding of individuality, the issue of individuation immediately arises: how can an individual be individualized? One answer is to interpret Leibniz's principle of identity of indiscernibles as a principle of individuation: as principle that uniquely singles out each individual. But whatever Leibniz's view on this principle ultimate was (for a recent discussion, see Rodriguez-Pereyra [2014]), this principle can also be understood in a metaphysically less inflationary way as just a logical principle, leaving aside the issue of individuation. (b) Weyl also briefly gestures at the way in which the issue of individuation has been approached empirically by quantum physics, an approach that emphasizes precisely the lack of identity of elementary particles, such as photons and electrons. (c) An additional, separate although related, issue concerns the conservation of identity over time. When do objects remain the same in time and when do they change? Here the identity of the self provides a rich source of examples, also with a long and complex history both within Eastern and Western philosophy (Weyl was familiar with central aspects of both).

As noted above, the crucial feature of aggregates is that they form an equivalence relation. This is a point that Weyl highlights:

> Balls may be white, red or green; electrons may be in this or that position; animals in a zoo may be mammals or fish or birds or reptiles; atoms in a molecule may be H, He, Li, ... atoms. The universal expression for such 'equality in kind' is by means of a binary relation $a \sim b$ satisfying the axioms of equivalence: $a \sim a$; if $a \sim b$ then $b \sim a$; if $a \sim b$, $b \sim c$, then $a \sim c$. Various words are in use to indicate equivalence, $a \sim b$, of two arbitrary elements, a, b, under a given equivalence relation $\sim$: a and b are said to be the same *kind* or *nature*, they are said to belong to the same *class*, or to be in the same *state* (Weyl [1927/1963], p. 239).

Although one of the formulations of an equivalence relation that Weyl provides is cast set-theoretically in terms of the objects that "belong to the same *class*", there is no need to reify the relation in this way. Crucial for an equivalence relation is the fact that the relevant objects are related to one another by reflexivity, symmetry, and transitivity. Whether the objects in question form (or not) a particular set, an equivalence class, is not required for that. Once again, sets are not central in this context and can be ultimately dispensed with. It is telling that Weyl also presents equivalence relations in explicitly non-set-theoretic terms, as objects that are "said to be the same *kind*" or that are "in the same *state*". This clearly indicates his awareness of a non-set-theoretic understanding of equivalence relations. And it is ultimately central for his conception of an aggregate that the commitment to the identity of the objects in the aggregate is not forthcoming. This eventually means bypassing the need to invoke set theory altogether, although Weyl's own formulation of an aggregate is still cast in terms of sets.

According to Weyl:

> An *aggregate* S is a set of elements each of which is in a definite state; hence the term aggregate is used in the sense of 'set of elements with equivalence relation'. Let us assume that an element is capable of k distinct states $C_1, \ldots, C_k$. A definite *individual state* of the aggregate S is then given if it is known, for each of the n marks p, to which of the k classes the element marked p belongs. Thus there are k_n possible individual states of S (Weyl [1927/1963], p. 239).

Even though Weyl characterized an aggregate as a set, the use of set theory in this context is entirely dispensable. An aggregate is nothing more than objects related to one another by an equivalence relation; these objects are, in Weyl's own words, "in a definite state". The state of the aggregate then results from the particular states of the objects that form the aggregate.

The central feature of aggregates, understood non-set-theoretically, is that they do not require the identity of the objects that are related to one another in the aggregate. In fact, as the result of the use of an equivalence relation, the objects in an aggregate are such that any individual differences among them ultimately do not matter. There is a clear sense in which an equivalence relation is an expression of such an indifference, which is achieved by not introducing artificial differences among the objects in the aggregate (such as by labeling them). Then only the cardinality of the aggregates is specified. As Weyl notes:

> If, however, no artificial difference between elements are introduced by their labels p and merely the intrinsic differences of state are made use of, then the aggregate is completely characterized by assigning to each class C_i ($i = 1, \ldots, k$) the number n_i of elements of S that belong to C_i. These numbers, the sum of which equals n, describe what may conveniently be called the *visible or effective state* of the system S. Each individual state of the system is connected with an effective state, and two individual states are connected with the same effective state if and only if one may be carried into the other by a permutation of the labels [...] (Weyl [1927/1963], pp. 239–240).

The significance of not introducing artificial differences *via* labeling is that one can then treat the objects in the aggregate for what they are without attempting to distinguish them individually. But this raises the issue, which underlies Weyl's discussion, of what does it take, in general, for an object to be an individual?

13.3.2 Individuals

I offer here a minimal formulation of individuals. Arguably, there are two basic requirements: one involves having well-defined identity conditions (without which it would not be possible for individuals to be singled out), and the other concerns persistence conditions (without which an individual would not persist over time). The requirements, formulated in generic terms, are:

(I_1) *Identity conditions*: an individual has (clearly determined) identity conditions. In terms of them, it is specified the ways in which an individual differs from other objects and the ways in which it is the same.

(I_2) *Persistence conditions*: an individual persists over time (despite changes in some of its properties). In terms of these conditions, it is specified what it takes for an individual to remain an individual while some of its features shift.

Of course, the requirements are only formulated in generic terms. But once these conditions have been fully specified, the identity and persistence of individuals can be determined. It is important, though, to recognize that not all individuals meet both conditions. We can call *episodic individuals* those that satisfy (I_1), but not (I_2). They are individuals, but do not last very long. *Robust individuals*, in turn, satisfy both (I_1) and (I_2). Their persistence conditions ensure that they remain beyond episodic moments.

To satisfy the identity conditions (I_1) is generally not difficult. Identity is understood here in a basic, non-metaphysically loaded, way: an equivalence relation for which substitutivity holds. No assumption is made, regarding identity, about there being essential properties, quiddities, or thisness of the objects involved. These are metaphysically contentious traits that are not part of identity *per se* but provide resources to interpret identity metaphysically. However, no such interpretation is, strictly speaking, required and none is assumed here.

In contrast with what happens with the satisfaction of identity conditions for individuals, there are at least two main ways of satisfying (I_2):

(I_2A) *Essential traits*: As long as essential (or necessary) features of an individual are preserved, the individual remains in existence. This requires, of course, the identification and articulation of such essential properties. As just noted, the proposal advanced here, just as, in my view, Weyl's own account, is not committed to such essentialism. After all, there is a more deflationary alternative readily available.

(I_2B) *Closest continuers*: Given an individual *i* that satisfies condition (I_1), at each moment in time, the closest continuer to individual *i* (the one that shares most properties with *i*) is taken to be *i* (for a thorough discussion, see Nozick [1981]). This is clearly a more deflationary alternative than essentialism since nothing like essences is required. One can interpret the closest continuer of a given individual as just the object that happens to be the one that shares more properties with that individual, and the existence of the object is just an empirical fact about the world. Alternatively, the identification of the closest continuer of a given individual *i* can be implemented pragmatically (still taking into account the available evidence), as the individual that, for all practical purposes and to the best of one's information, shares most properties with the individual *i*.

Interestingly, a deflationary version of the closest continuer theory was already formulated by Weyl as part of the development of his account of aggregates. As he notes:

> Whenever in reality identification of the same being at different times is carried out, it is of necessity based on the observable state. For a continuous flow of time and a continuous manifold of states, the underlying principle is by and large to be formulated as follows: suppose there exists at time t but one individual in a certain state C appreciably different from the states of all other individuals [thus an individual satisfies condition (I_1)]; if afterwards, especially if shortly afterwards, at a time t', one and only one individual is encountered in a state C' deviating by little from C, or 'typically similar' to C, then the presumption is justified that one is dealing with the *same* individual at both moments t and t' [thus an individual satisfies condition (I_2B)] (Weyl [1927/1963], p. 243).

Clearly, Weyl's preference for an empirically grounded closest continuer account of individuals is manifest. I fully agree with the importance of providing a metaphysically deflationary understanding of individuals, and that is a significant feature of the closest continuer account (at least on my reading of the proposal). This is especially so given the fact that the incursion through the perilous domain of individuals emerged in the context of a theory of aggregates. One of the chief motivations for the development of such a theory is precisely to make room for the theorizing of objects that lack well-defined identity conditions, namely, non-individuals. In fact, it is in contrast with individuals that an account of non-individuals is built.

13.3.3 Non-individuals

If quantum objects lack well-defined identity conditions, as the interpretation of non-relativist quantum mechanics favored by Weyl recommends, they are one of the primary examples of non-individuals. Similarly to what happens with individuals, there are two minimum requirements for non-individuals (the requirements are formulated in opposition to those of individuals):

(N_1) *Lack of identity conditions*: a non-individual does not have (clearly determined) identity conditions. This is the crucial aspect of a non-individual and the most salient feature to reckon with in the foundations of non-relativist quantum mechanics. To accommodate this feature is what prompted Weyl to articulate his account of aggregates.

(N_2) *Lack of persistence conditions*: a non-individual lacks persistence conditions over time. Nothing provides conditions in which a non-individual remains the same. In fact, it is unclear how any persistence conditions could be specified in the first place without the application of identity to at least the *kind* of non-individual under consideration. Particular electrons or protons are non-individuals and, as such, may not have well-determined identity conditions. But considered as kinds, they do have such conditions, given that electrons differ from protons. (There is no need to reify kinds to make this point: all that is needed are the properties that characterize the relevant objects: protons and electrons have different charges.)

Similarly to what happens with individuals, not every non-individual satisfies both conditions (N_1) and (N_2). *Regular* non-individuals satisfy (N_1). *Uber*-non-individuals satisfy both (N_1) and (N_2), that is, in addition to lacking identity conditions, they also do not have persistence conditions. They are extremely transitory things.

But how can a non-individual fail to satisfy condition (N_2)? That is, how could something that lacks identity conditions remain the same over time? After all the application of identity seems to be required if any persistence condition is to be in place. However, in principle, an object that does not have well-defined identity conditions may not, thereby, simply vanish immediately after coming into being. Perhaps nothing determines that such an object remains the same or not, but that does not entail that the object no longer exists. Of course, from an epistemic point of view, it is not required that one be able to determine whether an object remains the same or not for the object in question to continue to exist. And from a metaphysical perspective, perhaps nothing determines whether the object remains the same or not, but that does not entail that the object no longer exists. It is in this sense that non-individuals that lack well-defined identity conditions need not thereby also fail to have persistence conditions. At least, it is an open issue whether they have persistence conditions or not. To specify such conditions would require, it seems, the identity of the relevant kinds. Although an electron may not have identity conditions, it still remains an electron (rather than, say, a proton), despite the fact that nothing determines whether it is the same electron or not.

Weyl clearly recognizes that the objects that form an aggregate need not be individuals and thus the information about them would be incomplete if such objects were treated as individuals. After all, individuals typically have individuating conditions which non-individuals lack. As Weyl points out:

> If [...] at each moment attention is given to the visible state only, then the numbers [of things in a given state] $n_1(t)$, ..., $n_k(t)$ in their dependence on t contain the complete picture—however incomplete this information is from the 'individualistic' standpoint (Weyl [1927/1963], p. 242).

Weyl is also arguably aware of the two conditions just mentioned that are involved in the specification of non-individuals. He applies them to the components of an aggregate. First, he notes, the objects in the relevant state an aggregate is in lack identity conditions (that is, condition (N_1) of a non-individual is satisfied):

> For now we are told only how many elements, namely $n_i(t)$, are found in the state C_i at any time t, but no clues are available whereby to follow up the identity of the n individuals through time [since those things are non-individuals] (Weyl [1927/1963], p. 242).

Second, he continues, it is unclear whether the objects that form an aggregate remain the same over time (on his view, it is not even proper to ask whether they do). As he points out:

> We do not know, nor is it proper to ask, whether an element that is now in the state, say C_5, was a moment before in the state C_2 or C_6. The world is created, as it were, anew at every moment, no bond of identity joins the beings present at this moment with those encountered in the next. [...] This non-individualizing description is applicable even if the

total number $n_1(t) + \ldots + n_k(t) = n(t)$ of elements does not remain constant over time (Weyl [1927/1963], pp. 242–243).

Condition (N_2) is then also clearly satisfied. The close connection between aggregates and non-individuals is thereby clearly established.

13.4 Realism

It is important to highlight that the strategy sketched above of accommodating non-individuals in terms of an equivalence relation (formulated independently of set theory) is *neutral* regarding the ontological status of non-individuals. One can interpret the strategy realistically and claim that non-individuals exist. But one can also interpret the strategy anti-realistically, and be, for example, agnostic about the existence of non-individuals. Nothing in the proposed strategy settles the issue one way or another. This is a significant advantage of the view, since conceptualizing non-individuals should be independent of arguing whether they exist (or not).

The issue of realism about non-individuals emerges as part of the development of an interpretation of non-relativist quantum mechanics. Depending on what it is taken for the commitment to the existence of objects, different responses will emerge regarding non-individuals. Realists typically require at least support from suitably virtuous theories to conclude that certain objects (or structures) exist. This requires having a theory that posits the objects in question and satisfies familiar theoretical virtues: the theory is simple, unified, explanatory, and empirically adequate. (This is, in outline, what Jody Azzouni considers to be thin epistemic access; see Azzouni [2004].) Suppose, however, that there is no empirical, instrumental access to the objects in question, only the theoretical access just described. In light of the lack of access, some controversy usually emerges as to whether the relevant objects in fact exist or not. Some will insist they do, others understandably and legitimately raise doubts.

The issue about the existence of the objects normally will only be resolved once empirical, instrumental access to the relevant objects is established. To quality as epistemically adequate, the access needs to have a few features: it is robust (it obtains independently of particular beliefs about the objects in question); it can be refined (one can improve the access to the objects), and the access allows one to track the objects in space and time. (These are three of the four conditions for thick epistemic access; see Azzouni [2004]. I do not think that Azzouni's fourth condition is in fact relevant for instrumental access, so I did not include it here.)

Anti-realist views tend to challenge both the adequacy of these conditions and whether one is in a position to know (or have good reason to believe) that the conditions apply at all. With regard to the theoretical utility approach (thick epistemic access), anti-realists insist that simplicity, unification, and explanatory power are ultimately pragmatic virtues rather than epistemic ones; that is, they provide good reason to accept a theory (as a useful device to work with) but they need not settle the issue as to whether the theory is true or not (see van Fraassen [1980]). After all, a theory such as Newtonian physics clearly satisfies all three theoretical virtues: it is simple (it accounts for a variety of phenomena on the basis of gravity); it is unified (it accounts for both astronomical phenomena and phenomena near the surface of the earth), and it is explanatory (it explains the tides, planetary motion, etc.). The theory is also empirically adequate provided that it is restricted to domains in which strong gravitational fields or speeds that approach that of light are not involved. Despite that, Newtonian physics is false: it fails, for instance, to account for the perihelium of Mercury. Satisfaction of theoretical virtues is not truth conducive (see Bueno and Shalkowski [2019]).

Anti-realists typically grant that empirical adequacy is an epistemic virtue. Despite that, the fact that a theory is empirically adequate is not sufficient to establish the existence of unobservable entities given that empirically adequate theories can still be false.

With regard to the empirical, instrumental access to certain objects (thick epistemic access), anti-realists will typically grant the importance of such access but will note that without knowing (or, at least, having good reason to believe) that the conditions of robustness, refinement, and tracking are satisfied, one may still think that access to certain objects has been forged when, in fact, no such access is involved. This is similar to the concern that unless one knows (or has good reason to believe) that an instrument is reliable or has been properly calibrated, one is not in a position to know (or to have good reason to believe) that the information offered by the instrument is correct (see Bueno [2016]). Similarly, unless there are grounds to believe that the three conditions of thick access above have been satisfied, it is an open issue whether the objects in question exist or not. And depending on the objects under study, it may not be clear how to be in a position to know (or to have good reason to believe) that the conditions in question have indeed been satisfied.

Interestingly, this is precisely the situation with regard to non-individuals, such as electrons, that are involved in the foundations of non-relativist quantum mechanics. Realists argue that they satisfy both the theoretical considerations and the conditions for empirical, instrumental access. Hence, on their view, there are more than enough reasons to believe that these objects exist. Anti-realists, in turn, while not denying the existence of such non-individuals and the role they play in non-relativist quantum mechanics, question whether the considerations provided by realists to support the existence of non-individuals are known to apply. After all, despite the undeniable empirical success of quantum mechanics, it is not exactly clear that the theory is simple (its multiple, non-equivalent, and mutually inconsistent interpretations seem to question any such simplicity); the theory is not exactly unified (it fails to account for the behavior of astronomical objects), and it is not exactly clear how explanatory the theory ultimate is (given that unless an interpretation is provided, there is no answer to the question of what is going on beyond the appearances, and there is no agreement about which interpretation, if any, is ultimately correct).

As for the detection of non-individuals, once again, there is no doubt that quantum mechanics is extremely impressive. However, the point still stands that the interpretation of these results is not settled by the experiments alone (see Bueno [2018]), and there is much room for maneuver regarding what is really going on beyond the appearances, including, as it turns out, whether there are non-individuals at all. Given that non-relativist quantum mechanics is compatible with both interpretations that posit individuals and with those that posit non-individuals (see French and Krause [2006]), one cannot settle the issue of which of these interpretations (if any) is ultimately correct.

The result is that it is not clear that a commitment to non-individuals is required at this point. Of course, this does not entail that these objects do not exist, nor does it mean that careful consideration about what non-individuals are and how they are invoked in the foundations of quantum mechanics should not be pursued. On the contrary, a proper understanding of these issues will certainly illuminate one's comprehension of this crucial theory.

13.5 Conclusion

It is reassuring to see that even after so many decades, Weyl's contribution continues to motivate and inspire so much work. Far more, of course, needs to be said about the matters discussed in this article, but I hope enough was said to indicate the importance of Weyl's aggregates to the foundations of quantum mechanics and its philosophical understanding.

In fact, as argued above, Weyl's aggregates, suitably interpreted to avoid sets, provide a rich framework to characterize both individuals and non-individuals. It allows one to formulate a set-free and (in the case of non-individuals) identity-free account that can be used in the foundations of quantum mechanics. In future work, I intend to assess the overall feasibility of Weyl's approach and contrast it with attempts to make sense of the foundations of non-relativist quantum mechanics by jettisoning identity and revising both the underlying logic and the relevant set theory (such as the proposal developed in French and Krause [2006]). My impression is that Weyl's original approach has significant benefits. But this is something that needs to be left for another occasion.

Acknowledgements My thanks go to Jonas Arenhart, Chris De Ronde, Steven French, Roman Frigg, Roberto Giuntini, Miklós Redéi, Bryan Roberts, Simon Saunders, and especially Décio Krause, for extremely helpful discussions about the issues examined in this paper. Many thanks are also due to Alberto Cordero for his support and patience throughout the process of writing and completing this work.

References

Azzouni, J. (2004). *Deflating existential consequence: A case for nominalism*. New York: Oxford University Press.
Boolos, G. (1998). *Logic, logic, and logic*. Cambridge, MA: Harvard University Press.
Bueno, O. (2010). A defense of second-order logic. *Axiomathes, 20*, 365–383.
Bueno, O. (2014). Why identity is fundamental. *American Philosophical Quarterly, 51*, 325–332.
Bueno, O. (2015). Can identity be relativized? In A. Koslow & A. Buchsbaum (Eds.), *The road to universal logic* (Vol. II, pp. 253–262). Dordrecht: Birkhäuser.
Bueno, O. (2016). Epistemology and philosophy of science. In P. Humphreys (Ed.), *Oxford handbook in the philosophy of science* (pp. 233–251). Oxford: Oxford University Press.
Bueno, O. (2018). Can quantum objects be tracked? In O. Bueno, R. Chen, & M. Fagan (Eds.), *Individuation, process, and scientific practices*. New York: Oxford University Press.
Bueno, O., & Shalkowski, S. (2019). Troubles with theoretical virtues: Resisting theoretical utility arguments in metaphysics. *Philosophy and Phenomenological Research*. (forthcoming).
Bueno, O., Chen, R., & Fagan, M. (Eds.). (2018). *Individuation, process, and scientific practices*. New York: Oxford University Press.
Feferman, S. (1998). *In the light of logic*. New York: Oxford University Press.
Feferman, S. (2005). Predicativity. In S. Shapiro (Ed.), *Oxford handbook of philosophy of mathematics and logic* (pp. 590–624). Oxford: Oxford University Press.
French, S., & Krause, D. (2006). *Identity in physics: A historical, philosophical, and formal analysis*. Oxford: Clarendon Press.
Humphreys, P. (Ed.). (2016). *Oxford handbook in the philosophy of science*. Oxford: Oxford University Press.
Koslow, A., & Buchsbaum, A. (Eds.). (2015). *The road to universal logic* (Vol. II). Dordrecht: Birkhäuser.
Muller, F. (2011). Cantor-von Neumann set-theory. *Logique et Analyse, 213*, 31–48.
Nozick, R. (1981). *Philosophical explanations*. Cambridge, MA: Harvard University Press.
Rodriguez-Pereyra, G. (2014). *Leibniz's principle of identity of indiscernibles*. Oxford: Oxford University Press.
Sant'Anna, A., & Bueno, O. (2014). Sets and functions in theoretical physics. *Erkenntnis, 79*, 257–281.
Shapiro, S. (Ed.). (2005). *Oxford handbook of philosophy of mathematics and logic*. Oxford: Oxford University Press.
van Fraassen, B. C. (1980). *The scientific image*. Oxford: Clarendon Press.
van Heijenoort, J. (Ed.). (1967). *From Frege to Gödel: A source book in mathematical logic, 1879–1931*. Cambridge, MA: Harvard University Press.
von Neumann, J. (1925/1967). An axiomatization of set theory [in German]. *Journal für die reine und angewandte Mathematik 154*, 219–240. (English translation in van Heijenoort (ed.) [1967], pp. 393–413.)
Weyl, H. (1918/1987). *The continuum: A critical examination of the foundations of analysis*. (Translated from the German by Stephen Pollard and Thomas Bole. The first German edition was published in 1918.) New York: Dover.
Weyl, H. (1927/1963). *Philosophy of mathematics and natural science*. (Revised and augmented English edition, based on a translation by Olaf Helmer. The first German edition was published in 1927.) New York: Atheneum.
Weyl, H. (1928/1931). *The theory of groups and quantum mechanics*. (Translated from the second, revised German edition by H.P. Robertson. The first edition was published in 1928.) New York: Dover.
Zermelo, E. (1908/1967) Investigations in the foundations of set theory I" [in German]. *Mathematische Annalen 65*, 261–281. (English translation in van Heijenoort (ed.) [1967], pp. 199–215.)

Part V
Copenhagen Insights Revisited

Chapter 14
What Is Really There in the Quantum World?

Jeffrey Bub

Abstract The state of a classical system represents physical reality by assigning truth values, true or false, to every proposition about the values of the system's physical quantities. I present an analysis of the Frauchiger-Renner thought experiment (Frauchiger D, Renner R: Single-world interpretations of quantum mechanics cannot be self-consistent. arXiv eprint quant-ph/1604.07422, 2016), an extended version of the 'Wigner's friend' thought experiment (Wigner E: Remarks on the mind-body question. In: Good IJ (ed) The scientist speculates. Heinemann, London, 1961), to argue that the state of a quantum system should be understood as purely probabilistic and not representational.

14.1 Introduction

Quantum mechanics was born in 1925 with Heisenberg's seminal paper (Heisenberg 1925) 'On the quantum-theoretical re-interpretation of kinematical and mechanical relations.' Heisenberg thought that Bohr's atomic theory with its discrete electron orbits was not the right way to think about the structural features of atoms responsible for the emission and absorption spectra of gases. His proposal was to 're-interpret' classical mechanical quantities, like position, momentum, energy, angular momentum, as operations, subsequently in work with Born and Jordan (1925) and Born et al. (1925) represented by operators that act on and transform the states of quantum systems. Since operations needn't commute, the result was a noncommutative mechanics that explained the discrete frequencies of light emitted by atoms without appealing to electron orbits, and further elaborations of the theory explained other phenomena that couldn't be explained by classical physics.

Should the quantum state be understood as *representational* or *probabilistic*? Either response leads to conceptual puzzles. David Wallace asks the question in

J. Bub (✉)
University of Maryland, College Park, MD, USA
e-mail: jbub@umd.edu

A. Cordero (ed.), *Philosophers Look at Quantum Mechanics*,
Synthese Library 406, https://doi.org/10.1007/978-3-030-15659-6_14

(Wallace 2016) and takes the representational view: on the Everett interpretation the quantum state, or the wave function in configuration space, represents what's really there; the probabilistic role is recovered emergently and approximately.

In 'Two Dogmas About Quantum Mechanics' Bub and Pitowsky (2010), following Pitowsky (2007), Pitowsky and I argued that we should drop the 'dogma' that the quantum state is the Hilbert space analogue of the classical state in phase space, which represents physical reality by assigning truth values, true or false, to every proposition about the values of physical quantities. By contrast with the Everettian's many-worlds account of quantum mechanics, we proposed a single-world account along the following lines: Hilbert space encodes generic constraints on probabilistic correlations between measurement outcome events. As such, it is a probability theory of a new sort. We compared this to the way in which the geometry of Minkowski space-time in special relativity encodes generic constraints on spatio-temporal configurations of events. Both theories are 'principle' theories as opposed to 'constructive' theories in Einstein's sense (Einstein 1954). Hilbert space and Minkowski space-time don't provide representations of physical 'stuff.' Rather, they characterize the basic kinematic (or pre-dynamic) structure of their respective theories. Minkowski space-time provides the kinematic framework for the physics of a non-Newtonian, relativistic universe, in which there is no absolute spatial or temporal separation of events. Hilbert space provides the kinematic framework for the physics of an indeterministic, irreducibly stochastic universe, in which there are intrinsically random events. Just as Minkowski space-time imposes kinematic constraints on events to which a relativistic dynamics is required to conform, a quantum dynamics of matter and fields is required to conform, through its symmetries, to the kinematic structure of Hilbert space. See Michel Janssen (2009) for a defense of this view of special relativity contra Harvey Brown (2006).

By Gleason's theorem (1957), the non-Boolean subspace structure of Hilbert spaces of three or more dimensions uniquely characterizes the possible probability assignments to events (see Pitowsky 2003, 2007). These probabilities, expressed by the Born rule, are encoded in the quantum state, so the quantum state is a probability function, an assignment of probabilities to possible measurement outcome events. By a theorem of Wigner (1959) and Uhlhorn (1963), unitary (or anti-unitary) dynamical evolution is uniquely consistent with the Hilbert space structure. The unitary dynamics describes the evolution of probabilities and probabilistic correlations, not the evolution of events through time, or the evolution of truth values assigned to the corresponding propositions, as in classical mechanics.

A measurement in classical mechanics is a dynamical evolution in which an observable (a dynamical quantity) of a measured system becomes correlated with a 'pointer' observable of a system functioning as the measuring instrument. A 'measurement' in quantum mechanics is an entirely different matter: an experimental procedure in which an *indefinite* observable of a 'measured' system comes to have a *definite* value recorded in the change of a pointer observable of a macrosystem functioning as the 'measuring' instrument. The role of the measuring instrument is to define a chance set-up by selecting a particular set of measurement outcome events corresponding to a set of orthogonal eigenspaces in the Hilbert space of the 'measured' system. Putting it anthropomorphically: the 'measured' system is placed in an experimental situation where it is forced to 'choose' between alternative sequences of events, as a photon entering a beamsplitter is forced to 'choose' between alternative paths. The measurement outcome is *intrinsically random*, a genuinely stochastic event: there is no further story to be told about how a particular measurement outcome comes about dynamically (Colbeck and Renner 2011). Similarly, there is no further story to be told about Lorentz contraction or time dilation, once it is shown how to provide a dynamical account consistent with the kinematic constraints of Minkowski geometry. This transition, in which an indefinite observable becomes definite, has no counterpart in classical theory but makes sense only in a noncommutative framework, when the indefinite 'measured' observable does not commute with the observables with definite values. So there is no reason to accept John Bell's assertion (Bell 1990), that measurement in a fundamental theory of mechanics should always be open to a complete dynamical analysis, as applicable to quantum 'measurements.' This was the second dogma that we argued should be dropped.

Conditionalizing on a measurement outcome updates the probability assignment represented by the quantum state via the von Neumann Lüders rule and reflects the necessary information loss that occurs in a quantum 'measurement'—the notorious 'collapse' of the wave function representing the state. It is not a dynamical process. The unitary dynamics describes the change in probabilistic correlations between quantum 'measurements.' Just as Lorentz contraction is, ultimately, a kinematic effect in special relativity, the loss of information in a quantum 'measurement' is to be understood as a kinematic effect of the nonclassical quantum event space.

Frauchiger and Renner (2016), using an ingenious form of the Wigner's friend argument (Wigner 1961), have argued that single-world interpretations of quantum mechanics cannot be self-consistent. There are two rather different versions of the Frauchiger-Renner argument. In the following, I outline both versions and elaborate on the single-world account sketched above by showing how it avoids inconsistency in the Frauchiger-Renner scenarios.

Here is a 'cheat sheet' to orient the reader:

1. Alice can set up an experimental situation in which she measures a qubit observable with eigenstates $|0\rangle, |1\rangle$. On the single-world view outlined here, the outcome is an intrinsically random event, either definitely 0 or definitely 1, with certain probabilities depending on the set-up, specifically, in the Frauchiger-Renner example, 1/3, 2/3. At the end of the measurement, Alice assigns the state $|0\rangle$ or $|1\rangle$ to the qubit, depending on the outcome, which tells her the probabilities of possible outcomes of future measurements she might perform on the qubit. From an outside perspective, one could also assign the state $|0\rangle_Q|0\rangle_A$ or $|0\rangle_Q|0\rangle_A$ depending on the outcome to the composite system consisting of the qubit plus Alice's measuring instrument plus Alice's memory.
2. Frauchiger and Renner assume that all change is described by a unitary evolution, so Alice's experiment results in the entangled state

$$\frac{1}{\sqrt{3}}|0\rangle_Q|0\rangle_A + \sqrt{\frac{2}{3}}|1\rangle_Q|1\rangle_A \tag{14.1}$$

not one of the states $|0\rangle_Q|0\rangle_A$ or $|0\rangle_Q|0\rangle_A$. In their more elaborate argument considered in the following section, they show that a suitably powerful observer could measure a certain observable of a composite system consisting of a measured qubit plus Alice's measuring instrument plus Alice's memory and obtain an outcome, with finite probability given by the entangled state, that is inconsistent with the assumption that Alice's measurement resulted in one definite outcome. If one accepts that the correct quantum description of a situation in which Alice obtains a definite outcome for her measurement is described by an entangled state of the form (14.1) rather than $|0\rangle_Q|0\rangle_A$ or $|0\rangle_Q|0\rangle_A$, it is game over in favor of Frauchiger and Renner.

3. I allow that Alice could set up an experimental situation in which the final state is of the form (14.1), but I insist that she can also set up a different experiment that results in an intrinsically random outcome, where the final state is either $|0\rangle_Q|0\rangle_A$ or $|0\rangle_Q|0\rangle_A$. So something needs to be said about how to characterize the difference between the two situations: what does Alice do differently in the two experiments? A possible answer is that to obtain a definite measurement outcome Alice sets up the experiment in a way that allows environmental decoherence, but to obtain an entangled state of the form (14.1) she sets the experiment up to limit interaction with the environment. An alternative answer, implicit in Pitowsky's combinatorial approach to macroscopic objects in quantum mechanics (Pitowsky 2004, 2007), characterizes the difference on the basis of a structural or kinematic feature of Hilbert space for a many-particle system rather than decoherence, which is a dynamical process.
4. Given the distinction between these two experimental situations, I show that there is no inconsistency between the 'inner' description of events by Alice, and the 'outer' perspective of an observer for whom Alice and the system she measures is itself the object of a measurement. One could argue that the Frauchiger-Renner argument begs the question by denying the validity of this distinction from the outset.

14.2 The Frauchiger-Renner Argument

14.2.1 Original Version

Frauchiger and Renner derive a contradiction from a single-world view by considering an experiment where Alice prepares a qubit in a certain state and sends it to Bob, who measures a qubit observable. Alice and Bob are in separate closed laboratories. Wigner is outside these laboratories and is capable of measuring observables of any complexity involving all the systems in the laboratories, including the memories of

Alice and Bob (or one could consider Wigner and an equally powerful assistant making these measurements on the two laboratories). The idea is to derive a contradiction from the 'inner' perspective of Alice and Bob as observers who perform measurements and record outcomes, and the 'outer' perspective of Wigner who measures the observers Alice and Bob. It is assumed that Wigner's 'outer' perspective as an observer of observers can be described by a unitary evolution of the composite state of Alice, Bob, and their laboratories (i.e., no 'collapse'), and the contradiction follows from the assumption that this composite state is representational, in the sense that 0, 1 probability assignments by this state are not inconsistent with the possible events that are assumed to occur in the experiment, described from the 'inner' perspective of the observers who are being observed.

In effect, the argument is that a single-world interpretation of quantum mechanics is inconsistent if the observer is treated as a quantum system. Putting it differently, a single-world interpretation of quantum mechanics is consistent only if the observer in any application of the theory is left out of the quantum description and treated as a classical device than can register and record measurement outcomes as definite single-outcome events: *this* outcome, rather than *those* possible outcomes.

To begin the experiment, Alice tosses a biased quantum 'coin' and gets a definite outcome, heads or tails. The quantum coin is a qubit in the state

$$\frac{1}{\sqrt{3}}|h\rangle + \sqrt{\frac{2}{3}}|t\rangle \tag{14.2}$$

and to 'toss the coin' Alice measures the qubit observable with eigenvalues {heads, tails} = $\{h, t\}$, obtaining h with probability 1/3 and t with probability 2/3. If she gets h, she prepares a qubit in a state $|0\rangle$ and sends it to Bob. If she gets t, she prepares a qubit in the state $|+\rangle$ and sends it to Bob. where

$$|+\rangle = \frac{1}{\sqrt{2}}(|0\rangle + |1\rangle)$$

Bob measures the qubit observable with eigenvalues $\{0, 1\}$. Frauchiger and Renner describe the outcome of this sequence of events by the state

$$|\psi\rangle = \frac{1}{\sqrt{3}}|h\rangle_A|0\rangle_B + \sqrt{\frac{2}{3}}|t\rangle_A|+\rangle_B \tag{14.3}$$

$$= \frac{1}{\sqrt{3}}(|h\rangle_A|0\rangle_B + |t\rangle_A|0\rangle_B + |t\rangle_A|1\rangle_B) \tag{14.4}$$

Here the subscript A denotes the quantum coin, Alice (including Alice's memory, which is entangled with the coin state), and any relevant observables in Alice's laboratory that become entangled with Alice and the coin. Similarly, B denotes the qubit, Bob, and Bob's laboratory.

Now Wigner measures an A-observable, X, with eigenvectors

$$|\text{ok}\rangle_A = \frac{1}{\sqrt{2}}(|h\rangle_A - |t\rangle_A) \tag{14.5}$$

$$|\text{fail}\rangle_A = \frac{1}{\sqrt{2}}(|h\rangle_A + |t\rangle_A) \tag{14.6}$$

and a B-observable, Y, with eigenvectors

$$|\text{ok}\rangle_B = \frac{1}{\sqrt{2}}(|0\rangle_B - |1\rangle_B) \tag{14.7}$$

$$|\text{fail}\rangle_B = \frac{1}{\sqrt{2}}(|0\rangle_B + |1\rangle_B) \tag{14.8}$$

(or suppose Wigner measures X and his assistant measures Y if the laboratories are far apart).

The state $|\psi\rangle$ can also be expressed as

$$|\psi\rangle = \sqrt{\frac{2}{3}}|\text{fail}\rangle_A|0\rangle_B + \frac{1}{\sqrt{3}}|t\rangle_A|1\rangle_B \tag{14.9}$$

or as

$$|\psi\rangle = \frac{1}{\sqrt{3}}|h\rangle_A|0\rangle_B + \sqrt{\frac{2}{3}}|t\rangle_A|\text{fail}\rangle_B \tag{14.10}$$

It's easy to check that $|\psi\rangle$ has an overlap with $|\text{ok}\rangle_A|\text{ok}\rangle_B$. Specifically, the scalar product of $|\psi\rangle$ with $|\text{ok}\rangle_A|\text{ok}\rangle_B$ is $\frac{1}{2\sqrt{3}}$. So if Wigner measures X on Alice's laboratory and Y on Bob's laboratory, he will obtain the pair of outcomes ok_Aok_B with probability 1/12.

Wigner reasons as follows: If the outcome of my X measurement was ok_A, then Alice must have measured 'tails' for the quantum 'coin' and so prepared the qubit she sent to Bob in the state $|1\rangle$, because $|\text{ok}\rangle_A$ is orthogonal to the term $\sqrt{\frac{2}{3}}|\text{fail}\rangle_A|0\rangle_B$ in (14.9), and so the event corresponding to the state $|\text{fail}\rangle_A|0\rangle_B$ has zero probability. But if the outcome of my Y measurement was ok_B, then Alice must have measured 'heads' for the quantum 'coin' and so prepared the qubit she sent to Bob in the state $|0\rangle$, because $|\text{ok}\rangle_B$ is orthogonal to the term $\sqrt{\frac{2}{3}}|t\rangle_A|\text{fail}\rangle_B$ in (14.10), and so the event corresponding the state $|\text{t}\rangle_A|\text{fail}\rangle_B$ has zero probability. So the combined outcome ok_Aok_B leads to a contradiction on a single-world view, where the event 'heads' excludes the event 'tails.' Specifically, the assumption that there was one definite outcome to the quantum coin toss and a corresponding preparation of one definite qubit state contradicts a combined outcome for the two X, Y measurements that, according to quantum mechanics, occurs with 1/12 probability. So, on a single-world interpretation of quantum mechanics, the 'inner' and 'outer' descriptions by the quantum state of what happened are contradictory. On a many-worlds interpretation there is no contradiction, because 'heads' doesn't exclude 'tails' since 'heads' and 'tails' could both occur in different worlds.

14.2.2 Alternate Version

This version of the Frauchiger-Renner argument is due to Lluis Masanes (unpublished). My formulation is based on a talk by Matthew Pusey (2016) and email correspondence with Matthew Leifer.

Alice and Bob, who are in separate laboratories, measure observables A_1 and B_2 on two separated qubits in a Bell state. Alice's measurement is implemented by an interaction that entangles her measuring instrument and whatever else is considered part of the instrument in her laboratory, including Alice and the state of her memory, with the qubit in her laboratory, and Bob's measurement is implemented by a similar interaction with the qubit in his laboratory. Since unitary interactions are reversible, these measurement interactions can be undone by a suitably powerful Wigner, who is able to reverse the unitary transformations and measure observables of any complexity. Wigner implements a unitary transformation that reverses the interaction between Alice and her measuring instrument on the qubit in her laboratory, including the entanglement with her memory, and he does the same for Bob's measurement. After Wigner reverses Alice's and Bob's measurements, he measures the observables A_2 and B_2 on the two qubits, now in the restored Bell state. The observables measured are such that the expectation values $\langle A_1 B_1 \rangle$, $\langle A_1 B_2 \rangle$, $\langle A_2 B_1 \rangle$, $\langle A_2 B_2 \rangle$ violate the Clauser-Horne-Shimony-Holt version of Bell's inequality.

Suppose there are definite, agent-independent, non-perspectival facts of the matter about the outcomes of quantum measurements. If this experiment is performed many times, on each run of the experiment there is a fact of the matter about what Alice's and Bob's measurement outcomes were, even if all records of these outcomes are subsequently erased by Wigner, and there is a fact of the matter about what Wigner's measurement outcomes are. This means that on multiple runs of the experiment there exists a joint relative frequency distribution for the outcomes of all four measurements. Since this satisfies the axioms of probability theory, it follows that there exists a joint probability distribution over all four measurement outcomes.

Now, the expectation value $\langle A_2 B_2 \rangle$ for Wigner's measurements on the restored Bell state must agree with the prediction of quantum mechanics, because these measurements are the last step in the sequence and are not undone, and there remains an objective record of their outcomes. Also the expectation value $\langle A_1 B_1 \rangle$ for Alice's and Bob's measurements must agree with quantum mechanics. Wigner could have decided to delay or terminate the experiment after these measurements instead of undoing them. Then, clearly, the marginal for these two measurements should agree with the quantum prediction for the Bell state. (Agreement with the quantum prediction couldn't depend on whether or not Wigner subsequently undoes the measurements. For then no measurement would ever have to agree with the predictions of quantum mechanics, because any measurement might be undone by some sufficiently powerful agent at some point in the future, including the entanglement with the environment up to that point in time.) For the marginal $\langle A_1 B_2 \rangle$, since Wigner could have decided to delay or stop the experiment after Alice's measurement but reverse Bob's measurement and so restore Bob's half of the Bell state, the expectation value must also be as predicted by quantum mechanics, because A_1 and B_2 are measured on the original (restored) Bell state. By a similar argument, the marginal $\langle A_2 B_1 \rangle$ must agree with the quantum mechanical prediction.

It follows that all four marginals must agree with quantum mechanics, which means that they violate the Clauser-Horne-Shimony-Holt version of Bell's inequality (Clauser et al. 1969). But the marginals derivable from a joint distribution necessarily satisfy the Clauser-Horne-Shimony-Holt inequality. So we have a contradiction.

The contradiction is even more immediate for the superquantum correlation of a hypothetical nonlocal Popescu-Rohrlich box (PR box) (Popescu and Rohrlich 1994). A PR box has an Alice input and output, each 0 or 1, and a Bob input and output, also each 0 or 1. The inputs and outputs are correlated in the following way: if the inputs are both 1, the outputs are different; in the other three cases, the outputs are the same. The Alice and Bob parts of the box act separately like a random coin toss: a 0 input to one part of the box produces either output with equal probability, and similarly for a 1 input. The box is non-signaling: the marginal probabilities of 1/2 for inputs and outputs at either part of the box are independent of the input to the other part, or whether or not there is an input to the other part. The two parts of the box can be pulled arbitrarily far apart without affecting the correlation, which is independent of the time order of the two inputs, just as two qubits in an entangled Bell state can be separated arbitrarily far apart without affecting the quantum correlation, which is independent of the time order.

Now consider the Frauchiger-Renner scenario for a PR box rather than two qubits in a Bell state. Suppose Alice and Bob each input 0 into their part of the box. The outputs must be the same, so suppose they are also both 0. Wigner undoes these actions and their outcomes, restoring the box to its original state, and then inputs 1 into both inputs of the box. In this case, the outputs must be different, so suppose the outputs are 1 on the Alice side of the box and 0 on the Bob side. Now consider the Alice-Wigner and Wigner-Bob inputs and outputs. We immediately have a contradiction. Alice's input was 0 and her output was 0. Wigner's input on the Bob side of the box (after reversing the action of the box for Bob's input and output) was 1 and his output was 0, the same as Alice's. So far, this is in accord with the Popescu-Rohrlich correlation that the outputs should be the same for 0, 1 inputs. But Wigner's input on the Alice side of the box (after reversing the action of the box for Alice's input and output) was 1 and his output was 1, while Bob's input was 0 and his output was 0. Wigner's output should be the same as Bob's output in this case, but it's not. The argument is similar for any other combination of inputs and outputs: no quadruple of inputs and outputs can satisfy the correlation.

Evidently, undoing whatever happens when one part of a PR box produces an output for an input has to be impossible, and it's easy to see why. The production of an output for a given input at each part of a PR box is an intrinsically random event: the marginal probabilities remain 1/2, even conditional an any prior event, in any reference frame, before the PR box came into being (Bub 2016). The Popescu-Rohrlich correlation is possible if the box is required to produce outputs for a single pair of inputs, but not for multiple inputs. Once a part of a PR box produces an output for an input, it's done. There is no 'real factual situation' in Einstein's sense (Einstein 1949) for a PR box that would underwrite counterfactuals about what the outcome would be for an input other than the actual input. In other words, there is no hidden variable theory for a PR box, and one can show that a Bohmian hidden variable theory would violate the no-signaling constraint: access to such hidden variables would allow Alice and Bob to signal to each other instantaneously (Bub 2016).

One might argue that a PR box is a hypothetical device, and there is no reversible dynamics that would undo the production of an output for a given input: the only reversible transformations are local operations and permutations (Gross and Müller 2010). But there is a reversible dynamics in the case of quantum mechanics, and so a contradiction is unavoidable in a single-world interpretation of the theory.

14.3 Avoiding Inconsistency

Everettians and QBists avoid inconsistency because they accept the universality of unitarity but reject the assumption that there is a definite, agent-independent, non-perspectival fact of the matter about a measurement outcome in quantum mechanic. For Everettians, every possible outcome occurs in a measurement, in a different world. For QBists, each agent has his or her own individual perspective on reality. Quantum mechanics entails that we will end up in intersubjective agreement in the vast majority of scenarios we encounter, but not in Wigner's friend type scenarios like the Frauchiger-Renner scenario.

A single-world interpretation can avoid inconsistency by rejecting the two dogmas referred to in Sect. 14.1: insisting that the quantum state is not representational but a bookkeeping device for keeping track of probabilities and probabilistic correlations, and accepting the universality of unitarity for dynamical evolution between quantum 'measurements,' but denying that the updating of a quantum state by conditionalization after an indefinite observable has come to have a definite value is a dynamical process.

Consider a state of the form (14.4) in the original version of the Frauchiger-Renner argument:

$$\frac{1}{\sqrt{3}}(|h\rangle_a|0\rangle_b + |t\rangle_a|0\rangle_b + |t\rangle_a|1\rangle_b)$$

This state could represent an entangled state of two qubits, a and b, where $|h\rangle_a$ and $|t\rangle_a$ are eigenstates of an a-observable C, and $|0\rangle_b$ and $|1\rangle_b$ are eigenstates of a b-observable Q. The probabilistic information conveyed by this state is that a joint measurement of the observables C and Q on the qubits a and b would yield the joint outcomes $h, 0$ or $t, 0$ or $t, 1$ with equal probability 1/3. A joint measurement of observables X, with eigenstates $|\text{ok}\rangle_a$, $|\text{fail}\rangle_a$, and Y, with eigenstates $|\text{ok}\rangle_b$, $|\text{fail}\rangle_b$, that are linear superpositions of the eigenstates of C and of Q as defined in (14.5), (14.6), (14.7) and (14.8) with a for A and b for B, would yield the joint outcomes ok_aok_b with probability 1/12. Since the observables X, Y and C, Q are incompatible and represented by noncommuting operators, a measurement of X, Y provides no information about what the outcomes of C, Q measurements would have been or would be. It is well-known and understood that counterfactual inferences of this sort are illegitimate in quantum mechanics. If X and Y have definite values, then C and Q are indefinite and any assignment of definite values would be inconsistent (see Pitowsky 2007). One could, of course, first measure C and Q and obtain one of the joint pairs of outcomes $h, 0$ or $t, 0$ or $t, 1$, and then measure X and Y. In that case the joint outcomes ok_aok_b would be obtained with probability 1/4, but the measurement of X and Y would lead to a necessary loss of information about C and Q. There is no measurement in which a possible measurement outcome yields definite values for all four observables C, Q, X, Y. So something has gone wrong in the Frauchiger-Renner analysis of the scenario they describe by an entangled state of the same form.

There are two very different experimental situations that should be distinguished. You can send a photon through a beamsplitter with photon counters in the exit beams so that the photon is recorded in one of the counters, which are open to the environment allowing decoherence, in which case you can't undo the outcome. Or you can send a photon through a beamsplitter and then, with no counters in the exit beams so that there is no 'which way' information, reverse the unitary interaction occurring in the beamsplitter, bringing the two beams together to interfere and reproduce the photon in the initial state. The Frauchiger-Renner argument assumes that the two situations are equivalent for a suitably powerful Wigner. On the single-world view sketched in Sect. 14.1, this is not the case.

This distinction in the two experimental situations is the core idea in the proof of the 'no go' theorem for quantum bit commitment (Mayers 1997; Lo and Chau 1997). Bit commitment is a cryptographic protocol where one party, Alice, supplies an encoded bit to a second party, Bob. The information available in the encoding should be insufficient for Bob to ascertain the value of the bit, but sufficient, together with further information supplied by Alice at a subsequent stage when she is supposed to reveal the value of the bit, for Bob to be convinced that the protocol does not allow Alice to cheat by encoding the bit in a way that leaves her free to reveal either 0 or 1 at will.

To illustrate, suppose Alice commits to a bit by writing 0 or 1 on a piece of paper, which she locks in a safe. She hands the safe to Bob on Monday, but keeps the key. On Friday, she reveals the bit and hands the key to Bob, who can then unlock the safe and confirm that she actually made the commitment she claims to have made on Monday. The question is whether there exists a quantum analogue of this procedure that is unconditionally secure: provably secure according to quantum mechanics against cheating by either Alice or Bob. Bob can cheat if he can obtain some information about Alice's commitment before she reveals it (which would give him an advantage in repetitions of the protocol with Alice). Alice can cheat if she can delay actually making a commitment until the final stage when she is required to reveal her commitment, or if she can change her commitment at the final stage with a very low probability of detection.

Investigating the security of bit commitment is important because other cryptographic procedures can be built from a bit commitment protocol. There is no unconditionally secure bit commitment protocol in classical cryptography (although Adrian Kent (2005) has shown how to implement a secure classical bit commitment protocol by exploiting relativistic signaling constraints). The 'no go' quantum bit commitment theorem came as a surprise and was received with dismay by the quantum cryptography community. The theorem showed that Alice or Bob could always cheat without detection. The relevance of this for the Frauchiger-Renner argument is the difference between cheating and being honest in a quantum bit commitment protocol.

A quantum bit commitment protocol might involve several steps, where at each step Alice or Bob is required to make a choice between alternative actions, for example whether to perform one of a number of alternative measurements on a particle and return the particle to the other party after recording the outcome, or whether to implement one of a number of alternative unitary transformations on the particle before returning it.

To illustrate the difference between being honest and cheating, suppose the particle is a qubit and Alice is required to measure the observable with eigenvectors $\{|0\rangle_q, |1\rangle_q\}$. Alice can either be honest and perform the measurement and record the outcome, 0 or 1, before sending the qubit back to Bob, or she can cheat. Cheating involves entangling an ancilla with the qubit by a unitary interaction that produces the state $\frac{1}{\sqrt{2}}(|0\rangle_q|0\rangle_a + |1\rangle_q|1\rangle_a)$, where a here represents the ancilla state. Alice keeps the ancilla and sends the qubit to Bob. Bob can't detect this move: there is no information he can extract from the qubit that would allow him to tell whether the qubit is in one of the pure states $|0\rangle_q$ or $|1\rangle_q$, or whether it has been entangled with another system. The density operator available to Bob is the same in either case: an equal weight mixture of the states $|0\rangle_q$ and $|1\rangle_q$.

More realistically, Alice might be required to choose between two or more alternative measurements or other alternative actions. At the end of the commitment stage of the protocol, the composite system consisting of Alice's ancillas, the n particles that are passed in the communication channel between Alice and Bob, and Bob's ancillas will be represented by some composite entangled state $|0\text{-commit}\rangle$ or $|1\text{-commit}\rangle$, depending on Alice's commitment, on a Hilbert space $\mathcal{H}_A \otimes \mathcal{H}_B$, where $\mathcal{H}_A$ is the Hilbert space of the particles in Alice's possession at that stage (Alice's ancillas and the channel particles retained by Alice, if any), and $\mathcal{H}_B$ is the Hilbert space of the particles in Bob's possession at that stage (Bob's ancillas and the channel particles retained by Bob, if any). The density operators $W_B(0)$ and $W_B(1)$ characterizing the information available to Bob for the two alternative commitments are obtained by tracing the states $|0\text{-commit}\rangle$ and $|1\text{-commit}\rangle$ over $\mathcal{H}_A$. If these density operators are the same, then Bob will be unable to distinguish the 0-commitment from the 1-commitment without further information from Alice. In this case, the protocol is said to be 'concealing.'

What the proof establishes, by an application of the biorthogonal decomposition theorem, is that if $W_B(0) = W_B(1)$ then there exists a unitary transformation in $\mathcal{H}_A$ that will transform $|0\rangle$ to $|1\rangle$. That is, if the protocol is 'concealing' then it cannot be 'binding' on Alice: she can always make the 0-commitment and follow the protocol (with appropriate applications of the cheating strategy sketched above) to establish the state $|0\text{-commit}\rangle$. At the final stage when she is required to reveal her commitment, she can change her commitment if she chooses, depending on circumstances, by applying a suitable unitary transformation in her own Hilbert space to transform $|0\text{-commit}\rangle$ to $|1\text{-commit}\rangle$ without Bob being able to detect this move. So either Bob can cheat by obtaining some information about Alice's choice before she reveals her commitment, or Alice can cheat.

To return to the original Frauchiger-Renner scenario, suppose Alice and Bob perform the actions described. They can either do this in the 'honest' sense, or they can 'cheat' and 'keep the alternatives at the quantum level' by entangling the quantum coin and the qubit with ancillas in an appropriate way. If they perform the actions described in the 'honest' sense, the outcomes are $h, 0$ or $t, 0$ or $t, 1$ with probability 1/3, and the corresponding quantum states are $|h\rangle_A|0\rangle_B$ or $|t\rangle_A|0\rangle_B$ or $|t\rangle_A|1\rangle_B$. If they 'cheat,' the final state is (14.4):

$$\frac{1}{\sqrt{3}}(|h\rangle_A|0\rangle_B + |t\rangle_A|0\rangle_B + |t\rangle_A|1\rangle_B)$$

The subscript A here refers to the quantum coin and associated ancilla, and B to the qubit Alice sends to Bob and associated ancilla. As far as Wigner is concerned, the ancillas could be quantum systems of any complexity, and Wigner could also treat Alice and Bob as ancillas.

If Wigner measures the observables X and Y when Alice and Bob are 'honest,' he will obtain the pair of outcomes ok_Aok_B with probability 1/4 for any of the three possible outcomes of Alice and Bob's actions. As for the qubit example above, the X, Y measurements lead to a loss of information about the noncommuting observables C with eigenvectors $|h\rangle_A, |t\rangle_A$ and Q with eigenvectors $|0\rangle_B, |1\rangle_B$, so there is no contradiction. If Wigner measures the observables X and Y when Alice and Bob 'cheat,' he will obtain the pair of outcomes ok_Aok_B with probability 1/12. In this case, there are no measurement outcomes for Alice and Bob, so again there is no contradiction. In particular, the inference from the states (14.9) and (14.10) does not apply to a counterfactual situation in which C, Q do not have definite values.

A quantum state like (14.4) above is simply a probability assignment. The state tells Wigner what to expect for the outcomes of his measurements if Alice and Bob have 'cheated' and entangled ancillas with the quantum coin and the qubit Alice sends to Bob. It makes no difference to Wigner whether the ancillas are qubits or the entire laboratories of Alice and Bob, since we are assuming that he can manipulate the Hilbert spaces of these laboratories and their contents, including Alice and Bob. Wigner can choose to measure the observables X and Y, or he can choose to measure the noncommuting observables C and Q. If he measures C and Q, he will obtain outcomes corresponding to $h, 0$ or $t, 0$ or $t, 1$ with probability 1/3. His measurement procedure sets up an experimental situation in which these indefinite observables come to have definite values, corresponding to Alice and Bob and their measuring instruments recording definite values. What he cannot do is set up an experimental situation where X, Y and C, Q all come to have definite values.

For the second Frauchiger-Renner scenario, if Bob measures the observable B_1 in the 'honest' sense, so that an intrinsically random output has actually occurred, Wigner cannot undo the outcome. The conditionalized state is no longer the entangled state but a product state: Bob's measurement destroys the entanglement. So while Wigner can reverse any unitary interaction, he cannot undo an 'honest' measurement. If Alice also measures in the 'honest' sense, the expectation value $\langle A_1 B_1 \rangle$ will be as predicted by quantum mechanics for the Bell state, but not the expectation value $\langle A_1 B_2 \rangle$ for Alice's and Wigner's measurements, or the expectation value $\langle A_2 B_1 \rangle$ for Wigner's and Bob's measurements. Wigner's reversal would only be relevant if Alice and Bob 'cheated' and 'kept the alternatives at the quantum level.' But then there is no measurement outcome for A_1 or B_1. This blocks the argument that all four marginals $\langle A_1 B_2 \rangle$, $\langle A_2 B_1 \rangle$, and $\langle A_2 B_2 \rangle$ should accord with the prediction of quantum mechanics for the Bell state, and so there is no contradiction.

14.4 Conclusion

The Frauchiger-Renner argument is presented as a theorem that follows from three premises: QT (quantum theory: measurement outcomes forbidden by quantum theory cannot occur, even if the measured system is large enough to contain the observer), SW (single-world: measurements have definite, agent-independent, non-perspectival single outcomes), and SC (self-consistency: statements about measurement outcomes are logically consistent, even if they refer to the perspectives of different observers). My rebuttal of the Frauchiger-Renner argument accepts SW and SC, so it would appear that I must be denying QT by denying the universality of unitarity.

But consider in what sense I am denying universality. I accept a Wigner with unlimited capacity to perform unitary transformations on macroscopic systems of any size, from cats to galaxies. So I am not arguing that if systems are big enough then unitarity no longer applies. I am also not arguing that in addition to the unitary dynamics there is a 'collapse' dynamics that occurs during a quantum measurement. Rather, I am arguing that a single-world interpretation of quantum mechanics is consistent, provided the interpretation drops the two dogmas mentioned in

Sect. 14.1. Quantum mechanics is a probability theory for a world in which there are intrinsically random events in which indefinite observables become definite. Such events occur when a system is placed in an experimental situation defining a chance set-up for the indefinite alternatives, which we take as a 'measurement' of the observable. On this view, there is no inconsistency between an 'inner' and 'outer' perspective in assuming that something definite happened, provided the quantum state is understood as a bookkeeping device for keeping track of probabilities and probabilistic correlations rather than representing what's there, and that what happens when an outcome occurs in a quantum 'measurement' is an intrinsically random event, not described by the unitary dynamics of the theory.

In Pitowsky (2007) (see also Bub and Pitowsky 2010), Pitowsky distinguishes between a 'big' measurement problem and a 'small' measurement problem. The 'big' measurement problem is the problem of giving a dynamical account of how observables come to have definite values in quantum measurements. On the single-world view, Hilbert space provides the kinematic framework for the physics of an indeterministic, irreducibly stochastic universe, in which measurement outcomes are intrinsically random events, so the 'big' measurement problem is illusory.

I quote Pitowsky (2007):

> The BIG problem concerns those who believe that the quantum state is a real physical state which obeys Schrödinger's equation in all circumstances. In this picture a physical state in which my desk is in a superposition of being in Chicago and in Jerusalem is a real possibility; and similarly a superposed alive-dead cat. In fact the linearity of Schrödinger's equation implies that (decoherence notwithstanding) it is easy to produce states of macroscopic objects in superposition—which seems to contradict our experience, and sometimes, as in the cat case, does not even make much sense.
>
> In our scheme quantum states are just assignments of probabilities to possible events, that is, possible measurement outcomes. This means that the updating of the probabilities during a measurement follows the Von Neumann-Lüders projection postulate and not Schrödinger's dynamics. Indeed, the projection postulate is just the formula for conditional probability that follows from Gleason's theorem. So the BIG measurement problem does not arise. In particular, the cat in the Schrödinger thought experiment is not superposed, but is rather cast in the unlikely role of a particle spin detector. Schrödinger's equation governs the dynamics between measurements; it dictates the way probability assignments should change over time in the absence of a measurement. The general shape of the Schrödinger's equation is not a mystery either; the unitarity of the dynamics follows from the structure of $L(\mathcal{H})$ via a theorem of Wigner (1959) in its lattice theoretic form (Uhlhorn 1963). However, these remarks do not completely eliminate the measurement problem because in our scheme quantum mechanics is also applicable to macroscopic objects.

The last sentence concerns the 'small' measurement problem. This is the problem of accounting for why it is hard to observe macroscopic superposition or macroscopic entanglement. Pitowsky recognizes that decoherence is part of the explanation but proposes a more fundamental explanation that follows from the probabilistic structure. Entanglement witnesses are observables that distinguish between separable and entangled states. An entanglement witness for an entangled state is an observable whose expectation value lies in a bounded interval for any separable state, but is outside this interval for the entangled state. Pitowsky has shown that the measure of the set of entangled states for which the absolute value of the expectation value is greater than $C\sqrt{n\log n}$, where C is a positive universal constant, tends to zero as n tends to infinity for a large class of entanglement witnesses, and he conjectures that this is true in general. Setting aside cases where the entanglement witness is a thermodynamic observable corresponding to a global property of a macrosystem, the conjecture proposes that, in the cases where measurement of an entanglement witness requires many manipulations of individual particles, entangled states that can be distinguished from separable states become rarer and rarer as the number of particles increases. So quite apart from decoherence, which is a dynamical process, the kinematic structure of quantum mechanics entails that the combinatorial possibility of observing macroscopic entanglement in such cases is virtually impossible.

If Pitowsky's conjecture is true, a measuring instrument in quantum mechanics can be characterized as a many-particle system for which the set of entangled states that can be distinguished from separable states has measure zero, or close to zero. In this sense, a macrosystem is effectively a commutative or Boolean system, and as such can play the role of a measuring instrument in defining a chance set-up for the alternative values of an indefinite observable. (How it does so is another question.) This does not exclude the possibility of considering a suitably powerful Wigner capable of measuring observables of a system functioning as a measuring instrument. What I have shown is that a system can play the role of a measuring instrument for Alice and Bob, while also being the object of measurements by Wigner, without inconsistency between the 'inner' description of events by Alice and Bob and the 'outer' description by Wigner.

So on this view what is really there in a quantum world? Pitowsky put it clearly (Pitowsky 2007):

> Firstly, there are objects—particles about which the theory speaks—which are identified by a set of parameters that involve no uncertainty, and can be recorded in all circumstances and thus persist through time and context (Menahem 1988). Among them are the rest mass, electric charge, baryonic number, etc. The other part of quantum reality consists of events, that is, recordings of measurements in a very broad sense of the word.

Acknowledgements Thanks to Bill Demopoulos, Michel Janssen, Matthew Leifer, and Allen Stairs for illuminating discussions.

References

Bell, J. (1990). Against measurement. *PhysicsWorld, 8*, 33–40. Reprinted in A. Miller (Ed.), *Sixty-two years of uncertainty: Historical, philosophical and physical inquiries into the foundations of quantum mechanics* (pp. 17–31). New York: Plenum.

Born, M., & Jordan, P. (1925). Zur Quantenmechanik. *Zeitschrift fur Physik, 34*, 858–888.

Born, M., Heisenberg, W., & Jordan, P. (1925). Zur Quantenmechanik II. *Zeitschrift fur Physik, 35*, 557–615

Brown, H. R. (2006). *Physical relativity: Space-time structure from a dynamical perspective*. Oxford: Clarendon Press.

Bub, J. (2016). *Bananaworld: Quantum mechanics for primates* (p. 211). Oxford: Oxford University Press.

Bub, J., & Pitowsky, I. (2010). Two dogmas about quantum mechanics. In S. Saunders, J. Barrett, A. Kent, & D. Wallace (Eds.), *Many worlds? Everett, quantum theory, and reality* (pp. 431–456). Oxford: Oxford University Press.

Clauser, J. F., Horne, M. A., Shimony, A., & Holt, R. A. (1969). Proposed experiment to test local hidden-variable theories. *Pysical Review Letters, 23*, 880–884.

Colbeck, R., & Renner, R. (2011). No extension of quantum theory can have improved predictive power. *Nature Communications, 2*, 411. For two qubits in an entangled Bell state, Colbeck and Renner show that there can't be a variable, z, associated with the history of the qubits before the preparation of the entangled state in the reference frame of any inertial observer, that provides information about the outcomes of measurements on the qubits, so that Alice's and Bob's marginal probabilities conditional on z are closer to 1 than the probabilities of the Bell state. They show how the argument can be extended to any entangled state, and then to any quantum state.

Einstein, A. (1949). Autobiographical notes. In P. A. Schillp (Ed.), *Albert Einstein: Philosopher-scientist* (p. 85). Open Court: La Salle. But on one supposition we should, in my opinion, absolutely hold fast: the real factual situation of the system S_2 is independent of what is done with the system S_1, which is spatially separated from the former.

Einstein, A. (1954). What is the theory of relativity? *Ideas and opinions* (p. 228). New York: Bonanza Books. Reprinted from an article in the *London Times*, 28 Nov 1919.

Frauchiger, D., & Renner, R. (2016). Single-world interpretations of quantum mechanics cannot be self-consistent. arXiv eprint quant-ph/1604.07422.

Gleason, A. N. (1957). Measures on the closed subspaces of Hilbert space. *Journal of Mathematics and Mechanics, 6*, 885–893.

Gross, D., Müller, M., Colbeck, R., & Dahlsten, O. C. (2010). All reversible dynamics in maximally nonlocal theories are trivial. *Physical Review Letters, 104*, 080402.

Heisenberg, W. (1925). 'Über Quantentheoretischer Umdeutung kinematischer und mechanischer Beziehungen. *Zeitschrift fur Physik, 33*, 879–893.

Janssen, M. (2009). Drawing the line between kinematics and dynamics in special relativity. *Studies in History and Philosophy of Modern Physics, 40*, 26–52.

Kent, A. (2005). Secure classical bit commitment over finite channels. *Journal of Cryptology, 18*, 313–335; Unconditionally secure bit commitment. *Physical Review Letters, 83*, 1447–1450 (1999).

Lo, H.-K., & Chau, H. F. (1997). Is quantum bit commitment really possible? *Physical Review Letters, 78*, 3410–3413.

Mayers, D. (1997). Unconditionally secure quantum bit commitment is impossible. *Physical Review Letters, 78*, 3414–3417.

Menahem, Y. B. (1988). Realism and quantum mechanics. In A. van der Merwe (Ed.), *Microphysical reality and quantum formalism*. Dordrecht: Kluwer.

Pitowsky, I. (2003). Betting on the outcomes of measurements: A Bayesian theory of quantum probability. *Studies in History and Philosophy of Modern Physics, 34*, 395–414.

Pitowsky, I. (2004). Macroscopic objects in quantum mechanics: A combinatorial approach. *Physical Review A, 70*, 022103–1–6.

Pitowsky, I. (2007). Quantum mechanics as a theory of probability. In W. Demopoulos & I. Pitowsky (Eds.), *Festschrift in honor of Jeffrey Bub*. New York: Springer. arXiv e-print quant-ph/0510095.

Popescu, S., & Rohrlich, D. (1994). Quantum nonlocality as an axiom. *Foundations of Physics, 24*, 379.

Pusey, M. (2016). Is QBism 80% complete, or 20%? https://www.youtube.com/watch?v=_9Rs6l18MyY. Talk presented at a workshop: Information-theoretic interpretations of quantum mechanics, 11–12 June 2016, Western University, London.

Uhlhorn, U. (1963). Representation of symmetry transformations in quantum mechanics. *Arkiv Fysik, 23*, 307.

Wallace, D. (2016). What is orthodox quantum mechanics? arXiv eprint quant-ph/1604.05973.
Wigner, E. (1959). *Group theory and its applications to quantum mechanics of atomic spectra*. New York: Academic.
Wigner, E. (1961). Remarks on the mind-body question. In I. J. Good (Ed.), *The scientist speculates*. London: Heinemann.

Chapter 15
A Foundational Principle for Quantum Mechanics

Anton Zeilinger

Abstract In contrast to the theories of relativity, quantum mechanics is not yet based on a generally accepted conceptual foundation. It is proposed here that the missing principle may be identified through the observation that all knowledge in physics has to be expressed in propositions and that therefore the most elementary system represents the truth value of one proposition, i.e., it carries just one bit of information. Therefore an elementary system can only give a definite result in one specific measurement. The irreducible randomness in other measurements is then a necessary consequence. For composite systems entanglement results if all possible information is exhausted in specifying joint properties of the constituents.

15.1 Introduction

Quantum mechanics is magic.
Daniel M. Greenberger (Mermin 1985)

Physics in the twentieth century is signified by the invention of the theories of special and general relativity and of quantum theory. Of these, both the special and the general theory of relativity are based on firm foundational principles, while quantum mechanics[1] lacks such a principle to this day. By such a principle, I do not mean an axiomatic formalization of the mathematical foundations of quantum mechanics, but a foundational conceptual principle. In the case of the special theory, it is the Principle of Relativity, stating that all laws of physics must be the same in

This paper is dedicated to Daniel M. Greenberger on the occasion of his 65th birthday.

[1]Here "quantum theory," "quantum mechanics," and "quantum physics" are used interchangeably, all in a very broad sense.

A. Zeilinger (✉)
Institute of Experimental Physics, University of Vienna, Wien, Austria
e-mail: anton.zeilinger@univie.ac.at

A. Cordero (ed.), *Philosophers Look at Quantum Mechanics*,
Synthese Library 406, https://doi.org/10.1007/978-3-030-15659-6_15

all inertial reference frames, independent of their state of relative motion. In the case of the theory of general relativity, we have the Principle of Equivalences: (Einstein 1949) "In a gravitational field (of small spatial extension), things behave as they do in a space free of gravitation, if one introduces, in place of an "inertial system," a reference system that is accelerated relative to an inertial system." Both foundational principles are very simple and intuitively clear.

On these principles the two theories of relativity are built, which then lead to some surprising and in part even counterintuitive consequences, even as the theories themselves are based on such intuitively nearly obvious principles. I submit that it is because of the very existence of these fundamental principles and their general acceptance in the physics community that, at present, we do not have a significant debate on the interpretation of the theories of relativity. Indeed, the implications of relativity theory for our basic notions of space and time are broadly accepted.

In quantum mechanics, to the contrary, we do observe the presence of a broad discussion about the interpretation of the theory. In fact, we have a number of coexisting interpretations utilizing mutually contradictory concepts. (Zeilinger 1996) Possibly the coexistence of such a large number of philosophically quite different interpretations in itself contains an important message. I suggest that the message is that a generally accepted foundational principle for quantum mechanics has not yet been identified.

A few remarks are essential here in order to clarify what I mean by interpretation. As I analyzed earlier, (Zeilinger 1996) there exist at least two different levels of interpretation of a theory. On the first, basic, level, interpretation tells us how to verify the theoretical predictions. A huge set of operational and experimental rules and concepts connects the symbols used in the theory with observation. In the case of quantum mechanics, an essential ingredient at the basic level is the interpretation of the absolute square of the amplitude as a probability or probability density. On the second, the meta-level, less operational but conceptually more significant, interpretation means an analysis of what the theory implies for our general view of the world ("Weltbild"). It implies questions as to the meaning of the theory in a deeper sense.

It is my understanding that on the first, the operational, level, all interpretations of quantum mechanics essentially agree. They lead to the same experimental predictions. Suggestions actually to change quantum mechanics (Ghirardi et al. 1986) are not just interpretations but are really alternative theories. In view of the extremely high precision with which the theory has been experimentally confirmed, and in view of its superb mathematical beauty and symmetry, I consider a final success of such attempts to be extremely unlikely.

On the second level of interpretation, where we deal with questions of the meaning of the theory, the situation is complicated. Of the many interpretations, a very incomplete list includes the original Copenhagen Interpretation, (Bohr 1935) the Many-Worlds Interpretation, (Everett III 1957) the Transactional Interpretation, (Cramer 1986) Bohm's interpretation (Bohm 1952) in terms of a quantum potential, and, most recently, Mermin's Ithaca interpretation. (Mermin 1998) As I analyzed earlier, (Zeilinger 1996) these interpretations imply very different ideas about

Nature, about the world, or about our position in the world. While I personally prefer the Copenhagen interpretation because of its extreme austerity and clarity, the purpose of my present paper is not to compare and analyze these interpretations but to attempt to go significantly beyond them. In fact, I wish to suggest ideas for a foundational principle for quantum mechanics.[2]

15.2 Randomness

Die Schwäche der Theorie liegt ... darin, dass sie Zeit und Richtung des Elementarprozesses dem "Zufall" überläßt.[3]
Albert Einstein (Einstein 1917)

Our physical description of the world is represented by propositions. Any physical object can be described by a set of true propositions. A complete description of an object in general is a very long list of propositions. In everyday life and in classical physics, one regards these propositions as describing properties the object actually possesses, usually prior to and independent of observation. We now ask ourselves two questions. First, how do we arrive at such propositions and, second, how would we verify them? To answer the first question, we note that any such proposition is obtained through earlier observation. This need not be a single observation, and it need not be observations at the same time or the same place. To answer the second question, we note that any such proposition can be verified through future observation. We thus note that any properties we might assign to an object are arrived at only by observation and are tenable only as long as they do not contradict any further observation. In fact, the object therefore is a useful construct connecting observations.

We have knowledge, i.e., information, of an object only through observation. Thus, any concept of an existing reality has to be based on observations. Yet this does not imply – as tempting as such a conclusion might be – that reality is no more than a pure subjective human construct. From our observations we might mentally construct objects of reality. Predictions based on any such specific model of reality may then be checked by anyone. As a result we may arrive at intersubjective agreement on the model, thus lending a sense of objectivity to the mentally constructed objects.

What, then, is the role of physics? Using previously obtained information we wish to make predictions about the future. Again, our predictions might be formulated as predictions about some future properties of a system or object. Clearly, these predictions have implications for and indeed are propositions about

---

[2]A first, somewhat implicit, use of the principle was made in an analysis of two-photon entanglement and of quantum teleportation (see Ref. (Zeilinger 1997).

[3]The weakness of the theory lies ... in the fact, that it leaves time and direction of the elementary process to "chance." (translation by A.Z.).

specific future observations. It is, then, an important, though perhaps not the only, role of physics to connect past observation with future observation: or, more precisely, to make specific. but in general probabilistic, statements about results of future observations based on past observations. The connection might be very complicated. To express regularities and generalities in such connections is the point of laws of physics.

In quantum mechanics this expression is exactly what is achieved by the Schrödinger equation. The initial state $\psi\left(\vec{r}, t_i\right)$ at time t_i represents all our information as obtained by earlier observation, observation of any relevant features of our experimental setup. Using the Schrödinger equation, we can derive a time-evolved final state $\psi\left(\vec{r}, t_f\right)$ at some future time t_f. That state is just a short-hand way of representing the outcomes of all possible future observations. In general, those outcomes are probabilistic. By observations, we always mean observations of properties of our classical apparatus. It is not necessary to assume that the future properties of the classical apparatus can be predicted with certainty. Indeed, in general, quantum physics just gives the probabilities of observing specific future properties of the classical apparatus.

According to the standard understanding of quantum mechanics, the specific result is objectively random unless the quantum state is in an eigenstate of the projection operator describing the measurement. To illustrate that point, let us consider the state $|\psi\rangle$, which is an eigenstate of the projection operator $P_{op} = |\psi\rangle\langle\psi|$ with eigenvalue 1, that is, $|\psi\rangle = |\psi\rangle\langle\psi|\psi\rangle$. This simply means that the quantum system described by the state will be found with certainty to be in the state $|\psi\rangle$ if it is measured with the appropriate apparatus. What about other measurements?

Let us consider the specific case of a spin-1/2 particle with spin up along the z-axis, i.e., in the state $|\psi\rangle = |+z\rangle$. Then it follows immediately that the probability to find the particle's spin along a general direction at an angle θ with respect to the $+z$ direction is $P = \cos^2(\theta/2)$. Thus, specifically, for $\theta = 90^\circ$, we obtain $P = 1/2$, that is, the answer the experiment gives when we measure along that direction is completely random. Quantum mechanics does not provide any reason why in a specific run of the experiment the specific result observed is actually obtained. In essence, this is the famous measurement problem. Bell (Bell 1990) has expressed most clearly the misgivings of many about the measurement problem. His goal or hope was finally to "explain why events happen."

Here we turn the argumentation around. We will see that Bell's program is unachievable if we accept some very natural principles about the connection between information and elementary systems. In consequence, we will obtain new insight into the foundations of quantum mechanics.

In order to analyze the information content of elementary systems, we now decompose a system which may be represented by numerous propositions into constituent systems. It is natural to assume that each such constituent system will be represented by fewer propositions. How far, then, can this process of subdividing a system go? It is obvious that the limit is reached when an individual system finally represents the truth value to one single proposition only. Such a system we call an elementary system. We thus suggest a principle of quantization of information as follows.

An elementary system represents the truth value of one proposition.

To turn the principle around, the opposite would be absurd, namely, that the information content of a system would not scale with its size. We now note that the truth value of a proposition can be represented by one bit of information with "true" being identified with the bit value "1" and "false" being identified with the bit value 0. Thus, our principle becomes simply:

An elementary system carries 1 bit of information.

We remark that this might also be interpreted as a definition of what is the most elementary system. We stress, again, that by proposition we mean something which can be verified directly by experiment. In order to avoid misconceptions, I would like to underline that notions such as that a system "represents" the truth value of a proposition or that it "carries" one bit of information only implies a statement concerning what can be said about possible measurement results.

Let us again come back to our example above: the spin of a spin-1/2 particle represents the truth value of only one proposition.[4] In our case the true proposition is, "A measurement of spin along the z-axis will definitely give the result +." The spin of the particle carries the answer to one question only, namely, the question, What is its spin along the z-axis? Only if we actually perform a measurement in the z-basis can the measurement result be definite. Since this is the only information the spin carries, measurement along any other direction must necessarily contain an element of randomness. We remark that this kind of randomness must then be irreducible, that is, it cannot be reduced to hidden properties of the system, otherwise the system would carry more than a single bit of information. We have thus found a reason for the irreducible randomness in quantum measurement. It is the simple fact that an elementary system cannot carry enough information to provide definite answers to all questions that could be asked experimentally.

As discussed above, we know that in the case of a spin measurement, the degree of randomness depends on the relative orientation between the measurement direction and the direction along which our system is in an eigenstate. Clearly, from symmetry, the probability of finding a given spin value along the specified measurement direction must depend only on the angle between the measurement direction and the eigenstate direction. In a separate paper, it will be argued (Brukner and Zeilinger in press) that the most natural function describing this behavior consistent with the principle that the quantum system carries only one bit of information is the well-known sinusoidal dependence.

---

[4]Clearly, the state of an elementary particle is also characterized by other quantum numbers, it is in general an elementary system in more than one property. The cases of Hilbert spaces of higher dimension deserve separate analysis. E.g., a three-state system represents 1 trit of information. Here we restrict our analysis to two-state systems.

The extreme case is when the measurement direction is orthogonal to the eigenstate direction. Then, for the new measurement situation the system does not carry any information whatsoever, and the result is completely random. Or, in other words, the result is completely random because in such a measurement the elementary system carries no information whatsoever about the measurement result. We note that, most importantly, after the measurement the system is found in a new definite state. The information carried now by the system is not in any way determined by the information it carried before the measurement. Thus we conclude that the new information the system now represents has been spontaneously created in the measurement itself.

We finally remark that the viewpoint just presented lends natural support to Bohr's notion of complementarity. This notion is well-known, for example, for position and momentum or for the interference pattern and the path taken in a two-slit experiment; precise knowledge of one quantity excludes any knowledge of the other complementary quantity. In our case, measurements of a particle's spin along orthogonal directions are complementary, and the reason is, again, the fact that an elementary system carries only one bit of information.

15.3 Entanglement

It is wrong to think that the task of physics is to find out how Nature is. Physics concerns what we can say about Nature.
Niels Bohr (Petersen 1985)

Another fundamental feature of quantum mechanics is entanglement. (Zeilinger 1998) We now argue that entanglement follows from a slight generalization of our principle presented above. To do this we analyze how much information is contained in more complex objects, consisting of *N* elementary systems. Evidently, there are many ways in which the total information represented by a system can increase with its size. Here I argue for one specific solution to the question. Consider *N* elementary systems, which, by our principle above, therefore represent *N* independent individual propositions – evidently each system just one. Let us assume that these systems are completely separated initially. By complete separation I mean that we have no interaction between individual elementary systems and no additional information is contained in how the systems relate to each other. Then we have our principle of quantization of information generalized to:

N *elementary systems represent the truth values of* N *propositions.*
N *elementary systems carry* N *bits.*

Now let the initially separated systems interact with each other. It is then suggestive to assume that the information represented jointly by the *N* systems is conserved during the interaction process if there is no information exchange with the environment. That is, the interaction can neither increase the total amount of information represented by the total system nor reduce it. We remark that our principle does not make any statement of how the information contained in the N propositions (the N bits of information) is distributed over the N systems. After the interaction the N bits might still be represented by the N systems individually or, alternatively, they might all be represented by the N systems in a joint way, in the extreme with no individual system carrying any information on its own. In the latter case we have complete entanglement.

In order to analyze entanglement in view of the ideas just proposed above, let us consider two elementary systems carrying therefore two bits of information, i.e., representing the truth value of two propositions. For reason of simplicity, we consider two spin-1/2 particles. Which two propositions are possible to describe completely the system of our two particles?

A most simple case would be propositions which describe each one of the two particles separately. If, without loss of generality, we consider measurement of spin along the z-axis, then proposition 1 could simply be a statement about the spin of particle 1 along that axis, and proposition 2 could be a statement about the spin of particle 2 along that axis. Then four possibilities result:

$$\begin{aligned} |\,\psi\,\rangle_1 &= |+z\rangle_1\,|+z\rangle_2 \\ |\,\psi\,\rangle_2 &= |+z\rangle_1\,|-z\rangle_2 \\ |\,\psi\,\rangle_3 &= |-z\rangle_1\,|+z\rangle_2 \\ |\,\psi\,\rangle_4 &= |-z\rangle_1\,|-z\rangle_2 \end{aligned} \tag{15.1}$$

where, e.g., $|+z\rangle_1$ represents the state of particle 1 along the $+\,z$-axis. Evidently, the four resulting possibilities are rather trivial and are also present in classical mechanics. In our new language this is the case where each spin itself represents one proposition on its own. Actually, the four states (1) are the representation of the four possible two-bit combinations (true–true, true–false, false–true, false–false) of the truth values of the propositions, "The spin of particle 1 is up along z" and "The spin of particle 2 is up along z."

Instead of choosing propositions which describe the individual members of the system, we could alternatively choose propositions which describe results of joint observations. Consider, e.g., the proposition "The two spins are the same along z." Then, clearly, we have two possibilities: the two spins could be either both up along z or both down along z, i.e.,

$$\text{either}\quad |+z\rangle_1\,|+z\rangle_2 \quad\text{or}\quad |-z\rangle_1\,|-z\rangle_2 \tag{15.2}$$

If this is all we know,[5] then the system is incompletely described. This is necessarily the case because we have exhausted only one possible proposition, i.e., one bit of

[5]Just to stress our point again: By "are the same along z," we mean something like "Should they be measured along z, they would be found to be identical," and analogously for propositions about individual systems. This does not imply that the system necessarily "has" the measured property before the measurement.

information. A trivial way to describe the system completely is also to specify the spin af an individual member of the system, i.e., to assume that the system represents the truth value of a proposition like, "Spin 1 is up along the z-axis." Then we have two propositions and, actually, another proposition immediately follows, namely, "Spin 2 is up along the z-axis" and this case reduces to the one just discussed.

Yet we still have another, very different, possibility to complete the description of the system as started using the above proposition, "The two particles have the same spin along the z-axis." Instead of choosing as the second proposition one about the properties of an individual, we could choose another proposition also describing joint properties of the system. This could, e.g., be a proposition stating, for some other chosen direction, that the two spins are also equal along the new direction. Consider specifically the proposition, "The two spins are equal along the x-axis." Then we know that either both are up along x or both are down along x should they be measured along x. Quantum mechanically, the situation is

$$\text{either} \quad |+x\rangle_1\,|+x\rangle_2 \quad \text{or} \quad |-x\rangle_1\,|-x\rangle_2 \tag{15.3}$$

How can both (2) and (3) be true for the same two particles? They can if we note that these two propositions together, namely, that the two spins are equal along the z-axis and they are equal along the x-axis, now uniquely (up to a trivial phase factor) determine the entangled quantum state

$$\begin{aligned} |\phi^+\rangle &= \tfrac{1}{\sqrt{2}}(|+z\rangle_1|+z\rangle_2+|-z\rangle_1|-z\rangle_2) \\ &= \tfrac{1}{\sqrt{2}}(|+x\rangle_1|+x\rangle_2+|-x\rangle_1|-x\rangle_2) \end{aligned} \tag{15.4}$$

That state does not contain any information about the individuals; all information is contained in joint properties. In fact, now there cannot be any information carried by the individuals because the two bits of information are exhausted by defining that maximally entangled state, and no further possibility exists also to encode information in individuals. As an example to exhibit the richness of our approach, let us consider an alternative choice for the second proposition, the first one still being equality along the z-direction. Let us assume that the second proposition is now,

"The two spins are different along x." Then the two spins are

$$\text{either} \quad |+z\rangle_1\,|-x\rangle_2 \quad \text{or} \quad |-x\rangle_1\,|+x\rangle_2 \tag{15.5}$$

It can easily be seen that now the entangled quantum state representing the situation is

$$\begin{aligned} |\phi^-\rangle &= \tfrac{1}{\sqrt{2}}(|+z\rangle_1|+z\rangle_2-|-z\rangle_1|-z\rangle_2) \\ &= \tfrac{1}{\sqrt{2}}(|+x\rangle_1|-x\rangle_2+|-x\rangle_1|+x\rangle_2) \end{aligned} \tag{15.6}$$

This means that our first proposition determines which of the two terms appear in the entanglement when represented in the z-basis, and the second proposition fixes their relative phase. (Zeilinger 1997) As above, where the two propositions were used to determine properties of the individuals, we again obtain four orthogonal states

$$
\begin{aligned}
|\phi^+\rangle &= \tfrac{1}{\sqrt{2}}\left(|+z\rangle_1|+z\rangle_2 + |-z\rangle_1|-z\rangle_2\right)\\
|\phi^-\rangle &= \tfrac{1}{\sqrt{2}}\left(|+z\rangle_1|+z\rangle_2 - |-z\rangle_1|-z\rangle_2\right)\\
|\psi^+\rangle &= \tfrac{1}{\sqrt{2}}\left(|+z\rangle_1|-z\rangle_2 + |-z\rangle_1|+z\rangle_2\right)\\
|\psi^-\rangle &= \tfrac{1}{\sqrt{2}}\left(|+z\rangle_1|-z\rangle_2 - |-z\rangle_1|+z\rangle_2\right)
\end{aligned}
\tag{15.7}
$$

These four Bell states (Braunstein et al. 1992) are now the representation of the four possible two-bit combinations (true–true, true–false, false–true, false–false) of the truth values of the propositions, "The two spins are equal along z" and "The two spins are equal along x."

Note that we have sketched a natural understanding of quantum entanglement as a consequence of our fundamental principle. Also note that we do not make any statement as to the relative time ordering of the observations on the two systems, their relative space arrangement, and the like. Thus nonlocality comes in naturally.

We might finally remark that, from our viewpoint, quantum teleportation (Bennett et al. 1993) also obtains a very natural interpretation. All that changes by Alice's observation is the set of propositions describing possible results without any information actually transmitted to Bob as a consequence of her measurement alone!

It might amuse Dan Greenberger that this procedure can be continued to more and more elementary quantum systems. As a very specific result, three-particle entangled states, so-called GHZ states, (Greenberger and Zeilinger 1989) can be described by three elementary propositions. For example, consider the eight possible three-particle GHZ states first introduced by Mermin, (Mermin 1990)

$$
\begin{aligned}
&\tfrac{1}{\sqrt{2}}\left(|+++\rangle + |---\rangle\right)\\
&\tfrac{1}{\sqrt{2}}\left(|+++\rangle - |---\rangle\right)\\
&\tfrac{1}{\sqrt{2}}\left(|++-\rangle + |--+\rangle\right)\\
&\tfrac{1}{\sqrt{2}}\left(|++-\rangle - |--+\rangle\right)\\
&\tfrac{1}{\sqrt{2}}\left(|+-+\rangle + |-+-\rangle\right)\\
&\tfrac{1}{\sqrt{2}}\left(|+-+\rangle - |-+-\rangle\right)\\
&\tfrac{1}{\sqrt{2}}\left(|+--\rangle + |-++\rangle\right)\\
&\tfrac{1}{\sqrt{2}}\left(|+--\rangle - |-++\rangle\right)
\end{aligned}
\tag{15.8}
$$

where we use the abbreviation $|+\ +\ +\rangle = |+z\rangle_1|+z\rangle_2|+z\rangle_3$, and similarly for the other terms.

It is clear that in order to define all states uniquely, we need all eight combinations of the three propositions from true–true–true to false–false–false. It is also obvious that we cannot start as simply as before in the case of two particles by just taking as proposition 1, "The three spins are equal along z." This is because such a statement distinguishes only the first two states from the remaining six states, the latter not being distinguishable by just two bits, i.e., by the truth values of just two propositions.

Here I leave it as a puzzle what these three propositions are. I hope that Danny Greenberger, who is a passionate solver of Sunday crossword puzzles, will enjoy solving this small birthday present puzzle. And I am sure not only that he will have the solution in no time but also that he will immediately obtain many possible generalizations to more particles.

15.4 Comments

The principle given above is basic and elementary enough that it actually can serve as a foundational principle of quantum mechanics, that is, that it finally helps to answer the question, "Why the quantum?" (Wheeler 1983) This optimism is supported by the observation presented above that the principle carries in its heart two elementary notions of quantum mechanics, namely, the randomness of individual events and entanglement. It is clear that it may be a matter of taste whether or not one accepts the suggested principle as self-evident, as I do. If not, then I simply propose to turn the reasoning around and, based on our known features of quantum physics, argue for the validity of the principle.

The most fundamental viewpoint here is that the quantum is a consequence of what can be said about the world. Since what can be said has to be expressed in propositions and since the most elementary statement is a single proposition, quantization follows if the most elementary system represents just a single proposition.

While I have given here, only in a very sketchy way, a few points of a new view of quantum mechanics, a number of other fundamental concepts follow and will be elaborated upon in future papers. (Brukner and Zeilinger 1999) This will also include a more detailed analysis of philosophical and interpretational consequences. Suffice it to say here that, in my view, the principle naturally supports and extends the Copenhagen interpretation of quantum mechanics. It is evident that one of the immediate consequences is that in physics we cannot talk about reality independent of what can be said about reality. Likewise it does not make sense to reduce the task of physics to just making subjective statements, because any statements about the physical world must ultimately be subject to experiment. Therefore, while in a classical worldview, reality is a primary concept prior to and independent of observation with all its properties, in the emerging view of quantum mechanics the notions of reality and of information are on an equal footing. One implies the other and neither one is sufficient to obtain a complete understanding of the world.

About 20 years ago, I first met Daniel M. Greenberger. For me, this was one of the most important encounters of my life. Not only have we become personal friends, but his openness and ready acceptance of unusual views were very crucial in forming my own view of science. Danny Greenberger is one of the few living physicists who considers it not only possible but highly likely that our present worldview of physics may be overthrown in the not-too-distant future. This is a very healthy attitude against becoming too complacent. In his mind, one of the most useless ideas is that a final Theory of Everything is just around the next corner. I do hope that my suggestion presented above is met by Danny's approval: Not necessarily approval of its contents but, hopefully, approval of the fact that it tries to open up a new avenue for the understanding of quantum mechanics.

Acknowledgments I wish to thank D. M. Greenberger, M. A. Horne, and A. Shimony for discussions over many years on the foundations of quantum mechanics.

My initial hesitant confidence in the ideas presented here was strengthened by progress in the analysis of some of their implications done in collaboration with Caslav Brukner. This work was supported by Austrian Science Foundation (FWF), Grant S6503, and by U.S. National Science Foundation Grant PHY 97-22614.

References

Bell, J. (1990). *Physics World, 3*, 33.

Bennett, H., Brassard, G., Crépeau, C., Josza, R., Peres, A., & Wootters, W. K. (1993). *Physical Review Letters, 70*, 1895; Bouwmeester, D., Pan, J. W., Mattle, K., Eibl, M., Weinfurter, H., & Zeilinger, A. (1997). *Nature 390*, 575.

Bohm, D. (1952). *Physics Review, 85*, 166.

Bohr, N. (1935). *Physics Review, 48*, 696.

Braunstein, S. L., Mann, A., & Revzen, M. (1992). *Physical Review Letters, 68*, 3259.

Brukner, C., & Zeilinger, A. (1999). A quantum information invariant. In D. M. Greenberger, W. Reiter, & A. Zeilinger (Eds.), *Experimental and epistemological foundations of quantum mechanics* (Vienna Circle Yearbook, 1999). Dordrecht: Kluwer Academic.

Brukner, Č., & Zeilinger, A. (1999). Malus' law and quantum information. *Acta Physica Slovaca, 89*, 647–652.

Cramer, J. C. (1986). *Reviews of Modern Physics, 58*, 647.

Einstein, A. (1917). *Zeitschrift für Physik, 18*, 121.

Einstein, A. (1949). Autobiographical notes. In P. A. Schilpp (Ed.), *Albert Einstein: Philosopher-scientist* (Vol. 1, p. 65). Evanston: Open Court.

Everett, H., III. (1957). *Reviews of Modern Physics, 29*, 454.

Ghirardi, G. C., Rimini, A., & Weber, T. (1986). *Physical Review D, 34*, 470; Pearle, P. (1989). Physical Review A *39*, 2277.

Greenberger, M. A. H., & Zeilinger, A. (1989). Going beyond Bell's theorem. In M. Kafatos (Ed.), *Bell's theorem, quantum theory, and conceptions of the universe* (p. 69). Dordrecht: Kluwer Academic.

Mermin, N. D. (1985). *Physics Today, 38*(4), 38.

Mermin, N. D. (1990). *Physics Today, 43*(6), 9.

Mermin, N. D. (1998). *American Journal of Physics, 66*, 753.

Petersen, A. (1985). The philosophy of Niels Bohr. In A. P. French & P. J. Kennedy (Eds.), *Niels Bohr, a centenary volume* (p. 299). Cambridge, MA: Harvard University Press.

Wheeler, J. A. (1983). Law without law. In J. A. Wheeler & W. H. Zurek (Eds.), *Quantum theory and measurement* (p. 182). Princeton: Princeton University Press.

Zeilinger, A. (1996). On the interpretation and philosophical foundation of quantum mechanics. In Festschrift for K. V. Laurikainen & U. Ketvel et al. (Eds.), *Vastakohtien todellisuus*. Helsinki: Helsinki University Press. A copy can be found at our website www.quantum.at

Zeilinger, A. (1997). *Philosophical Transactions of the Royal Society of London. Series A, 355*, 2401.

Zeilinger, A. (1998). *Physica Scripta, T, 76*, 203.

Part VI
Calls to Reconceptualize QM

Chapter 16
A Reconstruction of Quantum Mechanics

Simon Kochen

Dedicated to the memory of Ernst Specker.

Abstract We show that exactly the same intuitively plausible definitions of state, observable, symmetry, dynamics, and compound systems of the classical Boolean structure of intrinsic properties of systems lead, when applied to the structure of extrinsic, relational quantum properties, to the standard quantum formalism, including the Schrödinger equation and the von Neumann–Lüders Projection Rule. This approach is then applied to resolving the paradoxes and difficulties of the orthodox interpretation.

16.1 Introduction

Almost a century after the mathematical formulation of quantum mechanics, there is still no consensus on the interpretation of the theory. This may be because quantum mechanics is full of predictions which contradict our everyday experiences, but then so is another, older theory, namely special relativity.

Although the Lorentz transformations initially gave rise to different interpretations, when Einstein's 1905 paper appeared it soon led to a nearly universal acceptance of Einstein's interpretation. Why was this? Einstein began with the new conceptual principle that time and simultaneity are relative to the inertial frame, dropping the classical assumption that they are absolute. By then using the linearity of transformations due to the local nature of special relativity and the experimental fact that the speed of light is constant, Einstein was able to derive the Lorentz

This work was partially supported by an award from the John Templeton Foundation.

S. Kochen (✉)
Mathematics Department, Princeton University, Princeton, NJ, USA
e-mail: kochen@math.princeton.edu

A. Cordero (ed.), *Philosophers Look at Quantum Mechanics*,
Synthese Library 406, https://doi.org/10.1007/978-3-030-15659-6_16

transformations. Furthermore, by introducing the natural classical notions of state, observable, and symmetry in the new setting, Einstein derived the new dynamical equations to replace the Newtonian equations. This manifestly consistent derivation allowed for a resolution of the apparent paradoxes which confounded the older ether theory, and led to the adoption of Einstein's interpretation by physicists.

In this paper, we shall endeavor to use Einstein's approach as a model for deriving and interpreting quantum mechanics. We also start with a new conceptual precept which replaces a classical premise. It is a basic assumption of classical physics that experiments measure pre-existing inherent observables and properties of systems, and any disturbance due to the interaction with the apparatus can be minimized or incorporated into its effect on the observables. By contrast, when we measure a particle's component of spin in a particular direction in a Stern-Gerlach experiment, it is the general belief that we are not measuring a pre-existing property. Rather, it is the interaction of the particle with the magnetic field, which is inhomogeneous in that direction, that creates the value of the spin. We shall say that such properties are *relational* or *extrinsic*, as opposed to the intrinsic properties of classical physics.

That quantum observables and properties take values only upon suitable interactions is, of course, not new to physicists. Bohr, the founder of the Copenhagen interpretation, wrote in Bohr (1937): "The whole situation in atomic physics deprives of all meaning such inherent attributes as the idealization of classical physics would ascribe to such objects." This is a radically new consequence of quantum physics that controverts one of the main conceptual assumptions of classical physics, that properties of a physical system are intrinsic.

The aim of this paper is to show that a mathematical formulation of this principle allows us to reconstruct the formalism of quantum mechanics. Let us give the basic idea in defining the structure of extrinsic properties, given in Sect. 16.2. Every experiment yields a σ-algebra of measured properties. For instance, in measuring an quantum observable with spectral decomposition $\sum a_i P_i$, the σ-algebra is generated by the projections P_i. It is shown in Sect. 16.2 that for quantum experiments the different measured σ-algebras cannot all be imbedded into a single σ-algebra. In the case of classical physics, on the other hand, the measured σ-algebras all sit inside the σ-algebra $B(\Omega)$ of intrinsic properties of the system, consisting of the σ-algebra generated by the open sets of the phase space Ω of the system.

To mathematically treat the extrinsic properties of quantum mechanics, we replace the encompassing σ-algebra $B(\Omega)$ of properties by a *σ-complex* Q, consisting of the union of all the σ-algebras of the system elicited by different decoherent interactions, such as measurements.

This change allows us to define in a uniform and natural manner the concepts of state, observable, symmetry, and dynamics, which reduce to the classical notions when Q is a Boolean σ-algebra, and to the standard quantum notions when Q is the σ-complex $Q(\mathcal{H})$ of projections of Hilbert space $\mathcal{H}$. Moreover, we use this approach to derive both the Schrödinger equation and the von Neumann-Lüders Projection Postulate. We also show on the basis of interferometry experiments why Q has the form $Q(\mathcal{H})$.

The most noteworthy feature of this reconstruction of quantum mechanics is that the classical definitions of the key physical concepts such as state, observable, symmetry, dynamics, and the combining of systems take on precisely the same form in the quantum case when they are applied to extrinsic properties.

In the standard formulation, these concepts take on a strikingly different form from the classical one. In particular, the definition of state as a complex function and the complex form of the Schrödinger equation, as opposed to the intuitive, real definitions of classical physics, led Bohr to speak of this formalism as only a symbolic representation of reality.

One purpose of this approach is to show that once the relational character of properties is accepted, the definitions of the basic concepts of quantum mechanics are as real and intuitive as is the case for classical mechanics. Of course, it is not our intention to dispense with the linear complex Hilbert space in treating problems in physics. The linearity of the Schrödinger equation is crucial for solving atomic problems. Our purpose in showing that our intuitive definitions of the notions are equivalent to the standard complex ones is rather to reduce the use of the complex Hilbert space to a technical computational tool, similar to the use of complex methods in classical electromagnetism and fluid mechanics.

At first sight the structure of a σ-complex Q is unusual. Operations between elements of Q are not defined unless they lie in the same Boolean σ-algebra within Q. That however is the whole point of this structure. Operations are only defined when they make physical sense. This points to the main difference of this approach to that initiated by Birkhoff and von Neumann (1936), and carried forward by Mackey (1963) and Piron (1976), among others. They define the logic of quantum mechanics to be a certain kind of lattice, consisting of the set of projection operators of Hilbert space. However, Birkhoff and von Neumann (1936) already raised the question:

> What experimental meaning can one attach to the meet and join of two given experimental propositions?

That question has never been adequately answered. Varadarajan, in his book (Varadarajan 1968) on the lattice approach to quantum mechanics, written some thirty years after the Birkhoff and von Neumann paper, writes:

> The only thing that may be open to serious question in this is [the] assumption ... which forces any two elements of $\mathcal{L}$ to have a lattice sum, ... We can offer no really convincing phenomenological argument to support this.

Replacing the structure of a complex Hilbert space by an equally mysterious structure of a lattice does not achieve the goal of a transparent foundation for quantum mechanics. What is perhaps surprising is that the far weaker structure of a σ-complex suffices to reconstruct the formalism of quantum mechanics. Our approach has nevertheless benefited from the lattice approach, especially as delineated in Varadarajan (1968), since theorems using lattices turned out often to have proofs using the weaker σ-complex structure.

One of the aims of a consistent, logical reconstruction of quantum mechanics is to resolve problematic questions and inconsistencies in the orthodox interpretation, such as the Measurement Problem, the Einstein-Podolsky-Rosen paradox, the Kochen-Specker paradox, the problem of reduction and the von Neumann-Lüders Projection Rule, and wave-particle duality. We discuss a resolution of these questions in the context of this reconstruction as they arise in this paper.

At various points in the paper we consider properties of systems as they are measured by experiments. We are not however espousing an operational view of quantum mechanics. We believe quantum mechanics describes general interactions in the world, independently of a classical macroscopic apparatus and observer. We do not subscribe to the Bohrian view that classical physics is needed to give meaning to quantum phenomena. The interactions we describe using a macroscopic apparatus could apply equally well to appropriate decoherent interactions between two systems in general. (See the discussion in Sect. 16.2). Nevertheless, we refer for the most part to experiments rather than general interactions in order to emphasize that the postulates have operational content and meaning. This has the merit of allowing those who prefer the operational approach to make sense of this reconstruction.

Another point is that since the properties that constitute a σ-complex correspond to the results of possible measurements, they refer to what in the orthodox interpretation are the properties that may hold as a result of reduction. We do not attempt to discuss the conditions under which reduction or decoherence occurs. There are discussions in the literature on the conditions under which reduction can occur. For instance, Bohm (2001) analyzes the strength of the inhomogeneity of the magnetic field for a successful reduction to occur in the Stern-Gerlach experiment. We consider these as interesting pragmatic questions which lie outside the purview of this paper.

We have not given a new axiomatization of quantum physics. In fact, there are no axioms in this paper, only definitions of the basic concepts, definitions which are identical with the classical ones. Rather, we have presented a framework that is common to all physical theories. It is the aim of every theory is to predict the probabilities of the outcomes of interactions between systems. Experiments are particular instances of such interactions. An experiment gives rise to a Boolean σ-algebra of events which reflects an isomorphic σ-algebra of properties of the system. The different possible experiments yield a family of σ-algebras, reflecting a family of σ-algebras properties of the system, whose union we call a σ-complex. This σ-complex helps determine the underlying theory, and conversely, a given theory determines the kind of σ-complex of properties that arises, but the general structure of a σ-complex as a union of σ-algebras is independent of any particular theory.

The main aim of the paper is to derive elementary quantum mechanics by applying the natural classical definitions of the physical concepts to extrinsic properties, and then use this derivation to resolve the standard paradoxes and problematic questions. We shall accordingly give only outlines of the proofs of the requisite theorems. To show that we have accomplished the goal of reconstructing the formalism, we shall use the textbook by Bohm (2001). This book has the advantage of explicitly introducing five postulates which suffice to treat the standard topics in quantum theory. We shall specify each of the Bohm postulates as we derive them in the paper.

To avoid repetition, we shall make the blanket assumption that the Hilbert space $\mathcal{H}$ that we deal with is a separable complex Hilbert space. The Appendix has a table which summarizes the reconstruction.

16.2 Properties

Scientific theories predict the probabilities of outcomes of experiments. We recall from probability theory that the individual outcomes of an experiment on a system form the *sample space* S. For instance, a Stern-Gerlach experiment which measures the z-components of spin for a spin 1 system has the sample space $S = \{s_{-1}, s_0, s_1\}$ corresponding to the three possible spots labeled s_{-1}, s_0, s_1 on the screen. An experiment to measure the temperature of water by a thermometer has (an interval of) the real line as sample space.

Out of the elementary outcomes, one forms an algebra of more complex outcomes, called *events*, consisting of a Boolean algebra B of subsets of the space S. The operations of B consist of union $a \vee b$, and complementation $a^{\perp}$, and all other Boolean operations, such as intersection $a \wedge b$, which are definable from them. If S is finite, then B consists of all subsets of S. If S is infinite, then the operation of countable union $\bigvee a_i$ of elements a_i of B, is added, and B is called a (Boolean) σ-algebra. (For the definition of and details about Boolean algebras see Koppelberg (1989).)

The algebra B of events, i.e. sets of outcomes, reflects the corresponding structure of properties of the system. For instance, in the above Stern-Gerlach experiment, the sets $\{s_{-1}\}$, $\{s_0\}$, and $\{s_1\}$ correspond to the properties $S_z = -1$, $S_z = 0$, and $S_z = 1$; the set $\{s_{-1}, s_1\}$ corresponds to the property $S_z = -1 \vee S_z = 1$ (where $\vee$ denotes 'or'), or equivalently, the property $\neg(S_z = 0)$ (where $\neg$ denotes 'not'), and so on. In this case, the Boolean algebra is clearly the eight element algebra. In the case of the above temperature measurement of the water, the elementary outcomes are open intervals of the real line, and the algebra of events is the σ-algebra of (*Borel*) sets generated by the intervals by complement and countable intersection.

Thus, for both classical and quantum physics, every experiment on a given system S elicits a σ-algebra of properties of S, which are true or false, i.e. have a truth value, for the system.

We come now to a crucial difference between the two theories. In classical physics, we assume that the measured properties of the system already exist prior to the measurement. It may be true that the interaction of the system with the apparatus disturbs the system, but this disturbance can be discounted or minimized. For instance, the thermometer may change the temperature of the water being measured, but this change can be accounted for, and there is no doubt that the water had a temperature prior to the measurement which is approximated by the measured value. The basic assumption is that systems have intrinsic properties, and the experiment measures the values of some them.

The family of intrinsic properties of a system form a Boolean algebra, and in the infinite case a σ-algebra. For classical physics, one introduces the phase (or state) space, with a canonical structure. The open sets of Ω generate a σ-algebra $B(\Omega)$ of *Borel* sets by complement and countable intersection. The algebra $B(\Omega)$ constitutes the σ-algebra of intrinsic properties of the system. Since the σ-algebras of measured properties are aspects of all the intrinsic properties of the system, these different σ-algebras must all be part of the σ-algebra $B(\Omega)$. Hence, the union $\cup B$ of all the σ-algebras arising from possible measurements is embeddable in $B(\Omega)$. In fact, if we assume that every property of the system is, in principle, experimentally measurable then the union $\cup B$ itself forms a σ-algebra.

In quantum mechanics, for measurements such as the Stern-Gerlach experiment, physicists do not believe that the value of the spin component S_z exists prior to the measurement. On the contrary, it is the interaction with the magnetic field, inhomogeneous in the z-direction, that results in a definite spot, say s_1, on the screen, reflecting the value, $S_z = 1$ of the spin of the particle.

This general conviction is, in fact, supported by a theorem, called the Kochen-Specker Paradox. This result showed that the spin component S_z cannot be an intrinsic property of a spin 1 particle. We recall that this result shows that there exist a small number of directions in space (33 suffice) such that any prior assignment of values to the squares of the components of spin in these directions contradicts the condition that $S_x^2 + S_y^2 + S_z^2 = 2$, for an orthogonal triple (x, y, z). Since the squares of the components of spin in orthogonal directions commute for a spin 1 system, we may measure them simultaneously for the triple (x, y, z). For instance, the measurement of the observable $S_x^2 - S_y^2$, with eigenvalues $1, -1, 0$ gives us the value 0 for S_x^2, S_y^2, or S_z^2, respectively, and 1 for the other two. We shall call such an experiment a *triple experiment on the frame* (x, y, z).

The operators S_x^2, S_y^2, S_z^2 generate an eight element Boolean algebra:

$$B_{xyz} = \{0, 1, S_x^2, S_y^2, S_z^2, 1 - S_x^2, 1 - S_y^2, 1 - S_z^2\}.$$

The 33 directions give rise to 40 orthogonal triples, and hence 40 Boolean algebras. It is important to note that the Boolean algebras have common sub-algebras. For instance, the algebra $B_{x'y'z}$ of the triple experiment on (x', y', z) has the Boolean algebra $B_z = (0, 1, S_z^2, 1 - S_z^2)$ in common with B_{xyz}.

The 40 Boolean algebras, and hence their union $\cup B_{xyz}$, cannot be embedded into a single Boolean algebra. We may see this directly from the fact that every Boolean algebra has truth values, i.e. a homomorphism onto the Boolean algebra $\{0, 1\}$, so that such an embedding would assign values to all the 40 Boolean algebras simultaneously, and hence to the 40 triples S_x^2, S_y^2, S_z^2, contradicting the Kochen-Specker theorem. (For a proof of this theorem, with the 40 triples, see Conway and Kochen 2009).

The conclusion is that, in general, quantum mechanical properties are not intrinsic to the system, but have truth values created by interactions with other systems. We shall call such interactive or relational properties *extrinsic*. The question now is: what mathematical structure captures the concept of extrinsic properties, to replace the Boolean σ-algebras that characterize intrinsic properties?

Such a structure must contain all the σ-algebras that are elicited by experiments. The minimal structure is then clearly the union $\cup B$, where B ranges over all the σ-algebras that arise in experiments. Intuitively, we may obtain such a structure by gluing together the σ-algebras at the "faces," i.e. the common sub-σ-algebras. This structure is the minimal one which contains all the σ-algebras arising from different experiments. We shall adopt it as embodying the idea of extrinsic properties. We now give the formal definition of this notion.

Definition[1] Let F be a family of σ-algebras. The σ-complex Q_F based on F is the union $\cup B$ of all σ-algebras B lying in F.

We shall generally leave the family F implicit, and simply refer to a σ-complex Q. We shall usually deal with σ-complexes that are closed under the formation of sub-σ-algebras. We can, in any case, always close a σ-complex by adding all its sub-σ-algebras.

The term σ-complex is based on the notion of a *simplicial complex* in topology. A simplicial complex is obtained by taking a family of simplices, which is closed under sub-simplices, and gluing together common simplicial faces. σ-complexes are not just analogous to simplicial complexes, but have a close correspondence, as we now outline. First recall that an *atom* of a Boolean algebra is an element x such that $y \leq x$ (i.e. $x \wedge y = y$) implies $y = 0$ or $y = x$. The atoms of a Boolean algebra in a closed Boolean complex define the vertices of a simplex, and the union of these simplices yield a simplicial complex. We may conversely define a Boolean complex from a simplicial complex. The graphs called K-S diagrams in the literature define simplicial complexes of the corresponding Boolean complexes. Strictly speaking, a simplicial complex is the family of simplices, and their union is called the carrier, so we should really call F the σ-complex. However, we shall find it convenient and harmless to conflate the two notions of σ-complex and its carrier.

Let $\mathcal{H}$ be a Hilbert space. Every set of pair-wise commuting projection operators closed under the operation of orthogonal complement $P^{\perp}(= 1 - P)$ and countable intersection $\bigwedge P_i$ forms a σ-algebra. We form the family of all such σ-algebras, and name their union, the σ-complex based on this family, $Q(\mathcal{H})$. The σ-complex $Q(\mathcal{H})$ is the structure in quantum mechanics that replaces the σ-algebra $B(\Omega)$ of Borel sets of the phase space Ω in classical mechanics.

We now summarize this discussion of properties in a form that will serve as a template for each of the other concepts we introduce in the later sections. We

[1] A Boolean σ-complex is a closely connected generalization of a *partial Boolean algebra* (introduced in Kochen and Specker (1967a), and further studied in Kochen and Specker (1964, 1967b)).

first give the classical form of the concept in terms of the σ-algebra $B(\Omega)$; then we generalize the concept by simply replacing the σ-algebra by a σ-complex Q; finally, we specialize to quantum mechanics by taking Q to be the σ-complex $Q(\mathcal{H})$. It then requires a theorem to show that the resulting concept is equivalent to the standard quantum definition on $\mathcal{H}$. Some of the classical concepts are defined in terms of the phase space Ω, rather than the σ-algebra $B(\Omega)$. We must then give an equivalent definition of the concept in terms of $B(\Omega\}$.

Classical Mechanics
The properties of a system form the σ-algebra $B(\Omega)$ of Borel sets of the phase space Ω of the system.

General Theory
The properties of a system form a σ-complex Q.

Quantum Mechanics
The properties of a system form the σ-complex $Q(\mathcal{H})$ of projections of the Hilbert space $\mathcal{H}$ of the system.

For a system S with a σ-complex Q, an appropriate interaction with another system, such as a measurement, or, more generally, a decoherent interaction, will elicit a σ-algebra B in Q of properties that have truth values. We shall call B *the (current) interactive algebra* for the system S in the interaction.

For instance, B_{xyz} is the interactive algebra in the triple experiment with the frame (x, y, z). Thus, a measurement of the observable $S_x^2 - S_y^2$ has the interactive algebra B_{xyz}. We may also consider an experiment for which the interactive algebra is $B_z = \{0, 1, S_z^2, 1 - S_z^2\}$. For instance, a variant of the Stern-Gerlach experiment with the magnetic field replaced by an inhomogeneous electric field measures the absolute value $|S_z|$ of S_z, since the electric field vector is a polar vector. For a spin 1 system this amounts to measuring S_z^2. Such an experiment is described in Wrede (1927).

In general, a measurement of the observable with discrete spectral decomposition $\sum a_i P_i$ has as interaction algebra the σ-algebra generated by the P_i's. The general case, where the observable contains a continuous spectrum, is described in Sect. 16.4.

In the triple experiment, the interaction algebra B_{xyz} of the measured system is reflected in the isomorphic eight element algebra of events consisting of the subsets of the three spots on the detection screen.

This isomorphism is, as we have seen, a general feature of a measurement, but it is also true for any appropriate decoherent interaction. If the state of the combined two interacting systems is $\sum a_i \phi_i \otimes \psi_i$ at the end of the interaction, then the interaction algebras of the systems are the two σ-algebras generated by the P_{ϕ_i} and the P_{ψ_i}, which are isomorphic. It is important to note that the macroscopic nature of the apparatus plays no role in the classical nature of the interaction algebras as Boolean σ-algebras. It simply follows from the nature that we attributed to extrinsic properties, that in every appropriate interaction they have the classical structure of a σ-algebra. As a consequence, we have no need to (and do not) subscribe to the Copenhagen interpretation, especially espoused by Bohr, that it is necessary to presuppose a classical physical description of the world in order explicate the quantum world. Quantum properties are not intrinsic, but the appropriate interaction elicits an interaction algebra with the classical structure of a σ-algebra.

16.3 States

16.3.1 Probability Measures

The theory of probability (following Kolmogorov) is based on a *probability measure*, a countably additive, [0,1]-valued measure, i.e. a function

$$p : B \to [0, 1]$$

with domain B a σ-algebra, such that $p(1) = 1$, and

$$p(\bigvee a_i) = \sum p(a_i) \text{ for pair-wise disjoint elements } a_1, a_2, \ldots \text{ in } B.$$

In the case of a measurement on a system S, the probability function p gives the probabilities of the σ-algebra of events, or equally of the measured properties of S. A physical theory predicts the probabilities of outcomes of any possible experiment, given the present state. This leads to the following concept of a state.

Classical Mechanics σ*-algebra* $B(\Omega)$
A *state* of a system with phase space Ω is a probability measure on the σ-algebra $B(\Omega)$.

General Theory σ*-complex* Q
A *state* of a system with a σ-complex of properties Q is a map $p : Q \to [0, 1]$ such that the restriction $p|B$ of p to any σ-algebra B in Q is a probability measure on B.

Quantum Mechanics σ*-complex* $Q = Q(\mathcal{H})$
Assume that $\mathcal{H}$ has dimension greater than two. There is a one-one correspondence between states p on $Q(\mathcal{H})$ and density operators (i.e. positive Hermitean operators of trace 1) w on $\mathcal{H}$ such that

$$p(x) = \text{tr}(wx) \text{ for all } x \in Q(\mathcal{H}).$$

That a density operator w defines a probability measure p on $Q(\mathcal{H})$ is an easy computation. The converse, that a state p defines a unique density operator w on $\mathcal{H}$, follows from a theorem of Gleason (1957). Gleason's theorem is the affirmative answer to a question of Mackey (1963), in which Mackey asked whether a state on the lattice of projections on $\mathcal{H}$ defines a unique density operator. A careful check of Gleason's proof of the theorem shows that, in fact, the stronger theorem stated above is true, and that the lattice operations on non-commuting projections are not needed for the proof.

As this result shows, the intuitive and plausible definition of classical states leads, with the change from intrinsic to extrinsic properties, to a similar characterization of quantum states.

16.3.2 Pure and Mixed States

The set of states on a σ-complex is closed under the formation of convex linear combinations: if $p_1, p_2, \ldots$ are states then so is $\sum c_i p_i$, for positive c_i, with $\sum c_i = 1$. The above one-one correspondence between states of $Q(\mathcal{H})$ and density operators is convexity-preserving. The extreme points of the convex set of states of a system are those that cannot be written as a non-trivial convex combination of states of the system.

Classical Mechanics σ*-algebra* $B(\Omega)$

A *pure state* of a system is an extreme point of the convex set of all states of the system.

For $B(\Omega)$, a pure state p has the form $p(s) = \begin{cases} 1 \text{ if } \omega \in s \\ 0 \text{ if } \omega \notin s \end{cases}.$

In other words, the classical pure states correspond to the points in Ω. Thus, the phase space Ω consists of the pure states, and so is also called the state space.

Thus, in the classical case all the properties of the system in a pure state are either true or false. As we would expect for intrinsic properties, measurements simply find out which measured properties are the case. The general states as mixtures of the pure states can then be interpreted as giving the probabilities of the properties which are true. These may be termed epistemic probabilities, based on the knowledge of the actual pure state that subsists.

General Theory σ*-complex* Q

A pure state of a system is an extreme point of the convex set of states of the system.

Quantum Mechanics σ*-complex* $Q = Q(\mathcal{H})$

There is a one-one correspondence between the pure states of a system and rays $[\psi]$ of unit vectors ψ in $\mathcal{H}$, such that $p(x) = \langle\psi, x\psi\rangle$.

For it is easily seen that the pure states correspond to one-dimensional projections P_ψ (with ψ in the image of P_ψ) and $p(x) = \text{tr}(P_\psi x) = \langle\psi, x\psi\rangle$. As in the classical case, the state space of the system consists of the pure states, and in this case corresponds to the projective Hilbert space of the rays of $\mathcal{H}$.

In the quantum case, even the pure states predict probabilities that are not 0 or 1, and so these are not the probabilities of properties that already subsist. This is, of course, what we should expect of extrinsic properties. A pure state simply predicts the probabilities of properties in possible future interactions, such as measurements. Mixed states are, as in the classical case, mixtures of the pure states. However, in this case there is no unique decomposition of a mixed case into pure states. This has led to a traditional difficulty in interpreting quantum mixed states. We shall postpone a discussion of our interpretation of mixed states until we have treated conditional probabilities in Sect. 16.8.

16.4 Observables

Some classical concepts such as observables are defined using the phase space Ω rather than the σ-algebra $B(\Omega)$. We can, in general, restate these definitions in terms of $B(\Omega)$. The reason for this is that the Stone Duality Theorem between Boolean algebras and spaces (and its extension by Loomis to σ-algebras) assures us that constructions on the algebras have their counterparts on the spaces and vice versa.

A classical observable is defined as a real-valued function $f : \Omega \rightarrow \mathbb{R}$ on the phase space Ω of the system. To avoid pathological, non measurable functions, f is assumed to be a Borel function, i.e. a function such that $f^{-1}(s) \in B(\Omega)$, for every set s in the σ-algebra $B(\mathbb{R})$ of Borel sets generated by the open intervals of $\mathbb{R}$.

The inverse function $f^{-1} : B(\mathbb{R}) \rightarrow B(\Omega)$ is easily seen to preserve the Boolean σ operations, i.e. to be a homomorphism. Moreover, as we see below, any such homomorphism allows us to recover the function f.

For our purposes, the advantage of using the inverse function is that it involves only the σ-algebra $B(\Omega)$ instead of the phase space Ω, allowing us to generalize the definition to a σ-complex.

Classical Mechanics σ-algebra $B(\Omega)$

An *observable* of a system with phase space Ω is a homomorphism $u : B(\mathbb{R}) \rightarrow B(\Omega)$, i.e. a map u satisfying

$$u(s^{\perp}) = u(s)^{\perp},$$
$$u(\bigvee s_i) = \bigvee u(s_i),$$

for all $s, s_1, s_2, \ldots inB(\mathbb{R})$.

There is a one-to-one correspondence between observables u and Borel functions $f : \Omega \rightarrow \mathbb{R}$ such that $u = f^{-1}$.

For given the map u we may define the Borel function f by the equation

$$f(x) = \inf\{y \mid y \in \mathbb{Q},\ x \in u((-\infty, y])\}.$$

The proof that f has the requisite properties is direct, using the denumerability of the rationals $\mathbb{Q}$ to apply the countable additivity of u. (See Varadarajan 1968, Theorem 14.)

General Theory σ*-complex* Q

An *observable* of a system with σ-complex Q is a homomorphism

$$u : B(\mathbb{R}) \to Q.$$

Note that the image of u lies in a single σ-algebra in Q.

Quantum Mechanics σ*-complex* $Q = Q(\mathcal{H})$

There is a one-one correspondence between observables $u : B(\mathbb{R}) \to Q(\mathcal{H})$ and Hermitean operators A on $\mathcal{H}$, such that, given u, $A = \int \lambda d P_\lambda$, where $P_\lambda = u((-\infty, \lambda])$.

Conversely, given a Hermitean operator A on $\mathcal{H}$, the spectral decomposition $A = \int \lambda d P_\lambda$ defines the observable u as the spectral measure $u(s) = \int_s d P_\lambda$, for $s \in B(\mathbb{R})$. This establishes the one-one correspondence.

It follows easily that if $u : B(\mathbb{R}) \to Q(\Omega)$ is an observable with corresponding Hermitean operator A, then, for the state p with corresponding density operator w, the expectation of u

$$\mathrm{Exp}_p(u) = \mathrm{tr}(Aw).$$

(See Postulates I and II of Bohm 2001.)

The theorem shows the close connection between the measurement of an observable and the interaction algebra of measured properties. For instance, for the case of a discrete operator A, the spectral decomposition $A = \sum a_i P_i$ defines the interaction algebra of measured properties generated by the P_i. Conversely, given the interaction algebra of measured properties, its atoms P_i allow us to define, for each sequence of real numbers a_i, the Hermitean operator $\sum a_i P_i$ which is thereby measured. In particular, we may in this way associate an observable with values 0 and 1 with every property in $Q(\mathcal{H})$. If A is a non-degenerate observable with eigenvalue λ belonging to eigenstate ϕ, we shall often speak of the property $A = \lambda$ to mean the projection P_ϕ which has image the ray of ϕ.

16.5 Combined Systems

An essential part of the formalism of physics is the mathematical description of the physical union of two systems. In this section we answer the question: what is the σ-complex of the union $S_1 + S_2$ of two systems with given σ-complexes Q_1 and Q_2?

In classical physics, given two systems S_1 and S_2 with the phase spaces Ω_1 and Ω_2, the phase space of the combined system $S_1 + S_2$ is the direct product space $\Omega_1 \times \Omega_2$, whereas for quantum systems with Hilbert spaces $\mathcal{H}_1$ and $\mathcal{H}_2$, the Hilbert space of $S_1 + S_2$ is the tensor product $\mathcal{H}_1 \otimes \mathcal{H}_2$. The direct and tensor products are very different constructions. The dimension of the direct product space is the sum of the dimensions of the two factor spaces, whereas the dimension of the tensor product is the product of the dimensions of the factor spaces. It is this difference that lies behind the promise of quantum computers.

We have nevertheless to combine these two operations via a single construction on the σ-complex Q. When $Q = B(\Omega)$, we may get a clue to the construction by means of Stone duality for Boolean algebras and Boolean spaces. The dual of the direct product of two Boolean spaces is the direct sum $B_1 \oplus B_2$ (also called the free product or co-product) of Boolean algebras. (See Koppelberg 1989, Chapter 4.) A similar duality extends to σ-algebras. (See Koppelberg 1989, Chapter 5.) We now use our general principle of defining a concept on a σ-complex by reducing it to the corresponding concept on its σ-algebras.

Classical Mechanics σ-algebra $B(\Omega)$

Given two systems S_1 and S_2 with σ-algebras $B(\Omega_1)$ and $B(\Omega_2)$, the combined system $S_1 + S_2$ has the σ-algebra $B(\Omega_1) \oplus B(\Omega_2)$. There is a unique space $\Omega_1 \times \Omega_2$ such that $B(\Omega_1) \oplus B(\Omega_2) \cong B(\Omega_1 \times \Omega_2)$.

The isomorphism is a well-known part of Stone Duality. For a proof see Koppelberg (1989, Chapters 4 and 5).

General Theory σ-complex Q

Given two systems S_1 and S_2 with σ-complexes Q_1 and Q_2, the combined system $S_1 + S_2$ has the σ-complex $Q_1 \oplus Q_2$, consisting of the closure (i.e. all the sub-σ-algebras) of the direct sums $B_1 \oplus B_2$ of all pairs of σ-algebras B_1 and B_2 in Q_1 and Q_2.

Quantum Mechanics σ-complex $Q = Q(\mathcal{H})$

Given the combined system $S_1 + S_2$ with the σ-complex $Q(\mathcal{H}_1) \oplus Q(\mathcal{H}_2)$, there is a unique Hilbert space $\mathcal{H}_1 \otimes \mathcal{H}_2$ such that $Q(\mathcal{H}_1) \oplus Q(\mathcal{H}_2) \cong Q(\mathcal{H}_1 \otimes \mathcal{H}_2)$. (See Postulate IVa of Bohm 2001.)

We give an outline of the proof when $\mathcal{H}_1$ and $\mathcal{H}_2$ have finite dimensions. It suffices to show that every element of $Q(\mathcal{H}_1 \otimes \mathcal{H}_2)$ lies in $Q(\mathcal{H}_1) \oplus Q(\mathcal{H}_2)$. The elements of $Q(\mathcal{H}_1) \oplus Q(\mathcal{H}_2)$ are generated by the one-dimensional projections $P_{\phi\otimes\psi}$, where $\phi \in \mathcal{H}_1$and $\psi \in \mathcal{H}_2$. We must show that if Γ is an arbitrary unit vector in $\mathcal{H}_1 \otimes \mathcal{H}_2$, then P_Γ lies in $Q(\mathcal{H}_1) \oplus Q(\mathcal{H}_2)$. One definition of the tensor product allows us to think of Γ as a conjugate-linear map from $\mathcal{H}_2$ to $\mathcal{H}_1$. (See Jauch 1968, for example.) The proof proceeds by induction on the rank of Γ as such a map. The maps of rank 1 are of the form $P_{\phi\otimes\psi}$, so the basis of the induction is true.

Now suppose Γ has rank n.

The proof is greatly simplified by choosing suitable orthonormal bases in $\mathcal{H}_1$ and $\mathcal{H}_2$ in which to expand Γ. We can construct bases $\{\phi_i\}$ and $\{\psi_i\}$ in $\mathcal{H}_1$ and $\mathcal{H}_2$ such that $\Gamma = \sum c_i\phi_i \otimes \psi_i$, with the c_i real. (Briefly, $\Gamma\Gamma^*$ and $\Gamma^*\Gamma$ have common strictly positive eigenvalues, say a_i, and respective eigenvectors ϕ_i and ψ_i; it follows that $\Gamma = \sum \sqrt{a_i}\phi_i \otimes \psi_i$. See Jauch (1968), for example.)

Let

$$\Theta = \begin{cases} -c_2\phi_1 \otimes \psi_1 + c_1\phi_2 \otimes \psi_2, & \text{for } n = 2 \\ c_1\phi_3 \otimes \psi_1 + c_2\phi_2 \otimes \psi_3 + c_3\phi_1 \otimes \psi_3 + \sum_{i>3} c_i\phi_i \otimes \psi_1, & \text{for } n > 2 \end{cases}$$

$$\Delta = c_1\phi_2 \otimes \psi_1 + c_2\phi_1 \otimes \psi_2 + \sum_{i\geq 3} c_i\phi_i \otimes \psi_2$$

Then Γ, Θ, and Δ are pairwise orthogonal unit vectors. Hence, P_Γ, P_Δ, and P_Θ mutually commute, and $P_\Gamma = (P_\Gamma \vee P_\Theta) \wedge (P_\Gamma \vee P_\Delta)$.

For $n = 2$, let $x_+ = c_2\Gamma + c_1\Theta$ and $x_- = c_1\Gamma - c_2\Theta$. For $n > 2$, let $x_\pm = \Gamma \pm \Theta$. Also, let $y_\pm = \Gamma \pm \Delta$. Then it is easily checked that the four vectors $x_\pm$ and $y_\pm$ are of rank $n-1$, and x_+ and x_- are orthogonal, as are y_+ and y_-. It follows that $[P_{x_+}, P_{x_-}] = [P_{y_+}, P_{y_-}] = 0$. Moreover, $P_\Gamma \vee P_\Theta = P_{x_+} \vee P_{x_-}$ and $P_\Gamma \vee P_\Delta = P_{y_+} \vee P_{y_-}$. Hence, $P_\Gamma = (P_{x_+} \vee P_{x_-}) \wedge (P_{y_+} \vee P_{y_-})$. Since P_{x_+}, P_{x_-}, P_{y_+}, and P_{y_-} inductively lie in $Q(\mathcal{H}_1) \oplus Q(\mathcal{H}_2)$ and each of the pairs (P_{x_+}, P_{x_-}), (P_{y_+}, P_{y_-}), and $(P_{x_+} \vee P_{x_-}, P_{y_+} \vee P_{y_-})$ lie in a common σ-algebra, it follows that P_Γ lies in $Q(\mathcal{H}_1) \oplus Q(\mathcal{H}_2)$. The proof provides an algorithm for constructing $x_\pm$ and $y_\pm$.

The uniqueness (up to isomorphism) is a routine consequence of the fact that $Q_1 \oplus Q_2$ is categorically a co-product (see Koppelberg (1989) for a proof in the σ-algebra case).

The infinite dimensional case is discussed in Sect. 16.10.

As an illustration we consider the simplest case of the tensor product $\mathcal{H}_1 \otimes \mathcal{H}_2$ of two-dimensional Hilbert spaces, which we may take to represent two spin $\frac{1}{2}$ particles. Each element of $Q(\mathcal{H}_1)$ (resp. $Q(\mathcal{H}_2)$) corresponds to the property $s_z \otimes I = \frac{1}{2}$ (resp. $I \otimes s_z = \frac{1}{2}$) for some direction z. For Γ in $\mathcal{H}_1 \otimes \mathcal{H}_2$ we shall identify P_{x_+}, P_{x_-}, P_{y_+}, and P_{y_-}.

We write the vector Γ in the diagonal form $c_1\phi_1 \otimes \psi_1 + c_2\phi_2 \otimes \psi_2$. Hence,

$$x_- = \phi_1 \otimes \psi_1, \; x_+ = \phi_2 \otimes \psi_2$$

$$y_+ = (\phi_1 + \phi_2) \otimes (c_1\psi_1 + c_2\psi_2), \; y_- = (\phi_1 - \phi_2) \otimes (c_1\psi_1 - c_2\psi_2)$$

Now ϕ_1 defines $s_z \otimes I = \frac{1}{2}$ for a direction z, and ψ_1 defines $1 \otimes s_w = -\frac{1}{2}$ in a direction w. Thus, $\phi_1 \pm \phi_2$ defines $s_x \otimes I = \pm\frac{1}{2}$ for a direction x orthogonal to z. Also, if we write $c_1 = \cos(\mu/2)$, then $c_1\psi_1 + c_2\psi_2$ defines $I \otimes s_u = \frac{1}{2}$ in a direction u at an angle μ from the w direction, and $c_1\psi_1 - c_2\psi_2$ defines $I \otimes s_v = \frac{1}{2}$ in a direction v at angle $-\mu$ from the w direction. It follows that

$$\begin{aligned} P_\Gamma &= (P_{x_+} \vee P_{x_-}) \wedge (P_{y_+} \vee P_{y_-}) \\ &= (s_z \otimes I = \frac{1}{2} \leftrightarrow I \otimes s_w = -\frac{1}{2}) \\ &\quad \wedge [(I \otimes s_u = \frac{1}{2} \rightarrow s_x \otimes I = \frac{1}{2}) \wedge (s_x \otimes I = \frac{1}{2} \rightarrow I \otimes s_v = \frac{1}{2})] \end{aligned}$$

In this manner every state in a combined system can be interpreted as a compound proposition about the factors.

A particularly interesting case is the singleton state $\Gamma = \sqrt{\frac{1}{2}}(\phi_z^+ \otimes \psi_z^- - \phi_z^- \otimes \psi_z^+)$, (with $s_z\phi_z^\pm = \pm\frac{1}{2}\phi_z^\pm$ and $s_x\psi_z^\pm = \frac{1}{2}\psi_z^\pm$) where

$$\begin{aligned}
P &= (P_\Gamma \vee P_\Theta) \wedge (P_\Gamma \vee P_\Delta) \\
&= (S_z = 0) \wedge (S_x = 0) \\
&= (P_{x_+} \vee P_{x_-}) \wedge (P_{y_+} \vee P_{y_-}) \\
&= (s_z \otimes I = \frac{1}{2} \leftrightarrow I \otimes s_z = -\frac{1}{2}) \wedge (s_x \otimes I = \frac{1}{2} \leftrightarrow I \otimes s_x = -\frac{1}{2}).
\end{aligned}$$

In Sect. 16.11 we shall apply this result to the EPR experiment.

This construction of the direct sum generalizes in an obvious way to the direct sum of an arbitrary number of σ-complexes, representing the union of several systems. The above theorems then generalize to:

$$\begin{aligned}
B(\Omega_1) \oplus B(\Omega_2) \oplus \cdots &\cong B(\Omega_1 \times \Omega_2 \times \cdots) \\
Q(\mathcal{H}_1) \oplus Q(\mathcal{H}_2) \oplus \cdots &\cong Q(\mathcal{H}_1 \otimes \mathcal{H}_2 \otimes \cdots).
\end{aligned}$$

These general sums are needed in discussing statistical mechanics. It is now routine to define symmetric and anti-symmetric direct sums of σ-complexes, yielding the corresponding symmetric and anti-symmetric tensor products of Hilbert spaces, needed to deal with identical particles. (See Postulate IVb of Bohm (2001). The spin-statistics connection that Bohm adds can also be added here.)

16.6 Symmetries

As Noether, Weyl, and Wigner showed, observables such as position, momentum, angular momentum, and energy arise from global symmetries of space and time, and the conservation laws for them arise from the corresponding symmetries of interactions. Other observables arise from local symmetries. In classical physics the symmetries appear as canonical transformations of phase space, and in quantum physics they appear as unitary or anti-unitary transformations of Hilbert space. For us they naturally appear as symmetries of a σ-complex.

Definition An automorphism of a σ-complex Q is a one-one transformation $\sigma : Q \to Q$ of Q onto Q such that for every σ-algebra B in Q and all $a, a_1, a_2, \cdots$ in B

$$\sigma(a^\perp) = \sigma(a)^\perp \text{ and } \sigma(\bigvee a_i) = \bigvee \sigma(a_i).$$

General Theory: σ-complex Q

A *symmetry* of a system with σ-complex Q is given by an automorphism of Q.

A symmetry σ defines a natural convexity-preserving map $p \to p_\sigma$ on the states of Q by letting $p_\sigma = p \circ \sigma^{-1}$, i.e. $p_\sigma(x) = p(\sigma^{-1}(x))$, for all $x \in Q$.

Quantum Mechanics *σ-complex* $Q = Q(\mathcal{H})$

There is a one-one correspondence between symmetries $\sigma : Q(\mathcal{H}) \to Q(\mathcal{H})$ and unitary or anti-unitary operators u on $\mathcal{H}$ such that $\sigma(x) = uxu^{-1}$, for all $x \in Q(\mathcal{H})$.

If a state p corresponds to the density operator w, then

$$p_\sigma(x) = p(\sigma^{-1}(x)) = \mathrm{tr}(wu^{-1}xu) = \mathrm{tr}(uwu^{-1}x),$$

so that the state p_σ corresponds to the density operator uwu^{-1}.

It is easy to check that unitary and anti-unitary operators define a symmetry on $Q(\mathcal{H})$. For the converse we use a well-known theorem of Wigner. (See Bargmann 1964.) The original theorem of Wigner posits a one-one map of the set of rays of $\mathcal{H}$ onto itself which preserves the inner product. Uhlhorn (1963) was able to weaken this to preserving the orthogonality of rays. As Bargmann states in Bargmann (1964), the proof he gives of Wigner's theorem may be easily modified to prove Uhlhorn's result. (For a proof see Varadarajan 1968.)

Now assume that σ is a symmetry of $Q(\mathcal{H})$. Then σ is a one-one map of the set of atoms, i.e. one-dimensional projections P_ψ, of $Q(\mathcal{H})$ onto atoms of $Q(\mathcal{H})$. In other words, rays $[\psi]$ of $\mathcal{H}$ are one-to-one mapped onto rays of $\mathcal{H}$. Moreover, since σ-algebras are mapped by σ to σ-algebras, the orthogonality of rays is preserved. The Uhlhorn version of Wigner's theorem then shows there is a unique (up to a multiplicative constant) unitary or anti-unitary map u on $\mathcal{H}$ such that $\sigma(x) = uxu^{-1}$.

In the case of classical physics, with $Q = B(\Omega)$, a symmetry is defined by a canonical transformation of the manifold. Every such transformation defines an automorphism of the σ-algebra $B(\Omega)$. However, the converse is not true. Although the automorphism still defines a continuous map from Ω to itself, the structure of a σ-algebra is too weak to recover the canonical structure. It is remarkable that the σ-complex structure is sufficient to allow one to define the symmetries of the Hilbert space. In that sense, quantum physics allows a more satisfactory reconstruction than classical physics. As Sect. 16.9 suggests, we may recover the classical canonical structure from the quantum structure in the limit of an increasing number of particles.

16.7 Dynamics

Now that we have shown that the symmetries of $Q(\mathcal{H})$ are implemented by symmetries of $\mathcal{H}$, we may use time symmetry to introduce a dynamics for systems.

To define dynamical evolution, we consider systems that are invariant under time translation. For such systems, there is no absolute time, only time differences. The change from time 0 to time t is given by a symmetry $\sigma_t : Q \to Q$, since the structure of the system of properties is indistinguishable at two values of time. We assume that if the state evolves first for a time t and then the resulting state for a time t', then this yields the same result as the original state evolving for a time $t + t'$. Moreover, we assume that evolution over a small time period results in small changes in the probability of properties occurring.

The passage of time is thus given by a continuous representation of the additive group $\mathbb{R}$ of real numbers into the group $\text{Aut}(Q)$ of automorphisms of Q under composition:
i.e. a map $\sigma : \mathbb{R} \to \text{Aut}(Q)$, such that

$$\sigma_{t+t'} = \sigma_t \circ \sigma_{t'}$$

and $p_{\sigma_t}(x)$ is a continuous function of t.

The image of σ is then a continuous one-parameter group of automorphisms on Q.[2]

We have seen that an automorphism σ corresponds to a unitary or anti-unitary operator. Anti-unitary operators actually occur as symmetries, for instance in time reversal. However, for the above representation only unitary operators u_t corresponding to the symmetry σ_t can occur, since $u_t = u_{t/2}^2$, which is unitary.[3]

It follows that the evolving state p_{σ_t} corresponds to the density operator $w_t = u_t w u_t^{-1}$. By Stone's Theorem,

$$u_t = e^{-\frac{i}{\hbar}Ht},$$

where $\hbar$ is a constant to be determined by experiment; so

$$w_t = e^{-\frac{i}{\hbar}Ht} w \, e^{\frac{i}{\hbar}Ht}.$$

Differentiating,

$$\partial_t w_t = -\frac{i}{\hbar}[H, w_t].$$

This is the Liouville-von Neumann Equation.

Conversely, this equation yields a continuous representation of $\mathbb{R}$ into $\text{Aut}(Q(\mathcal{H}))$. For $w = P_\psi$, a pure state, $w_t = P_{\psi(t)}$ and this equation reduces to the Schrödinger Equation:

$$\partial_t \psi(t) = -\frac{i}{\hbar} H \psi(t).$$

---

[2]The group $\text{Aut}(Q)$ may, in fact, be construed as a topological group by defining, for each $\epsilon > 0$, an ϵ-neighborhood of the identity to be $\{\sigma \mid |p_\sigma(x) - p(x)| < \epsilon \text{ for all } x \text{ and } p\}$. We may then directly speak of the continuity of the map σ, in place of the condition that $p_{\sigma_t}(x)$ is continuous in t.

[3]More precisely, we have a projective unitary representation of $\mathbb{R}$, but such a representation of $\mathbb{R}$ is equivalent to a vector representation. (See, e.g., Varadarajan 1968.)

(See Postulate Va of Bohm (2001). Postulate Vb is the Heisenberg form of the equation, and follows similarly.)

We stop here without specifying any further the form of the Hamiltonian H. This form depends upon calculating the linear and angular momentum observables as operators from the homogeneity and isotropy of space, using the corresponding unitary representations that we have used for time homogeneity. This a well-known part of quantum mechanics and need not be explored further here. (See Jauch 1968, for example.) We have treated the non-relativistic dynamical equation. The connection between automorphisms of $Q(\mathcal{H})$ and unitary operators given above allows to us to treat the relativistic dynamical equations in a similar manner, following Wigner's work. (See Varadarajan 1968.)

16.8 Reduction and Conditional Probability

16.8.1 Conditional States

With these results, which cover four of Bohm's five postulates, we can now recover much of quantum theory. So far however, we will never predict interference. The states we introduced are probability measures on Q, which for any experiment is a classical probability measure on the σ-algebra of properties being measured. In fact, the probability must be classical, since it is mirrored in the probability measure on the experiment's σ-algebra of events, which are generated by macroscopic spots on a screen.

How then does interference enter the picture? In dealing with experiments, we have omitted a key ingredient that is usually referred to as "the preparation of state." To calculate the probability $p(x)$ of a property holding at the end of an experiment, we need to know both the property x and the state p. In general, when we are presented with a particle to be measured, we do not know its state. One way to know the state is to prepare it by means of a prior interaction.

For instance, the book Feynman et al. (1966) by Feynman, Leighton, Sands introduces quantum mechanics via a spin 1 system by discussing the probability of, for instance, going to state $S_x = 1$, given that it is in state $S_z = 0$. The particle is prepared in state $S_z = 0$ by sending it through a Stern-Gerlach field in the z direction, and then filtering it through a one-slit screen to allow only the central beam through. If the system is not detected as hitting the filtering screen, then it is reduced to the state $S_z = 0$. If allowed to hit a final detection screen it is certain to register the central spot. But we are free to send it through another Stern-Gerlach field in the x direction to measure $S_x = 1$, say. This is a reduction by preparation of the original, possibly unknown, state to the state $S_z = 0$.

Some physicists think that reduction is a phenomenon unique to quantum mechanics that has no counterpart in classical mechanics, but this not the case. Consider a one slit experiment with bullets. If we shoot at a target, we get a probability distribution on the target that defines a mixed state for the bullet. Since the target screen can be placed anywhere from the gun to any distant point, the probability distribution is a function of time that gives a time evolution of this state, satisfying the classical Liouville equation for mixed states. If we now interpose a one-slit screen between the gun and the target screen, we find that after the evolution of the state p up to the one-slit screen, the bullet either has hit this screen, or if not, has passed through with a new state $p(\cdot \mid y)$, where y is the property that it has not hit the screen. This is classically called conditionalizing the state p to y. The new state $p(\cdot \mid y)$ is defined by $p(x \mid y) = p(x \wedge y)/p(y)$, as the frequency definition of probability can verify. This filtering to a new state is entirely similar to the filtering of a spin 1 system described earlier, and is the classical equivalent of reduction.

Now that we have the classical form of reduction as conditionalization, we can follow our prescription by generalizing from a σ-algebra to a σ-complex.

Classical Mechanics σ-algebra $B(\Omega)$

Let p be a state on the σ-algebra $B(\Omega)$ and $y \in B(\Omega)$ such that $p(y) \neq 0$. By a *state conditionalized on* y we mean a state $p(\cdot \mid y)$ such that for every x in $B(\Omega)$,

$$p(x \mid y) = p(x \wedge y)/p(y).$$

General Theory: σ-complex Q

Let p be state on a σ-complex Q and $y \in Q$ such that $p(y) \neq 0$. By a *state conditionalized on* y we mean a state $p(\cdot \mid y)$ such that for every σ-algebra B in Q containing y and every $x \in B$,

$$p(x \mid y) = p(x \wedge y)/p(y).$$

In the literature, there exist generalizations of probability measures and conditional probability to non-commutative algebras, and, in particular, to lattices of projections. (See Beltrametti and Cassinelli 1981.) In general, it is by no means clear that such a state $p(\cdot \mid y)$ either exists or is unique, as is obviously the case for classical mechanics. However, for the quantum σ-complex $Q(\mathcal{H})$ this can be proved:

Quantum Mechanics σ-complex $Q = Q(\mathcal{H})$

If p is a state on $Q(\mathcal{H})$ and $y \in Q(\mathcal{H})$ such that $p(y) \neq 0$, then there exists a unique state $p(\cdot \mid y)$ conditionalized on y. If w is the density operator corresponding to p, then $ywy/\operatorname{tr}(ywy)$ is the density operator corresponding to the state $p(\cdot \mid y)$.

To see that the operator $ywy/\operatorname{tr}(ywy)$ corresponds to the state $p(\cdot \mid y)$, note that if x lies in the same σ-algebra as y, then x and y commute, so

$$\operatorname{tr}(ywyx)/\operatorname{tr}(ywy) = \operatorname{tr}(wxy)/\operatorname{tr}(wy) = p(x \wedge y)/p(y) = p(x \mid y).$$

For uniqueness, it suffices to consider the case when $x \in B(\mathcal{H})$ is a one-dimensional projection. Let $p(\cdot \mid y)$ be a state conditionalized on y, and let v be the corresponding density operator. Let ϕ be a unit vector in the image of x. We can write $\phi = y\phi + y^{\perp}\phi$. Then

$$\begin{aligned} p(x \mid y) &= \operatorname{tr}(vx) = \langle \phi, v\phi \rangle \\ &= \langle y\phi, vy\phi \rangle + \left\langle y\phi, vy^{\perp}\phi \right\rangle + \left\langle y^{\perp}\phi, vy\phi \right\rangle + \left\langle y^{\perp}\phi, vy^{\perp}\phi \right\rangle. \end{aligned}$$

Now, $\operatorname{tr}(vy^{\perp}) = p(y^{\perp} \mid y) = p(y^{\perp} \wedge y)/p(y) = 0$, so $vy^{\perp}\phi = 0$. Hence,

$$p(x \mid y) = \langle y\phi, vy\phi \rangle = \|y\phi\|^2 \operatorname{tr}(vP_{y\phi}) = \|y\phi\|^2 p(P_{y\phi})/p(y),$$

since $P_{y\phi} \leq y$. If $p'(\cdot \mid y\}$ is another state conditionalized on y, then

$$p'(x \mid y) = \|y\phi\|^2 p(P_{y\phi})/p(y) = p(x \mid y),$$

proving uniqueness.

The change from w to $ywy/\operatorname{tr}(wy)$ in state preparation or measurement is the general formula for the reduction of state given by the von Neumann-Lüders Projection Rule. In the orthodox interpretation this rule is an additional principle that is appended to quantum mechanics. Here it appears as the unique answer to conditionalizing a state to a given property. (See Postulate IIIa of Bohm 2001.)

The natural definition of applying a symmetry σ to a conditionalized state $p(\cdot \mid y)$ is given by

$$p_{\sigma}(x \mid y) = p(\sigma^{-1}(x) \mid \sigma^{-1}(y)).$$

16.8.2 Classical and Quantum Conditional Probability

In the well-known paper (Finkelstein 1963), Feynman writes that the basic change from classical to quantum mechanics lies in the revision in the probability rule called the Law of Alternatives,
$p(a \mid c) = \sum_i p(a \mid b_i)p(b_i \mid c)$ for disjoint b_i, to the quantum law that $\langle \alpha \mid \beta \rangle = \sum_i \langle \alpha \mid \beta_i \rangle \langle \beta_i \mid \gamma \rangle$, giving an additional interference term.

We agree that this is an important difference in the two theories. However, we shall derive it from what we consider the more basic difference, that between intrinsic and extrinsic properties.

Let $y_1, y_2, \cdots$ lie in a σ-algebra with $y_i \wedge y_j = 0$ for $i \neq j$, and let $y = \bigvee y_i$. Then

Classical Mechanics

$$
\begin{aligned}
p(x \mid y) &= p(\bigvee (x \wedge y_i))/p(y) \\
&= \sum (p(x \wedge y_i)/p(y_i)) \cdot (p(y_i)/p(y)) \\
&= \sum p(x \mid y_i) p(y_i \mid y),
\end{aligned}
$$

The Law of Alternatives in classical probability theory.

On the other hand, by Sect. 16.8, we have

Quantum Mechanics

$$
\begin{aligned}
p(x \mid y) &= \mathrm{tr}(ywyx)/\,\mathrm{tr}(wy) \\
&= \mathrm{tr}(\bigvee_{i,j} y_i w y_j x)/\,\mathrm{tr}(wy) \\
&= \sum \mathrm{tr}(y_i w y_i x)/\,\mathrm{tr}(wy) + \sum_{i \neq j} \mathrm{tr}(y_i w y_j x)/\,\mathrm{tr}(wy) \\
&= \sum p(x \mid y_i) p(y_i \mid y) + \sum_{i \neq j} \mathrm{tr}(y_i w y_j x)/\,\mathrm{tr}(wy).
\end{aligned}
$$

This shows that in condionalizing for the extrinsic properties of quantum mechanics an interference term must be added to the classical law of alternatives.

16.8.3 Conditionalizing on Several Properties

There is a different kind of preparation of state, one which leads to a mixed state. This occurs when, instead of all but one of the beams being blocked, as in Sect. 16.8, the beams are allowed to pass through the filter, while being registered. For instance, Feynman et al. (1966) describes a version of the two-slit experiment in which the particle scatters high frequency photons that register which slit the particle passed through. In this case, the property y_1 of passing through slit 1 is true or the property y_2 of passing through slit 2 is true, so that the state of the particle is either the conditional state $p(\cdot \mid y_1)$ or the state $p(\cdot \mid y_2)$.

If we consider an ensemble of particles, then each of the particles in the ensemble will be in the state $p(\cdot \mid y_i)$ with probability $p(y_i)$, for $i = 1, 2$, so that the ensemble is in the mixed state $p(y_1)p(\cdot \mid y_1) + p(y_2)p(\cdot \mid y_2)$. Thus, by registering the results of passage through each of the two slits, we restore the classical Law of Alternatives.

For a single particle, the same mixed state describes its predicted state upon passage through the registering two-slit screen. However, upon actual passage through the registered slits, the state is either $p(\cdot \mid y_1)$ or $p(\cdot \mid y_2)$. We may say that even after the passage, the state of the particle for an experimenter who is not aware of the registered result the state remains the mixed state. In this regard, the mixture has a similar interpretation as in the classical case, viz., the ignorance interpretation of mixtures.

A measurement of an observable is the most familiar example of conditionalizing with respect to several properties. If the observable has a spectral decomposition $\sum a_i P_i$, then measuring the observable amounts to registering the values of the properties given by the P_i. The interaction algebra B is the σ-algebra generated by the P_i.

We now formulate this notion of conditioning with respect to several conditions. Given a system with σ-complex Q and disjoint elements $y_1, y_2, \ldots$ in a common σ-algebra in Q, with $\bigvee y_i = 1$, and a state p, we define the state conditionalized on $y_1, y_2, \ldots$ to be $p(\cdot \mid y_1, y_2, \ldots) = \sum p(y_i)p(\cdot \mid y_i)$. We shall also write this more succinctly as $p(\cdot \mid B)$, the state conditionalized on the interaction algebra B, the σ-algebra generated by the y_i.

For quantum mechanics, with $Q = Q(\mathcal{H})$, if w is the density operator corresponding to the state p:

$$p(\cdot \mid B) = \sum \mathrm{tr}(wy_i)(y_i w y_i / \mathrm{tr}(wy_i)) = \sum y_i w y_i,$$

so that for each x the probability $p(x \mid B) = \sum \mathrm{tr}(y_i w y_i x)$. This gives the state of an ensemble without selection. (See Postulate IIIb of Bohm 2001.)

The natural definition for applying a symmetry to the conditioned state is given by

$$p_\sigma(x \mid B) = p(\sigma^{-1}(x) \mid \sigma^{-1}B).$$

Note that the non-uniqueness of the decomposition of a degenerate density operator into pure states causes no problems in this interpretation. This is because mixed states arise as mixtures of given pure states in the conditionalization from an experiment or the evolution of the mixture. The σ-algebra B generated by the $y_1, y_2, \ldots$ is simply the current interaction algebra of the σ-complex, and is always given to us as part of the interaction.

The fact that degenerate density operators do not have a unique decomposition into pure states has led some to put mixed and pure states on an equal footing, and to deny them the role as mixtures. This puts the cart before the horse, and ignores the historical development of the concept of mixed states. Mixtures of pure states were in long use in quantum mechanics (as well as in classical statistical mechanics) when von Neumann introduced the invariant formulation of a mixed state as a density operator. The use of the density operator has the advantage of allowing the introduction of the abstract notion of mixed state, without requiring the explicit mention of any basis of pure states, which could be recovered in the non-degenerate case. For us, however, in any interaction (and subsequent evolution) the interaction algebra is always given, which yields a unique decomposition of the mixed state as a mixture of pure states even in the degenerate case.

16.9 Reconstructing the σ-Complex $\mathcal{Q}(\mathcal{H})$

We saw in Sect. 16.2 that if we restrict ourselves to classical experiments, then the σ-complex of interaction algebras can be imbedded into a σ-algebra. On the other hand, the 40 quantum triple experiments yield a σ-complex that cannot be so imbedded. Thus, increasing the set of experiments has changed the structure of the σ-complexes of systems. It may then be possible that a sufficiently comprehensive family of experiments may force the structure of the σ-complex $\mathcal{Q}$ to be isomorphic to $\mathcal{Q}(\mathcal{H})$. In this section we shall see that this is indeed the case.

The result is based on the paper Reck et al. (1994). The interactions arise from a composition of interferometers. First, Mach-Zender interferometers together with beam splitters allow one to construct $\mathcal{Q}(\mathcal{H}_2)$, where $\mathcal{H}_2$ is a two-dimensional Hilbert space. A standard theorem, which allows one to decompose n-dimensional unitary operators as a product of two-dimensional ones, is then used to treat the σ-complex of higher dimensional Hilbert spaces.

We outline the construction in Reck et al. (1994) (from which the diagrams below are copied). The experimental realization of a general two-dimensional unitary matrix is obtained by a Mach-Zender interferometer consisting of two mirrors, two 50–50 beam splitters, an ω-phase shifter, and a ϕ-phase shifter at one output port:

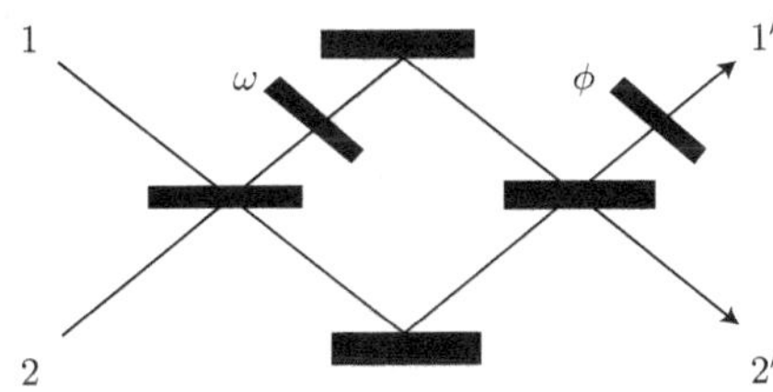

This device transforms the input state with modes (k_1, k_2) into the output state with modes (k_1', k_2'), which are related by the unitary matrix:

$$\begin{pmatrix} k_1' \\ k_2' \end{pmatrix} = \begin{pmatrix} e^{i\phi} \sin\omega & e^{i\phi} \cos\omega \\ \cos\omega & -\sin\omega \end{pmatrix} \begin{pmatrix} k_1 \\ k_2 \end{pmatrix}.$$

We can then realize all 2-dimensional unitary matrices by varying the phase shifters.

To treat $n \times n$ unitary matrices, the authors in Reck et al. (1994) show how to eliminate the off-diagonal element u_{jk} of a unitary matrix U by multiplying U by the matrix T_{jk} which is obtained from the $n \times n$ identity matrix I by replacing the (jj), (jk), (kj), (kk) entries by the entries of a matrix of the above 2-dimensional unitary form. This inductively results in the product

$$UT_{nn-1}T_{nn-2} \cdots T_{32}T_{31}T_{21} = D$$

where D is a diagonal unitary matrix with diagonal entries of modulus 1. Hence

$$U = DT_{21}^{\dagger}T_{31}^{\dagger}T_{32}^{\dagger}\cdots T_{n1}^{\dagger}T_{n2}^{\dagger}\dots T_{nn-1}^{\dagger}.$$

We now combine copies of the above interferometers so that the outputs of one are the inputs of the succeeding one, corresponding to the above product of the $T_{jk}^{\dagger}$ matrices, followed by n phase shifters to account for the matrix D. The result is a device which realizes the matrix U. For instance, for $n = 3$, we have:

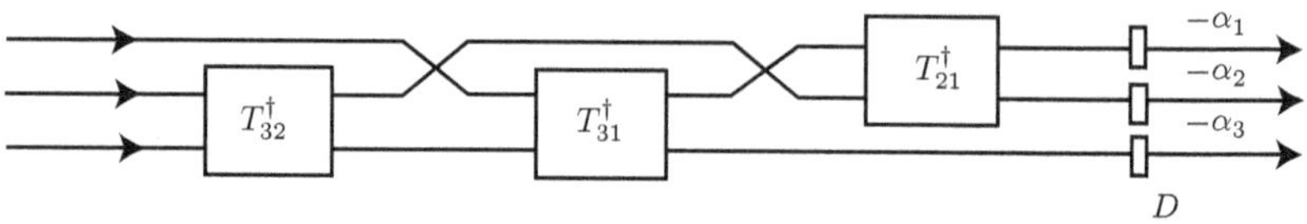

(Each box represents an interferometer of the above type.)

To realize an n-dimensional Hermitean matrix A, we use additional beam splitters to superpose those beams that correspond to the same eigenspace of A, and then add detectors for the resulting beams. The use of beam splitters to superpose beams is well-known. (See e.g. Zukowski et al. 1997.)

This is a précis of the construction in Reck et al. (1994). It allows us to realize every element of $Q(\mathcal{H})$, where $\mathcal{H}$ is an n-dimensional complex Hilbert space. What is significant is that we can also realize the σ-complex structure of $Q(\mathcal{H})$. To see this it suffices to consider the two Boolean operations of complementation $x^{\perp}$ and join $x \vee y$. The output for a projection x consists of two beams, labeled the 1-beam and the 0-beam according to the eigenvalues of x. The operation of complementation $x^{\perp}$ requires only a transposition of the 1 and 0 labels. The join $x \vee y$ of two projections corresponds to superposing the two 1-beams of x and y. These two operations suffice to define all the Boolean operations, and therefore the σ-complex structure of $Q(\mathcal{H})$. Note that this realization of the σ-complex of properties via the different σ-algebras generated by the outcomes of interferometer experiments follows the general prescription given in Sect. 16.2 for defining the σ-complex of properties of a system by means of the different σ-algebras of events defined by the experimental outcomes.

It is instructive to contrast the simple experimental counterparts to the σ-complex structure with the lattice structure of the set of projections. We know of no corresponding experimental realization to the lattice join (or meet) of two non-commuting projections. This is due to the difficulty of relating the eigenspaces of two non-commuting operators to the eigenspaces of their sum (or, for projections, to their union), while for commuting operators there is a simple relation. It is this difficulty that is alluded to in our earlier quotations from Varadarajan (1968) and Birkhoff and von Neumann (1936) in the introduction.

We have seen that if we can in principle form arbitrarily large networks of interferometers, then we can realize the σ-complex $Q(\mathcal{H})$ for Hilbert spaces of all finite dimensions. The single minimal space $\mathcal{H}$ for which $Q(\mathcal{H})$ realizes *all* the interferometer experiments, and hence contains all finite dimensional Hilbert spaces, is an infinite dimensional separable pre-Hilbert space, i.e. an inner product space $\mathcal{H}$, whose completion forms a separable Hilbert space $\mathcal{H}_\omega$. To see this note that $\mathcal{H}$ may be construed as the space of all complex sequences $\{a_i\}$ that are non-zero for only finite many i, with inner product $\langle\{a_i\},\{b_i\}\rangle = \sum a_i b_i$.

Thus in the infinite dimensional case we must add ideal elements which are limits of sequences of realized elements. We cannot expect to realize $Q(\mathcal{H}_\omega)$ via experiments without adding limits since the world itself may be finite. This is similar to the use of probability in physica as an ideal limit of relative frequency for longer and longer sequences of experiments. Of course, even the above realization of $Q(\mathcal{H})$ in the finite dimensional case is an idealization, since it requires ω-phase shifters for arbitrary real ω, in $[0, 2\pi]$.

We may now extend the result $Q(\mathcal{H}_1) \oplus Q(\mathcal{H}_2) \simeq Q(\mathcal{H}_1 \otimes \mathcal{H}_2)$ of Sect. 16.5 to the infinite dimensional case.

The fact that $\mathcal{H}$ is the *minimal* space such that $Q(\mathcal{H})$ is realized by the above interferometry experiments highlights the open-ended nature of our reconstruction. If we restrict ourselves to experiments of classical physics, then the σ-complex reduces to a σ-algebra, and the concepts lead to classical physics. If we add the forty triple experiments, the resulting σ-complex cannot be imbedded into a σ-algebra. If we allow for the interferometry experiments of this section, then Q must take the form $Q(\mathcal{H})$. It thus suffices to consider these interferometry experiments to realize the structure of quantum physics. We may then apply the resulting theory to general interactions.[4] As we have emphasized throughout the paper, the special nature of experiments, with the macroscopic apparatus, plays no role in the theory. Any appropriate decoherent interaction gives rise to isomorphic σ-algebras for the two systems. Experiments do play the pragmatic role of allowing us to become cognizant of a sufficient number of interactions to help determine the theory.

It is possible that other experiments may require a different realization of the σ-complexes. For instance, if we consider systems which satisfy *superselection rules* (see e.g. Beltrametti and Cassinelli 1981), then the σ-complex Q has a non-trivial σ-algebra which is common to all the σ-algebras B in Q. In this case Q is not of the form $Q(\mathcal{H})$, but is a sub-σ-complex of $Q(\mathcal{H})$. $\mathcal{H}$ takes the form of a direct sum $\oplus\mathcal{H}_i$ of Hilbert spaces with the pure states forced to lie in a factor $\mathcal{H}_i$.

16.10 From Quantum Physics to Classical Physics

With the description in Sect. 16.5 of the σ-complex of combined systems, it is possible to treat the statistics of a large number of particles such as macroscopic bodies. This is, of course, a major subject in quantum statistics, and we shall not

[4]Historically, of course, it was not such interferometry experiments, but rather spectroscopic experiments that lead Schrödinger to his equation.

venture there. However, we wish to say a few words on how the σ-complex of quantum mechanics tends to a classical σ-algebra with an increasing number of particles, so that the quantum system becomes effectively classical.

We shall adapt a remark in Finkelstein (1963) for this purpose. Let S be an ensemble of n non-interacting copies of a system S_i, $i = 1, 2, \ldots, n$, with σ-complex $Q(\mathcal{H}_i)$. Then S has the σ-complex

$$Q(\mathcal{H}_i) \oplus Q(\mathcal{H}_2) \oplus \cdots \oplus Q(\mathcal{H}_n) \simeq Q(\mathcal{H}_1 \otimes \mathcal{H}_2 \otimes \cdots \otimes \mathcal{H}_n).$$

Suppose each S_i is in the pure state ϕ. Then S is in the state $\Phi = \phi \otimes \phi \otimes \cdots \otimes \phi$. Consider the observable $\mathbf{A}$ of S which is the average of the same observable A of each S_i:

$$\mathbf{A} = (A \otimes I \otimes \cdots \otimes I + I \otimes A \otimes \cdots \otimes I + \cdots + I \otimes I \otimes \cdots \otimes A)/n.$$

We recall that the uncertainty ΔR of an operator R is the square root of the variance: $(\Delta R)^2 = \mathrm{Exp}((R - \mathrm{Exp}\, R)^2)$. Hence,

$$\begin{aligned}(\Delta \mathbf{A})^2 &= \langle \Phi, (\mathbf{A} - \mathrm{Exp}\,\mathbf{A})\Phi \rangle \\ &= (1 - 1/n)\, \langle \phi, (A - \mathrm{Exp}\, A)\phi \rangle^2 + (1/n) \left\langle \phi, (A - \mathrm{Exp}\, A)^2 \phi \right\rangle \\ &= (\Delta A)^2/n.\end{aligned}$$

Hence, if

$$\mathbf{B} = (B \otimes I \otimes \cdots \otimes I + I \otimes B \otimes \cdots I + \cdots + I \otimes I \otimes \cdots \otimes B)/n$$

is another such averaged observable, then for the commutator $[A, B]$ we have

$$\Delta[\mathbf{A}, \mathbf{B}] = \Delta[A, B]/n.$$

Thus, $\lim_{n \to \infty} \Delta[\mathbf{A}, \mathbf{B}] = 0$. It follows that the averaged observables of S all commute in the limit, and so the σ-complex of S becomes essentially a σ-algebra for very large n, as in a macroscopic body.

This calculation was made under the assumption that S is an ensemble of non-interacting replicas of one particle. In a real body the states and observables need not be identical. Without going into details, it is possible to give conditions on the allowed variation of the states of the particles and the averaged observables so that $\Delta[\mathbf{A}, \mathbf{B}]$ still tends to zero with increasing n. In any case, the result is at least suggestive that in a real body, the σ-complex of S will be very close to a σ-algebra.

The change in dynamics accompanying the move from the Hilbert space $\mathcal{H}$ to the phase space Ω has been well-studied. In essence, the quantum bracket $\frac{i}{\hbar}[X, Y]$ is replaced by the Poisson bracket $\{X, Y\}$, so that the von Neumann-Liouville equation $\partial_t w_t = -\frac{i}{\hbar}[H, w_t]$ is replaced by the classical Liouville equation $\partial_t f_t = -\{H, f_t\}$. (See Faddeev and Yakubovskii 2009, for example.) We saw in Sects. 16.6 and 16.7 that the lack of sufficient structure of a σ-algebra did not allow us to derive the classical dynamics from the automorphisms of $B(\Omega)$, whereas we could do so in the quantum case $Q(\mathcal{H})$. We can see now how it is possible to recover the classical dynamical equation by an excursion into the quantum structure $Q(\mathcal{H})$.

16.11 Interpreting and Resolving Quantum Paradoxes

16.11.1 The K-S Paradox and the Projection Rule

We have already applied this reconstruction to treat several issues in the interpretation of the formalism. One of these, the Kochen-Specker Paradox, which showed that the assumption that all properties are intrinsic leads to a contradiction, was the motivation for introducing the σ-complex of extrinsic properties. Conversely, assuming the relational nature of properties resolves this paradox. Another issue, discussed in Sect. 16.8, is the nature of reduction and the von Neumann-Lüders Projection Rule, which here appears as the counterpart to classical conditionalizing, not as an ad hoc addition to quantum theory. We now consider a number of other controversial questions from the literature.

16.11.2 Wave-Particle Duality

We discuss wave-particle duality in the context of the two-slit experiment. Let y_1 and y_2 be the projections of position in the regions of the two slits δ_1 and δ_2. Then $y_1 \vee y_2$ is the projection of position for the union $\delta_1 \cup \delta_2$. Let x be the property of position in a local region Δ on the detection screen.

If passage through each of the two slits is registered, then the Law of Alternatives of Sect. 16.8 tells us that $p(x|y_1 \vee y_2) = p(x|y_1)p(y_1|y_1 \vee y_2) + p(x|y_2)p(y_2|y_1 \vee y_2)$, which, in the case of symmetrical positioned slits, is proportional to the sum $p(x|y_1) + p(x|y_2)$ of the probabilities of passage through the individual slits, just as in the classical case.

In the case where the passage through the two slits by the quantum particle is not registered, we have shown in Sect. 16.8 that there is an additional interference term

$$[tr(y_1 w y_2)x) + tr(y_2 w y_1 x)]/tr(w(y_1 \vee y_2)).$$

Note that if x and y_1 and y_2 commute, this interference term vanishes. This happens if the detector is right next to the two-slit screen. If the detector is a distance from the two-slit screen, then the particle undergoes free flight evolution σ_t, so $\sigma_t(y_i) = u_t y_i u_t^{-1}$ no longer commutes with x, giving rise to the non-zero interference term.

An explanation of the interference effect that is often given is that the particle is, or acts as, a pair of waves emanating from the slits, which exhibit constructive and destructive interference effects. This was, of course, the explanation for Young's original experiment with the classical electromagnetic field. For individual quantum particles however, it leads to the paradoxical effect that the wave suddenly collapses to a local region at the detection screen.

The explanation given here is a different one. A system forms a localized particle if there is a position operator for the system, so that a measurement of position detects the system at a localized region in space. Until the position is measured the position has no value, since position in a region is an extrinsic property. We may view the two-slit screen as a preparation of state for the particle, for which the position is conditionalized, or reduced, to the region $\delta_1 \cup \delta_2$. This reduction is not a position measurement, since $\delta_1 \cup \delta_2$ is not a localized region (as it would be for a single-slit screen). It is only at the detection screen, where the particle, in interaction with the screen, is reduced to the local region Δ, that its position has a value.

The question of why the particle shows the interference effects of a wave is answered in Sect. 16.7, where the evolution of the quantum particle was defined by a trajectory in the space Aut(Q). This yielded the Schrödinger equation, which is a wave equation. On the other hand, a trajectory in the phase space of a classical particle passing through a two-slit screen is governed by the classical Liouville equation, without any wave properties. Thereby, the wave-like properties of a quantum particle are explained by the extrinsic character of its properties.

16.11.3 The Measurement Problem

The Measurement Problem refers to an inconsistency in the orthodox interpretation of quantum measurement. The interpretation assumes that an isolated system undergoes unitary evolution via Schrödinger's equation. We quote from Bohm (2001, Chapter XII):

> If time evolution is a symmetry transformation, then the mathematical structure (in particular the algebraic relations) of the algebra of observables does not change in time; this means that the physical structure is indistinguishable at two different points in time. Our experience shows that there are physical systems that have this property and in fact it is this property that defines the isolated systems. Thus isolated physical systems do not age, an absolute value of time has no meaning for these systems, and only time differences are accessible to measurement. Irreversible processes do not take place in isolated physical systems defined as above.

Accordingly, in the orthodox interpretation, for a measurement of an observable A of a system S by an apparatus T, the total system $S+T$, which is assumed to be isolated, undergoes unitary evolution.

We outline the standard description of an ideal measurement. Suppose the spectral decomposition of an observable is $A = \sum a_i P_i$, where each P_i is a one-dimensional projection with eigenstate ϕ_i. The apparatus is assumed to be sensitive to the different eigenstates of A. Hence, if the initial state of S is ϕ_k and the apparatus T is in a neutral state ψ_0, so that the state of $S+T$ is $\phi_k \otimes \psi_0$, then the system evolves into the state $\phi_k \otimes \psi_k$, where the ψ_i are the states of the apparatus co-ordinate corresponding to the states ϕ_i of the system. By linearity, if S is in the initial state $\phi = \sum a_i \phi_i$, then $S+T$ evolves into the state $\Gamma = \sum a_i \phi_i \otimes \psi_i$. The intractable problem for the orthodox interpretation is that the completed measurement gives a particular apparatus state ψ_k, indicating that the state of S is ϕ_k, so that the state of the total system is $\phi_k \otimes \psi_k$, in contradiction to the evolved state $\sum a_i \phi_i \otimes \psi_i$. We may also see the reduction from the viewpoint of the conditionalization of the states. If the state p of $S+T$ just prior to measurement is P_Γ, then after the measurement it is the conditionalized state

$$\begin{aligned} p(\,\cdot\,|P_{\phi_k} \otimes I \wedge I \otimes P_{\psi_k}) &= (P_{\phi_k} \otimes I \wedge I \otimes P_{\psi_k}) P_\Gamma (P_{\phi_k} \otimes I \wedge I \otimes P_{\psi_k}) / \\ &\qquad tr((P_{\phi_k} \otimes I \wedge I \otimes P_{\psi_k}) P_\Gamma)) \\ &= P_{\phi_k \otimes \psi_k}. \end{aligned}$$

Hence, the new conditionalized state of $S+T$ is the reduced state $\phi_k \otimes \psi_k$.

The orthodox interpretation then has to reconcile the unitary evolution of $S+T$ with the measured reduced states of S and T. The present interpretation stands the orthodox interpretation on its head. We do not begin with the unitary development of an isolated system, but rather with the results of a measurement, or, more generally, of a decoherent interaction. In fact, the original motivation for forming a σ-complex of properties was via the set of measured, and hence reduced, properties which form the current interaction algebra. For us, it is the conditions under which dynamical evolution occurs that is to be investigated, rather than the reduced state. We cannot take for granted what is assumed in the orthodox interpretation, as in the above quotation, that an isolated system evolves unitarily. So we must answer the question whether in a measurement the σ-complex structure of $S+T$ undergoes a symmetry transformation at different times of the process. As Sect. 16.7 showed, this is formalized as the condition for the existence of a representation $\sigma : \mathbb{R} \to \mathrm{Aut}(Q)$.

It is easy to see, however, that in the process of a completed measurement or a state preparation there are two distinct elements of $Q(\mathcal{H})(= Q(\mathcal{H}_1 \otimes \mathcal{H}_2))$ at initial time 0 which end up being mapped to the same element at a later time t. We have seen that an initial state $\phi \otimes \psi_0$ results in a state $\phi_k \otimes \psi_k$, for some k. However, $\phi_k \otimes \psi_0$ also results in the state $\phi_k \otimes \psi_k$. If we choose the state ϕ to be distinct from ϕ_k, then the two elements $P_{\phi \otimes \psi_0}$ and $P_{\phi_k \otimes \psi_0}$ of $Q(\mathcal{H})$ both map to the same element $P_{\phi_k \otimes \psi_k}$. However, any automorphism σ_t is certainly a one-to-one map on Q, so the measurement process cannot be described by a representation $\sigma : \mathbb{R} \to \mathrm{Aut}(Q)$, and hence a unitary evolution.

In our interpretation, the Measurement Problem is thus resolved in favor of reduction rather than unitary evolution. The point can be made intuitively that points of absolute time do exist in a measurement and also in state preparation, namely the point (or, better, small interval) of time at which reduction takes place. If for instance, we consider a Stern-Gerlach experiment with a state preparation in which a filter registers the passage of a particle through one of several slits, before the particle reaches a detection screen, then the interval of time of passage through the slit, in which the state of the particle is reduced, is such an absolute point of time: the state after passing through the slit is the conditionalized state, whereas before it is not.

Time and its passage is a problematic concept in physics, so to reinforce the point we shall give another example, in which time homogeneity is tied to spatial symmetry. Consider a particle resulting, say, from decay in which its state has spherical symmetry. Assume that the particle is initially at the center of a spherical detector system. During the passage of the particle until it hits the detector, the combined system of particle and detector is spherically symmetric and time homogeneous. At the moment of registering the impact on a local region of the detector, the system loses both its isotropy in space and its time symmetry. If it is difficult to argue against this breaking of space symmetry in favor of a particular direction, it seems to us to be equally hard to gainsay the breaking of time symmetry at the moment this non-isotropy occurs.

For a composite system it is not only outside forces that can break symmetry, but internal interactions. As opposed to the quotation of Bohm (2001) above, we believe that symmetry-breaking processes do take place in isolated compound systems with internal decoherent interactions during reduction of state. To argue that nevertheless symmetry has not been broken for the combined system is to favor the theoretical formalism ahead of the facts on the ground. It is notable that with this interpretation the system consisting of the universe as a whole, for which there are no external systems, acquires reduced or, as we say, conditionalized states as a result of the interactions of component systems.

Note that our alternative term *interactive property* is more appropriate here than *extrinsic property*. The reduction of the state to $\phi_k \otimes \psi_k$ happens for the composite system $S_1 + S_2$ because of the interaction of the component systems S_1 and S_2 which are internal to $S_1 + S_2$ rather than an interaction of $S_1 + S_2$ with an external system.

16.11.4 The Einstein-Podolsky-Rosen Experiment

We shall discuss the EPR phenomenon in the Bohm form of two spin $\frac{1}{2}$ particles in the combined singlet state Γ of total spin 0. Suppose that in that state the two particles are separated and the spin component s_z of particle 1 is measured in some direction z. That means that the observable $s_z \otimes I$ of the combined system is being measured.

Let $P_z^{\pm} = \frac{1}{2}I \pm s_z$. We have the spectral decomposition

$$s_z \otimes I = \tfrac{1}{2}P_z^+ \otimes I + (-\tfrac{1}{2})P_z^- \otimes I,$$

so the interaction algebra $B = \{0, 1, P_z^+ \otimes I, P_z^- \otimes I\}$. We expand the singlet state

$$\Gamma = \sqrt{\tfrac{1}{2}}(\phi_z^+ \otimes \psi_z^- - \phi_z^- \otimes \psi_z^+),$$

where $P_z^{\pm}\phi_z^{\pm} = \phi_z^{\pm}$ and $P_z^{\pm}\psi_z^{\pm} = \psi_z^{\pm}$. Thus, if particle 1 has spin up, the state $p(\cdot \mid P_z^+ \otimes I)$ of the system is, by Sect. 16.8, given by

$$p(\cdot \mid P_z^+ \otimes I) = (P_z^+ \otimes I)P_\Gamma(P_z^+ \otimes I)/\operatorname{tr}((P_z^+ \otimes I)P_\Gamma) = P_{\phi_z^+ \otimes \psi_z^-}.$$

This is, of course, equivalent to projecting the vector Γ into the image of $P_z^+ \otimes I$:

$$P_z^+ \otimes I(\Gamma) = \sqrt{\tfrac{1}{2}}(\phi_z^+ \otimes \psi_z^-).$$

Similarly, if particle 1 has spin down the state $p(\cdot \mid P_z^- \otimes I)$ is given by the vector

$$P_z^- \otimes I(\Gamma) = \sqrt{\tfrac{1}{2}}(\phi_z^- \otimes \psi_z^+).$$

This shows that if s_z is measured for particle 2, it is certain to have opposite value of s_z for particle 1. It does *not* mean that after s_z is measured for particle 1, then s_z has a value for particle 2. The properties $I \otimes P_z^+$ and $I \otimes P_z^-$ do not lie in the interaction algebra $B = \{P_z^+ \otimes I, P_z^- \otimes I, 0, 1\}$, and so have no value. The spin components are extrinsic properties of each particle, which do not have values until the appropriate interaction. To claim otherwise is to revert to the classical notion of intrinsic properties.

This is a necessary consequence of our interpretation, but it also follows from a careful application of standard quantum mechanical principles. For after the measurement of s_z on particle 1 gives a value of $\frac{1}{2}$, the state of the combined system is $\phi_z^+ \otimes \psi_z^-$, which is an eigenstate of $I \otimes s_z$. Born's Rule implies that an eigenstate of an observable will yield the corresponding eigenvalue as value only if and when that observable is measured.

The situation is entirely similar to the unproblematic triple experiment. A triple experiment on the frame (x, y, z) yields the interaction algebra B_{xyz}. If $S_z^2 = 0$, then $S_x^2 = S_y^2 = 1$. If (x', y', z) is another frame, then it is also the case that $p(S_{x'}^2 = 1 | S_z^2 = 0) = 1$, so that $S_{x'}^2$ is certain to have the value 1 if the triple experiment on the frame (x', y', z) is performed. But $S_{x'}^2$ does not have a value unless and until that experiment is carried out since $S_{x'}^2 = 1$ does not lie in the interaction algebra B_{xyz}.

We have not in this discussion mentioned a word about special relativity. Indeed, the spin EPR phenomenon has nothing to do with position or motion and is independent of relativistic questions. However, EPR with space-like separated particles has been used to put in question the full Lorentz invariance of quantum mechanics. This is replaced by a weaker notion that EPR correlations cannot be used for faster than light signaling. We believe that Lorentz invariance is a fundamental symmetry principle, which gives rise to basic observables, and is not simply an artifact of signaling messages between agents.

The relativistically invariant description of the EPR experiment is that if experimenters A_1 and A_2 measure particles 1 and 2, and the directions of spin in which they are measured are the same, then an experimenter B in the common part of the future light cones of A_1 and A_2 will find that the spins are in opposite directions.

Now that we have studied what EPR actually says, we shall treat the question of how correlations can exist between the different directions of spins of two particles when such spins cannot simultaneously have values.

To set the stage for EPR, we again first consider the triple experiment. For a spin 1 particle the proposition $S_z^2 = 1$ defines the same projection in $Q(\mathcal{H})$ as the proposition

$$S_x^2 = 0 \leftrightarrow S_y^2 = 1 \tag{16.1}$$

If we perform the (x, y, z) triple experiment with interaction algebra B_{xyz} and find that $S_z^2 = 1$, then we can check that either $S_x^2 = 0$ and $S_y^2 = 1$ or $S_x^2 = 1$ and $S_y^2 = 0$, so that (16.1) is true. However, for the orthogonal triple (x', y', z)

$$S_{x'}^2 = 0 \leftrightarrow S_{y'}^2 = 1 \tag{16.2}$$

is the same projection as (16.1) and so is also true. But $S_{x'}^2$ and $S_{y'}^2$ do not lie in the interaction algebra B_{xyz}, and so have no truth value unless and until the (x', y', z) triple experiment is performed. Thus, the correlation (16.2) is true without its component properties $S_{x'}^2$ and $S_{y'}^2$ having truth values.

Now consider the EPR experiment. We have seen in Sect. 16.5 that $S = 0$ is the same projection as $(S_z = 0) \wedge (S_x = 0)$, and $S_z = 0$ and $S_x = 0$ are in turn respectively the same projections as

$$s_z \otimes I = \frac{1}{2} \leftrightarrow I \otimes s_z = -\frac{1}{2} \tag{16.3}$$

and

$$s_x \otimes I = \frac{1}{2} \leftrightarrow I \otimes s_x = -\frac{1}{2}. \tag{16.4}$$

If the projections $S_z = 0$ and $S_x = 0$ are true, then so are the correlations (16.3) and (16.4) since they define the same projections. As in the triple experiments, we see that these correlations subsist simultaneously, even though the spins s_z and s_x for each particle cannot have values simultaneously. Thus, the existence of seemingly paradoxical EPR correlations in different directions can be understood via the logic of extrinsic properties.

In summary, the extrinsic properties of a σ-complex may have relations subsisting among its elements because of general laws of physics, such as conservation laws, which are timeless and independent of particular interactions. The σ-complex structure accommodates such relations in the form of compound formulas such as (16.3) and (16.4), which are true, even when the constituent parts do not have truth values. This fact allows us to interpret the EPR phenomenon in a fully relativistically invariant way. For extrinsic properties a compound property may have truth values even when the component parts do not.

16.12 On the Logic of Quantum Mechanics

As we have stressed throughout this paper, the major transformation from classical to quantum physics in this approach lies not in modifying the basic classical concepts such as state, observable, symmetry, dynamics, combining systems, or the notion of probability, but rather in the shift from intrinsic to extrinsic properties.

Now properties, whether considered as predicates or propositions, are the domain of logic. Boolean algebras correspond to propositional logic and σ-algebras to predicate logic. Hence the change to a σ-complex of extrinsic properties should entail a new logic of properties. At first sight however, it would appear that the logic of extrinsic properties as elements of a σ-complex Q is no different than classical propositional logic, since these elements can only be compounded when they lie in the same σ-algebra in Q. This is far from the case; in fact, the difference in logic plays an important role in resolving some of the quantum paradoxes. The underlying reason is that a compound property such as $x \vee y$ may be lie in an interaction algebra and so have a truth value, even though neither x nor y lie in the algebra, and have no truth value.

The logic of extrinsic properties has been systematically studied in Kochen and Specker (1964, 1967b), where a complete axiomatization of the propositional calculus of extrinsic properties is given. Here we shall confine ourselves to pointing out some uses of this logic that appeared in this paper.

1. The simplest such case is $x \vee x^{\perp}$, which equals 1 in Q, and so is always true, even though x may have no truth value.[5] Thus, for a spin $\frac{1}{2}$ particle, $s_z = \frac{1}{2} \vee s_z = -\frac{1}{2}$ is true simultaneously for all directions z, though s_z may have no value.

[5]This is reminiscent of Aristotle's famous sea battle in *De Interpretatione*: "A sea battle must either take place tomorrow or not, but it is not necessary that it should take place tomorrow neither is it necessary that it should not take place, yet it is necessary that it either should or should not take place tomorrow."

2. In the two-slit experiment (Sect. 16.11), we saw that it is this lack of truth value that leads to the interference pattern at the detector screen. The source of the interference pattern is not some non-classical probability, but rather the applications of classical Kolmogorov axioms of probability to the logic of extrinsic properties. The conditional probability $p(x|y)$ is the probability of x given that y has happened and so has a truth value. Therefore the probability $p(x|y_1 \vee y_2)$ implies that $y_1 \vee y_2$ is true. However, neither y_1 nor y_2 has happened. We should not expect the classical Law of Alternatives connecting $p(x|y_1 \vee y_2)$ to $p(x|y_1)$ and $p(x|y_2)$ to be valid unless y_1 and y_2 are events that have happened. In that case the Law of Alternatives is in fact valid in quantum mechanics.
3. In the EPR experiment, the singleton state $S = 0$ implies that $s_z \otimes I = \frac{1}{2} \leftrightarrow I \otimes s_z = -\frac{1}{2}$ is true for any direction z. In fact, as shown in Sect. 16.5 the element $S = 0$ equals

$$(s_z \otimes I = \frac{1}{2} \leftrightarrow I \otimes s_z = -\frac{1}{2}) \wedge (s_x \otimes I = \frac{1}{2} \leftrightarrow I \otimes s_x = -\frac{1}{2}).$$

Thus, the correlation exists in both the z and x directions even though the spins cannot simultaneously have values in these directions. Section 16.5 shows how general superpositions of states of combined systems may be reformulated as compound statements of this quantum logic.
4. The K-S Paradox in Sect. 16.2 can be stated as a proposition that is classically true but false in quantum mechanics. To see this, let $+$ denote exclusive disjunction. Then $x + y + z + x \wedge y \wedge z$ is true if and only if exactly one of x, y, and z is true.

   The statement $\bigvee_{i \leq 40}(x_i + y_i + z_i + x_i \wedge y_i \wedge z_i)^{\perp}$, where (x_i, y_i, z_i) range over the orthogonal triples of the 40 triple experiments of Sect. 16.2 is classically true, but false under a substitutions $x_i \mapsto S^2_{x_i}$, $y_i \mapsto S^2_{y_i}$, $z_i \mapsto S^2_{z_i}$.

   For two spin $\frac{1}{2}$ particles there is a K-S Paradox which yields a much simpler such proposition in four dimensional Hilbert space:

$$[(x \leftrightarrow y) \leftrightarrow (z \leftrightarrow w)] \leftrightarrow [(x \leftrightarrow z) \leftrightarrow (y \leftrightarrow w)].$$

   This classically true proposition is false under the substitution

$$x \mapsto s_z \otimes I = \frac{1}{2},\ y \mapsto I \otimes s_z = \frac{1}{2},\ w \mapsto s_x \otimes I = \frac{1}{2},\ z \mapsto I \otimes s_x = \frac{1}{2}.$$

   For details, see Conway and Kochen (2002). Kochen and Specker (1967a) Theorem 4 shows that every K-S Paradox corresponds to a classically true proposition which is false under a substitution of quantum properties.

Appendix: Summary Table of Concepts

	General mechanics	Classical mechanics	Quantum mechanics
Properties	σ-complex $Q = \cup B$, with B a σ-algebra	σ-algebra $B(\Omega)$	σ-complex $Q(\mathcal{H}\}$
States	$p : Q \to [0, 1]$ $p \mid B$, a probability measure	$p : B(\Omega) \to [0, 1]$ a probability measure	$w : \mathcal{H} \to \mathcal{H}$ Density operator $p(x) = \mathrm{tr}(wx)$
Pure states	Extreme point of convex set	$\omega \in \Omega$	1 dim operator i.e. unit $\phi \in \mathcal{H}$ $p(x) = \langle x, x\phi\rangle$
Observables	$u : B(\mathbb{R}) \to Q$ homomorphism	$f : \Omega \to \mathbb{R}$ Borel function	$A : \mathcal{H} \to \mathcal{H}$ Hermitean operator
Symmetries	$\sigma : Q \to Q$ automorphism	$h : \Omega \to \Omega$ canonical transformation	$u : \mathcal{H} \to \mathcal{H}$ unitary or anti-unitary operator $\sigma(x) = uxu^{-1}$
Dynamics	$\sigma : \mathbb{R} \to \mathrm{Aut}(Q)$ representation	Liouville equation $\partial_t \rho = -[H, \rho]$	von Neumann -Liouville equation $\partial_t w_t = -\frac{i}{\hbar}[H, w_t]$
Conditionalized states	$p(x) \to p(x \mid y)$ for $x, y \in B$ in Q $p(x \mid y) = p(x \mid y)/p(y)$	$p(x) \to p(x \mid y)$ $= p(x \wedge y)/p(y)$	$w \to ywy/\mathrm{tr}(wy)$ von Neumann -Lüders Rule
Combined systems	$Q_1 \oplus Q_2$ direct sum of σ-complexes	$\Omega_1 \times \Omega_2$ direct product of phase spaces	$\mathcal{H}_1 \otimes \mathcal{H}_2$ tensor product of Hilbert spaces

References

Bargmann, V. (1964). Note on Wigner's theorem on symmetry operations. *Journal of Mathematical Physics, 5*, 862.

Beltrametti, E. G., & Cassinelli, G. (1981). *The logic of quantum mechanics*. Reading: Addison-Wesley.

Birkhoff, G., & von Neumann, J. (1936). The logic of quantum mechanics. *Annals of Mathematics, 37*, 823.

Bohm, A. (2001). *Quantum mechanics: Foundations and applications*. New York: Springer

Bohr, N. (1937). Causality and complementarity. *Philosophy of Science, 4*, 289.

Conway, J., & Kochen, S. (2002). The geometry of the quantum paradoxes. In R. A. Bertlmann & A. Zeilinger (Eds.), *Quantum [un]speakables* (p. 257). Berlin: Springer.

Conway, J., & Kochen, S. (2009). The strong free will theorem. *Notices American Mathematical Society, 56*, 226.

Faddeev, L. D., & Yakubovskii, O. A. (2009). *Lectures on quantum mechanics for mathematics students*. Providence: American Mathematical Society.

Feynman, R. P. (1948). Space-time approach to non-relativistic quantum mechanics. *Reviews of Modern Physics, 20*, 36.

Feynman, R. P., Leighton, R. B., & Sands, M. (1966). *The Feynman lectures on physics* (vol. 3). Reading: Addison-Wesley.

Finkelstein, D. (1963). The logic of quantum physics. *Transactions of the New York Academy of Sciences, 25*, 621.
Gleason, A. M. (1957). Measures on the closed subspaces of a Hilbert space. *Journal of Mathematics and Mechanics, 6*, 885.
Jauch, J. M. (1968). *Foundations of quantum mechanics*. Reading: Addison-Wesley.
Kochen, S., & Specker, E. P. (1964). The calculus of partial propositional functions, methodology and philosophy of science, Congress at Jerusalem (vol. 45).
Kochen, S., & Specker, E. P. (1967a). The problem of hidden variables in quantum mechanics. *Journal of Mathematics and Mechanics, 17*, 59.
Kochen, S., & Specker, E. P. (1967b). Logical structures arising in quantum mechanics. The theory of models, symposium at Berkeley (vol. 177).
Koppelberg, S. (1989). *Handbook of boolean algebras* (vol. 1). Amsterdam: North-Holland.
Mackey, G. W. (1963). *Mathematical foundations of quantum mechanics*. Amsterdam: Benjamin.
Piron, C. (1976). *Foundations of quantum physics*. Reading: Benjamin.
Reck, M., Zeilinger, A., Bernstein, H. J., & Bertani, P. (1994). Experimental realization of any discrete unitary operator. *Physical Review Letters, 73*, 58.
Uhlhorn, U. (1963). Representation of symmetric transformations in quantum mechanics. *Arkiv Fysik, 23*, 307.
Varadarajan, V. S. (1968). *The geometry of quantum theory*. Princeton: Van Nostrand.
Wrede, E. (1927). Uber die Ablenkung von Molekularstrahlen elektrischer Dipolmolekule iminhomogenen elektrischen Feld. *Z. Phys. A, 44*, 261.
Zukowski, M., Zeilinger, A., & Horne, M. A. (1997). Realizable higher-dimensional two-particle entanglements via multiport beam splitters. *Physical Review A, 55*, 2564.

Chapter 17
What is Orthodox Quantum Mechanics?

David Wallace

Abstract What is called "orthodox" quantum mechanics, as presented in standard foundational discussions, relies on two substantive assumptions—the projection postulate and the eigenvalue-eigenvector link—that do not in fact play any part in practical applications of quantum mechanics. I argue for this conclusion on a number of grounds, but primarily on the grounds that the projection postulate fails correctly to account for repeated, continuous and unsharp measurements (all of which are standard in contemporary physics) and that the eigenvalue-eigenvector link implies that virtually all interesting properties are maximally indefinite pretty much always. I present an alternative way of conceptualising quantum mechanics that does a better job of representing quantum mechanics as it is actually used, and in particular that eliminates use of either the projection postulate or the eigenvalue-eigenvector link, and I reformulate the measurement problem within this new presentation of orthodoxy.

17.1 Introduction: The Orthodox View of Orthodoxy

"Orthodox" or "standard" quantum mechanics, as typically presented in textbook philosophy-of-physics discussions,[1] consists of these components:

The structural core: This has three parts:

1. **States:** The possible states of a quantum system are represented by normalised vectors in some complex Hilbert space.

[1] See, e.g., Albert (1992), Barrett (1999), Bub (1997), and (Penrose 1989, ch. 5–6).

D. Wallace (✉)
Department of Philosophy, University of Southern California, Los Angeles, CA, USA
e-mail: davidmwwallace@outlook.com

A. Cordero (ed.), *Philosophers Look at Quantum Mechanics*,
Synthese Library 406, https://doi.org/10.1007/978-3-030-15659-6_17

2. **Observables:** To any physical quantity used to describe the system (often called an 'observable') is associated a self-adjoint operator on that same Hilbert space.
3. **Dynamics:** The state of a quantum system evolves over time according to the *Schrödinger equation*:

$$\frac{\mathrm{d}}{\mathrm{d}t} |\psi(t)\rangle = -\frac{i}{\hbar} \widehat{H} |\psi(t)\rangle \tag{17.1}$$

where $\widehat{H}$ is the self-adjoint operator corresponding to the system's energy.

(This can be generalised in certain respects, in particular by allowing quantum states to be mixed rather than pure, and in fact I think this generalisation does a better job of capturing real-world quantum mechanics than the pure-state version (cf. Wallace 2013) but for simplicity I use the pure-state version in this paper.)

The Born (probability) rule: Suppose some quantity O has associated operator $\widehat{O}$, which can be written

$$\widehat{O} = \sum_i o_i \widehat{\Pi}(i) \tag{17.2}$$

where the o_i are the distinct eigenvalues of the operator and $\widehat{\Pi}(i)$ projects onto the subspace of states with eigenvalue o_i. (Recall that any self-adjoint operator can be so written—this is the 'spectral resolution' of the operator.) Then if O is measured on a quantum system with state $|\psi\rangle$, then:

1. The only possible outcomes of the measurement are the eigenvalues o_i of the operator;
2. The probability of the measurement giving result o_i is

$$\Pr(O = o_i) = \langle\psi| \widehat{P}(i) |\psi\rangle . \tag{17.3}$$

The projection postulate (aka the collapse law): Suppose some quantity O, as above, is measured on a quantum system in state $|\psi\rangle$. Then the measurement induces a stochastic transition on the state, so that:

1. Immediately after the measurement, the system is in one of the states

$$|\psi_i\rangle = \frac{\widehat{\Pi}(i) |\psi\rangle}{\| \widehat{\Pi}(i) |\psi\rangle \|} . \tag{17.4}$$

2. The probability that the system transitions into state $|\psi_i\rangle$ is given by

$$\Pr(|\psi\rangle \to |\psi_i\rangle) = \langle\psi| \widehat{P}(i) |\psi\rangle . \tag{17.5}$$

(The projection law thus restricts the generality of the Schrödinger equation: systems evolve under it *only when a measurement is not taking place*.)

The eigenvector-eigenvalue link (E-E link): Given an quantity O as above:

1. A system in state $|\psi\rangle$ possesses a definite value of O if and only if $|\psi\rangle$ is an eigenstate of $\widehat{O}$, $\widehat{O}\,|\psi\rangle = o_i\,|\psi\rangle$.
2. In this case, the definite value is the associated eigenvalue o_i.

Given one additional assumption—that if a measurement of O returns value o_i, the measured system actually has value o_i of O—the Born rule can be derived from the projection postulate and the eigenvalue-eigenvector link. For if O is measured on a system in state $|\psi\rangle$, by the projection postulate it will transition into an eigenstate of $\widehat{O}$, with the probability of transitioning given by (17.5); after the collapse, it will have a definite value of O by the eigenvalue-eigenvector link; if measurement simply reports that definite value, the Born rule follows.

In any case, it is standard in foundations of quantum mechanics to treat both the projection postulate and the E-E link as core components of orthodox QM. Interpretations of QM like Everett's and Bohm's, for instance, are specifically described as 'no-collapse' interpretations in view of the fact that they drop the collapse law from the postulates of QM; discussions of the ontology of the GRW collapse theory (Albert and Loewer 1996) talk of the need to abandon the E-E link; attempts at interpretation-neutral discussions of the ontology of QM (e.g., Skow 2010; Darby 2010; Bokulich 2014; Wolff 2015; Wilson 2016) typically take the E-E link as a starting point.

Furthermore, typical statements of the quantum measurement problem typically take the E-E link, and/or the projection postulate, as central. The measurement problem is the problem of *macroscopic indefiniteness*, of quantum states that describe macroscopic systems in states that are indefinite with regard to ordinary properties such as the location of pointers or the heartbeats of cats. Or—if macroscopic indefiniteness is to be removed via the projection postulate—it is the problem of *dynamical ill-definedness*, of the lack of any well-defined recipe as to when collapse occurs (it is easy to show that collapse cannot be a consequence or special case of the Schrödinger equation applied to a complex measuring system).

The purpose of this paper, by contrast, is to argue that orthodox quantum mechanics in fact consists only of the structural core and the Born rule. The projection postulate, and the eigenvector-eigenvalue link, are at best parts of a proposed interpretation of QM that goes beyond orthodoxy, at worst unmotivated distractions. As such, in formulating (as opposed to solving) the quantum measurement problem, we should begin with just the structural core and the Born rule. We might introduce one or both as part of a *solution* to the measurement problem, but we confuse the dialectic by taking them as initial common ground.

To be clear what I mean: I will not (here) argue that the *best* or *right* way to interpret QM, or to solve the quantum measurement problem, involves abandoning collapse and/or the E-E link. I will argue that QM as actually practiced by physicists—and what does "orthodox" QM mean, if not that?—already proceeds without either.

In Sect. 17.2 I provide some evidence that physicists in practice do not seem to make use of a collapse rule. I strengthen this in Sects. 17.3 and 17.4 by arguing that the collapse rule is incapable of handling two standard kinds of experimental setup: those involving repeated measurements, and those involving continuous observation. In Sect. 17.6 I point out that the rapid spreading of wavepackets under the Schrödinger equation means that the E-E link makes the ridiculous claim that essentially any system, including macroscopic systems, is maximally indefinite in position, and hence that the E-E link does not have the resources to say when systems are actually localised; in Sect. 17.7 I deploy a result of Hegerfeldt to show that this generalises to pretty much any observable.

For the remainder of the paper, I explore what 'orthodox quantum mechanics' is, shorn of the E-E link and the projection postulate. (This part of the paper draws on some parts of Wallace (2016), albeit deployed in a rather different context). I consider (in Sect. 17.8) a view which treats preparation and measurement as primitive, but ultimately reject it (in Sect. 17.9) on the grounds both that it too struggles with continuous and repeated measurements, and that it cannot handle applications of QM where results from QM are integrated into larger pieces of historical science. With this as a starting point, I finally suggest (Sect. 17.10) that orthodoxy should be understood as an inchoate attitude to the quantum state, where its dynamics are always unitary but where it is interpreted either as physically representational or as probabilistic, according to context. In the concluding section I reflect on the right formulation of the measurement problem given this conception of what 'orthodox QM' actually is.

17.2 Against Collapse: Indirect Evidence

The projection postulate appears in Dirac (1930) and von Neumann (1955), the first two codifications of the axioms of QM. It continues to be widespread, though not universal, in first courses on QM to this day: an unscientific perusal of my shelf reveals that collapse is included in about half of the books there that present QM from scratch. (The Born rule, of course, appears in all of them.) But for all this (I will argue) it plays no real role in applications of quantum mechanics in physics. It is rather hard to prove a negative, but here I give some suggestive reasons to think that physical practice abjures collapse.

Firstly, collapse is conspicuously absent from *second* courses in QM, and in particular in courses on relativistic QM. This ought to strike a student as peculiar (it certainly struck the author, as an undergraduate, as peculiar): collapse, as formally defined in QM, is a *global* phenomenon, applying to the whole quantum state and so affecting, simultaneously, systems spatially far from one another. In relativity, this notion of simultaneity is frame-dependent (or simply meaningless, depending how you think about conventionality of simultaneity, but in any case problematic). One would expect, if collapse is really part of orthodox QM, that the first chapter of any relativistic QM textbook would start with a careful discussion of exactly how the collapse postulate is to be applied in the relativistic context. I have not once seen any such textbook so much as consider the question.

Again, to be clear: the point is not that collapse is *unsatisfactory* in the relativistic regime. Of course it is; the tension between relativity and QM has been known at least since the EPR paper (Einstein et al. 1935). But relativistic QM textbooks contain, not an unsatisfactory collapse rule, but no collapse rule at all. One concludes that the theory must be applicable without any mention of collapse. And indeed it is: the name of the game in relativistic QM is to calculate probability distributions over physical quantities—most often, over the various energies, momenta and particle numbers of the decay products of some scattering experiment—and for this, only the Born rule is required; collapse plays no part.

Secondly, the theoretical physics community has been worrying for forty years about the so-called "black hole information loss paradox" originally identified by Hawking (1976). (See, e.g., Page (1994) and Belot et al. (1999) and references therein, though the debate continues in lively fashion to this day.) At its heart, the paradox is simply that black hole decay is *non-unitary* and as such can't be described within the Schrödinger-equation framework. But state-vector collapse is also non-unitary! So if the collapse law is part of orthodox QM, quantum-mechanical dynamics were never unitary in the first place: they were an alternating series of unitary and non-unitary processes. So why be so desparate to preserve unitary in the exotic regime of black hole decay, when it is ubiquitous in far more mundane cases? One has the clear impression that (at least this part of) the theoretical physics community does *not* in fact think that dynamics is non-unitary in any other contexts in physics, rendering black hole decay uniquely problematic. Tempting though it might be for this advocate of the Everett interpretation to claim that the community has adopted the many-worlds theory *en masse*, a more mundane account is simply that (what they regard as) *orthodox* QM does not include the collapse postulate.[2]

Thirdly, modern quantum field theory largely abandons Hamiltonian methods in favour of the path-integral approach. But in that approach it is not even clear how collapse is to be defined (and, again, textbook presentations never seem to mention the issue), and yet the theory still seems to produce empirically successful predictions.

Finally, and as an admittedly crude indicator, searching the archives of *Physical Review* for projection postulate, wave-function collapse and the like turns up only a few hundred references, nearly all of which turn out to be (a) foundational discussions, (b) discussions of proposed alternatives to quantum theory, or (c)

[2]Of course, plenty of people working on black hole decay *are* fairly explicit advocates of the Everett interpretation, and I have argued elsewhere that quantum cosmology generally is *tacitly* committed to the Everett interpretation, but it's clear that the majority of the community embrace Mermin's "shut up and calculate" approach (Mermin 2004).

theoretical quantum-computation discussions. (For comparison, searches for terms like state vector or Hilbert space or Schrodinger equation typically turn up several tens of thousands of references.)

17.3 Against Collapse: Inadequacy for Repeated Measurements

The case of repeated measurements—when some quantity is measured on a quantum system and then, a short while later, measured again—has actually been used, since Dirac, as an argument *for* state-vector collapse. The argument goes like this: repeated measurements must give identical results; so if a measurement of O gives outcome o_i, then a subsequent measurement of O immediately afterwards must also give outcome o_i, The only way this is compatible with the Born rule is if the state of the system immediately before this second measurement is an eigenstate of $\widehat{O}$ with eigenvalue o_i—so to get repeated measurements right, wavefunction collapse is a requirement.

...which would be all very well, if repeated measurements *did* give identical results. But:

- photon detectors typically absorb photons: immediately after a measurement on a photon, the photon no longer exists;
- The Stern-Gerlach apparatus detects an atom's spin by slamming it very hard into a screen; this process is in no way guaranteed to preserve that atom's spin.
- More generally, measuring something by slamming it very hard into something else is probably the single most commonly used tool in the experimental physicist's toolbox.

In fact, 'non-disturbing' measurements, in which repeated measurements indeed give the same results, are decidedly uncommon in quantum mechanics and require some skill to set up (see Home and Whitaker (1997) for discussion). So a collapse rule explicitly designed to ensure that repeated measurements give the same results is in flat conflict with a lot of observed physics.

By contrast, quantum mechanics *without* collapse has no trouble with repeated measurements—non-disturbing or otherwise. The familiar trick, following von Neumann's original prescription, is to include the measurement device in the physical analysis. Suppose for simplicity that $\widehat{O}$ is non degenerate,

$$\widehat{O} = \sum_i o_i \left|o_i\right\rangle \left\langle o_i\right|, \tag{17.6}$$

and suppose that the measurement device has some observable M corresponding to the possible measurement outcomes. In von Neumann's original version, M is the position of the centre of mass of some pointer; here for convenience I take $\widehat{M}$ too as being discrete and nondegenerate,

$$\widehat{M} = \sum_i m_i \, |m_i\rangle \, \langle m_i| \, . \tag{17.7}$$

Then the measurement interaction is assumed to have form

$$|o_i\rangle \otimes |m_0\rangle \rightarrow |\varphi_i\rangle \otimes |m_i\rangle \, . \tag{17.8}$$

Applying this measurement process to a system initially in state

$$|\psi\rangle = \sum_i \lambda_i \, |o_i\rangle \tag{17.9}$$

and a measurement device initially in state $|m_0\rangle$ gives the outcome

$$|\psi\rangle \otimes |m_0\rangle \rightarrow \sum_i \lambda_i \, |\varphi_i\rangle \otimes |m_i\rangle \, . \tag{17.10}$$

Applying the Born rule to a measurement *of M* now tells us that the probability of getting m_i is $|\lambda_i|^2$—exactly what the Born rule requires for a measurement of O on the original system, and it is for exactly this reason that this process indeed qualifies as a measurement.

There is no requirement here that $|\varphi_i\rangle = |o_i\rangle$ or even that the distinct $|\varphi_i\rangle$ are orthogonal—indeed, the measurement process could perfectly well dump the measured system in some fixed post-measurement state $|\varphi_0\rangle$ (as in the case of photon absorption) in which case the measurement process is

$$|\psi\rangle \otimes |m_0\rangle \rightarrow |\varphi_0\rangle \otimes \sum_i \lambda_i \, |m_i\rangle \, . \tag{17.11}$$

The 'non-disturbing' measurements are then the ones where indeed $|\varphi_i\rangle = |o_i\rangle$. In these cases, but only these, if we bring in a second copy of the measurement device and repeat the measurement interaction, we get

$$|\psi\rangle \otimes |m_0\rangle \otimes |m_0\rangle \rightarrow \left(\sum_i \lambda_i \, |o_i\rangle \otimes |m_i\rangle\right) \otimes |m_0\rangle \rightarrow \sum_i \lambda_i \, |o_i\rangle \otimes |m_i\rangle \otimes |m_i\rangle \, . \tag{17.12}$$

Applying the Born rule in this case to a joint measurement of $M \times M$, we find that indeed, when measurements are non-disturbing the probability is 100% that two successive measurements give the same result.

If there is a lesson to learn from repeated measurements it is that the Born rule, by itself, does not define *transition* probabilities, but only probabilities at an instant (an issue I return to in Sect. 17.10). But "wave functions collapse on measurement" does not solve this problem satisfactorily, and indeed gives flatly incorrect results.

17.4 Against Collapse: Inadequacy for Continuous Measurement

Continuous measurements—where a system is constantly observed to see if, or how quickly, it undergoes some change—are commonplace in physics. For instance, radioactive decay measurements—where a Geiger counter is placed near some radioactive substance, and the rate of decay is recorded—are among the most straightforward demonstrations of quantum mechanics' probabilistic nature. Yet they fit strikingly badly into the wavefunction-collapse framework.

It is not that the *physics* of (for instance) radioactive decay is problematic, at least phenomenologically. (Actually calculating decay rates *ab initio* is another matter: the nucleus is a complex and strongly bound system, and hard to treat analytically.) The idea is that—if the decay rate is $1/\tau$—then an undecayed particle, in state $|\text{undecayed}\rangle$, evolves over some short time interval δt like

$$|\text{undecayed}\rangle \to \left(1 - \frac{\delta t^2}{2\tau}\right)|\text{undecayed}\rangle + \sqrt{\frac{1}{\tau}\delta t}\,|\text{decay products}\rangle\,. \tag{17.13}$$

Meanwhile, the decay-product state's own evolution over time, which can be represented as

$$|\text{decay products}\rangle \to \widehat{U}(t)\,|\text{decay products}\rangle \equiv |\text{decay products};t\rangle \tag{17.14}$$

explores a very large region of Hilbert space and, in particular, satisfies

$$\langle\text{decay products}|\text{decay products};t\rangle \simeq 0 \tag{17.15}$$

for $t_0 < t < T$, where T is the (extremely large) Poincaré recurrence time for the system and $t_0 \ll \tau$ (i.e., the rate at which the radioactive products evolve away from their original state is much quicker than the particle's decay rate). Under these assumptions we can deduce

$$\widehat{U}(t)\,|\text{undecayed}\rangle = \mathrm{e}^{-t/2\tau}\,|\text{undecayed}\rangle + \int_0^t \mathrm{d}\xi\,\frac{\mathrm{e}^{-\xi/2\tau}}{\sqrt{\tau}}\,|\text{decay products};(t-\xi)\rangle \tag{17.16}$$

at least for $T \gg t \gg \tau$.

Applying the Born rule to this system gives exactly the results we would expect: at time t, the probability of the system being undecayed is $|\mathrm{e}^{-t/2\tau}|^2 = \mathrm{e}^{-t/\tau}$. And no assumption of wavefunction collapse is required to derive this probability. But suppose we make that assumption anyway: when, in that case, is the wavefunction supposed to collapse?

One possibility would be to model the continuous process of measurement by a frequent but discrete series—applying the projection postulate every Δ seconds—

and then taking $\Delta \to 0$. As long as Δ is long enough—technically speaking: as long as $\Delta \gg t_0$—this iterated collapse will leave the probabilities unaffected. But it is the content of the quantum Zeno paradox (Misra and Sudarshan 1977) that as $\Delta \to 0$, the evolution of the system is entirely halted: in this limit, the state of the particle remains $|\text{undecayed}\rangle$ forever.

Misra and Sudarshan did assume (at least for the purposes of their paper) that observation required collapse, and so that continuous observation required continuous collapse; hence "paradox". But it is the collapse postulate, not anything about continuous observation per se, that delivers this impossible result. Modelling of measurement as a physical process, as per the previous section, reveals (cf. Home and Whitaker 1997) that:

- The Zeno 'paradox' is a real (and empirically confirmed) physical effect: if a discrete measurement process is carried out repeatedly (and, crucially, if the time taken to carry out every individual measurement is short compared to the timescales on which the measured system evolves), then the rate of evolution of the system really is reduced by the measurements, and tends to zero as the frequency of repetition tends to infinity.
- A continuous observation can also be modelled as a physical process, and in this case the relevant variable is the response speed of the measurement device compared to the timescale on which the measured system evolves. Again, when the former is much faster than the latter, evolution is heavily suppressed. But an observation can be 'continuous' even while its response time is relatively slow: in the case of a Geiger counter, the response time is so slow compared to the evolution timescales of decay that Zeno slowing is negligible. (The relevant system timescale is not the decay rate, but the evolution time of the decay products, i.e. t_0.)

This is not to say that the Zeno effect is *entirely* non-paradoxical, even when understood without the distorting reference to collapse. Paradoxical (though non-contradictory) consequences arise when the measurement process involves energy exchange between measurement device and system only when the system is in a state distinct from its original state, so that the presence of the measurement device appears to halt the system's evolution even though the two are not interacting. This is related to the phenomenon of interaction-free measurement, as seen in the Elitzur-Vaidman bomb problem (Elitzur and Vaidman 1993). For further discussion see Home and Whitaker, *ibid*, or (for an unapologetically pro-Everettian perspective) see Wallace (2012, pp. 390–3).

17.5 Against Collapse: Inadequacy for Unsharp Measurement

The view that measurements are represented by collections of mutually orthogonal projectors is now thirty years out of date. Quantum measurement theory now regards the "projection-valued measurements" (PVMs) that can be so represented merely as a special case of a more general framework: "positive-operator-valued measurements" (POVMs).[3] In the POVM framework, measurements are represented by collections $\{\widehat{M}_i\}$ of self-adjoint operators that (i) are positive (that is, have no negative eigenvalues, or equivalently, satisfy $\langle\psi|\,\widehat{M}_i\,|\psi\rangle \geq 0$ for any state $|\psi\rangle$); (ii) sum to unity, $\sum_i \widehat{M}_i = \widehat{1}$.

For instance, consider measuring a particle's phase-space position: that is, consider simultaneously measuring its position and its momentum. Within the PVM framework, this is impossible: position and momentum do not commute. But in modern measurement theory, this simply tells us that we cannot make a simultaneous *sharp* measurement of position and momentum. We can measure both provided we are prepared to accept a little noise in the measurement process, and for macroscopically large systems the noise can be very small indeed—which is reassuring, since manifestly we *do* simultaneously measure the position and momentum of macroscopic bodies.

A phase-space POVM (in, for simplicity, one spatial dimension) can be defined by starting with some state $|\varphi\rangle$ that is a wavepacket approximately localised around position and momentum zero (say, a Gaussian), so that

$$|\varphi(p,q)\rangle = \exp(-i\widehat{X}p)\exp(+i\widehat{P}q)\,|\varphi\rangle \tag{17.17}$$

is the same state translated so as to be localised around position q and momentum p. Then the family of operators

$$\widehat{M}_{p,q} = \frac{1}{2\pi}\,|\varphi(p,q)\rangle\,\langle\varphi(p,q)| \tag{17.18}$$

is a POVM and can be used to represent the unsharp phase-space measurement. It will give probability distributions over position and momenta separately which are smearings-out of the sharp results obtained from the Born rule, with the level of smearing depending on the width of the wavepacket in position and momentum space and becoming negligible in both cases for macroscopically large systems.

Similarly, the POVM framework can handle fuzzy measurements of a single quantity, as might be appropriate when the measurement device is imperfect. Given an observable O corresponding to an operator $\widehat{O}$ with spectral resolution (17.2), suppose that $f^1, \ldots f^N$ are N functions from the spectrum of $\widehat{O}$ to the nonnegative reals, satisfying

$$\sum_{k=1}^{N} f^k(o_i) = 1 \tag{17.19}$$

for all i. Then the family of operators $f_k(\widehat{O})$ is a POVM. If $f_k(o_i) = \delta_i^k$, this just reduces to a sharp measurement of O, but more general measurements of O can be represented by more general choices of the f^k.

---

[3] For more detail on the physics of this section, see, e.g., Busch et al. (1996).

The POVM generalisation of traditional measurement theory is by now routine, and mathematically speaking is a straightforward generalisation of the Born rule. But it has no associated collapse law, and so it is opaque how to apply collapse in POVM contexts. In addition, POVMs are not associated with the spectral decompositions of the operators representing physical quantities, so to deduce what POVM is being applied, we need to model the measurement process as a unitary interaction with the measurement device, and then in due course apply the Born rule with respect to a macroscopic quantity pertinent to the measurement device.

The lessons of continuous, repeated and unsharp measurements are the same: in any measurement processes more complicated than a simple, non-repeated discrete measurement, reliably getting the physics right requires treating the system's behaviour unitarily, and if necessary physically modelling the measurement process. Collapse is *at best* an unreliable shorthand. And of course, it is *only* in "measurement processes more complicated than a simple, non-repeated discrete measurement" that the collapse rule could play any role in physical practice anyway. If we measure the system once and immediately discard it, the Born rule is all we need.

I conclude that the collapse postulate plays, and can play, no real part in actual applications of quantum mechanics.

17.6 Against the Eigenvalue-Eigenvector Link: Problems for Position

Consider a mass-m point particle—either a fundamental particle, or, more typically, the centre-of-mass degree of freedom of some rigid body like a dust mote or a table. Restricting it, for simplicity, to one dimension, its most significant observables are position and momentum, corresponding to operators $\widehat{X}$ and $\widehat{P}$ respectively, obeying the commutation relation $\left[\widehat{X}, \widehat{P}\right] = i\hbar$.

It is a standard result of quantum mechanics[4] that:

- $\widehat{X}$ and $\widehat{P}$ have continuous spectra (reflecting the fact that these are not quantised quantities, that any position or momentum is a possible result of a measurement), and can be expressed as

$$\widehat{X} = \int_{-\infty}^{+\infty} \mathrm{d}x \, x \, |x\rangle \langle x| \quad \text{and} \quad \widehat{P} = \hbar \int_{-\infty}^{+\infty} \mathrm{d}k \, k \, |k\rangle \langle k| \, . \tag{17.20}$$

- Any quantum state can be expressed in the position basis as

$$|\psi\rangle = \int_{-\infty}^{+\infty} \mathrm{d}x \, |x\rangle \langle x|\psi\rangle \equiv \int_{-\infty}^{+\infty} \mathrm{d}x \, \psi(x) \, |x\rangle \tag{17.21}$$

[4] And, like most 'standard results of quantum mechanics', there are some tacit additional mathematical assumptions required. See Ruetsche (2011, ch.3) for details.

where $\psi(x) \equiv \langle x|\psi\rangle$ is the *position-space wavefunction* (or often just *wavefunction*) of the state.

- Similarly, any quantum state can be expressed in the momentum basis as

$$|\psi\rangle = \int_{-\infty}^{+\infty} \mathrm{d}k\ |k\rangle\ \langle k|\psi\rangle \equiv \int_{-\infty}^{+\infty} \mathrm{d}k\ \hat{\psi}(k)\ |k\rangle \tag{17.22}$$

where $\psi(k) \equiv \langle k|\psi\rangle$ is the *momentum-space wavefunction* of the state.

- The position and momentum bases are related by

$$|k\rangle = \frac{1}{\sqrt{2\pi}} \int_{-\infty}^{+\infty} \mathrm{d}x\ \mathrm{e}^{ikx}\ |x\rangle \tag{17.23}$$

from which it follows that the position and momentum representations are Fourier transforms of one another:

$$\hat{\psi}(k) = \frac{1}{\sqrt{2\pi}} \int_{-\infty}^{+\infty} \mathrm{d}x\ \mathrm{e}^{-ikx}\ \psi(x). \tag{17.24}$$

Suppose we apply the E-E link to the position of the particle. For the particle to *definitely have position* x, it would need to be an eigenstate of $\widehat{X}$—that is, it would need to be in state $|x\rangle$. That isn't possible: because the spectrum of the position operator is continuous, the eigenstates of position are so-called 'improper eigenstates'—at least as QM is normally used, they do not represent an actually-attainable state of a quantum system. ('Legal' quantum states are normalised—$\langle\psi|\psi\rangle = 1$—whereas the norm $\langle x|x\rangle$ is infinite, or at any rate undefined.)

So: no system has a *perfectly* definite position. This is not in itself problematic. It is a standard result of functional analysis that functions $f(\widehat{X})$ may be defined by

$$f(\widehat{X}) = \int_{-\infty}^{+\infty} \mathrm{d}x\ f(x)\ |x\rangle\ \langle x|\,. \tag{17.25}$$

In particular, if Σ is some compact (i.e., closed and bounded) subset of the real numbers, and if Λ_Σ is defined by

$$\begin{aligned} \Lambda_\Sigma(x) &= 1 \ \text{ if } x \in \Sigma \\ &= 0 \text{ otherwise} \end{aligned} \tag{17.26}$$

—that is, if Λ_Σ represents the property of being in Σ—then

$$\Lambda_\Sigma(\widehat{X}) = \int_\Sigma |x\rangle\ \langle x| \tag{17.27}$$

projects, according to the E-E link, onto all and only those states which are definitely located in Σ. In the position representation, this is all and only states whose wavefunction vanishes outside Σ. This suggests that if we want to represent a reasonably-well-localised particle, we should choose one with a wavefunction confined to some reasonably small Σ. (And similarly *mutatis mutandis* if we want to consider systems localised in *momentum*.)

The first thing to say about this is that it is not how physicists *in fact* represent localised particles. The standard strategy in physics is to represent a particle localised at some point x_0 by a Gaussian, i.e. a state with wavefunction

$$\psi(x) = \mathcal{N} \exp(-(x - x_0)^2/2L^2). \tag{17.28}$$

$\psi(x)$ is very small when $|x - x_0|/L \gg 1$, so a state like this, if its position is measured, is nearly certain to be found within a few multiples of L from x_0—L is the "effective width" of the state, the size of the region in which it is 'effectively localised' in physics parlance. But $\psi(x) \neq 0$ for every value of x there is—so according to the E-E link, the particle is completely delocalised, no matter how small L might be.

Perhaps not too much should be made of this. Physicists use Gaussians because they are mathematically very convenient, rather than from some deep commitment to what 'true' localisation is like. Perhaps we should think of the Gaussian as just a very convenient approximation to a 'really' localised state, with the latter having a wavefunction with genuinely compact support.

But suppose that a quantum system, at some initial time, *does* have such a wavefunction—say, ψ_0, which is localised inside some compact region Σ. If we represent that same system instead in the momentum-space representation—that is, with its momentum-space wavefunction, which is the Fourier transform $\hat{\psi}$ of ψ—then we find that $\hat{\psi}$ does not itself have compact support. (This is a consequence of the classical Paley-Wiener theorem,[5] which says *inter alia* that the Fourier transform of a compactly supported L^2 function is holomorphic.) Via the $E - E$ link, this tells us that any particle whose *position* is not completely indefinite has a completely indefinite *momentum*.

That might itself be worrying: we might have hoped, given the uncertainty principle, that a particle could definitely have both (a) a position within some region of width L and (b) a momentum within some region of width $\hbar/L$, but we won't get that from the $E - E$ link. But worse is to come: for consider the *time evolution* of this system. We might expect that a particle whose momentum is completely indefinite will spread out instantaneously over all of space—and indeed, this is exactly what happens. Even 'confining' the system inside some potential well will not prevent its spreading out, for a quantum system will 'tunnel' through any potential barrier unless it is infinite, and so unphysical.

So: no body can be localised to any degree at all for more than an instant, at least on the E-E definition of 'localised'. (I have argued for this on intuitive physical grounds but will supply a mathematical proof in the next section) No assumption

[5] See, e.g., Rudin (1991, pp. 196–202).

about the 'microscopic' nature of the body in question has been made: the argument applies as readily to chairs, tables and planets as to electrons or atoms, and so chairs, tables and planets, according to the E-E link, have at almost all times a completely indefinite location. If we assume the projection postulate, of course, a system *is* localised immediately after a position measurement—but the operative word is 'immediately'. An arbitrarily short time after the measurement, delocalisation is complete.

I conclude that the E-E link is of no use in understanding what it really is for a physical system to be localised to any degree.[6] A fortiori, it cannot be being used in physics to do useful work in our understanding of localisation. As we shall see, this is not a feature unique to spatial localisation.

17.7 Against the Eigenvector-Eigenvalue Link: Problems for Basically Any Quantity

Hegerfeldt's theorem (Hegerfeldt 1998a,b) is as follows:

Hegerfeldt's theorem: Suppose that the spectrum of the Hamiltonian of some quantum system is bounded below (something that holds of essentially any physically reasonable Hamiltonian) and let $|\psi(t)\rangle$ be some dynamical history of that system (i.e., some solution to the Schrödinger equation). Then if $\widehat{\Pi}$ is any projection operator,[7] *exactly one* of the following holds:

1. $\langle\psi(t)|\,\widehat{\Pi}\,|\psi(t)\rangle \neq 0$ for all times t except for some nowhere dense set of times of measure zero (i.e., for all times except some set of isolated instants);
2. $\langle\psi(t)|\,\widehat{\Pi}\,|\psi(t)\rangle = 0$ for all times t.

Hegerfeldt proved the theorem as part of an investigation into localisation in relativistic quantum mechanics[8] but in fact it causes severe difficulties for the E-E link in general. For consider again our operator

$$\widehat{O} = \sum_i o_i \widehat{\Pi}_i \tag{17.29}$$

(which includes, as a special case, the sort of discretisations of position we considered previously.) The observable O corresponding to $\widehat{O}$ will according

[6]The line of argument here has some resemblance to that used by Albert and Loewer (1996) to argue that the E-E link should be rejected in the GRW theory in place of a "fuzzy link". But Albert and Loewer attributed the problem to the Gaussian collapse function used in the GRW theory, whereas as we have seen, the problem arises even in the absence of any collapse event, as a consequence of ordinary Schrödinger dynamics.

[7]In fact, it suffices for $\widehat{\Pi}$ to be a positive operator.

[8]See Halvorson and Clifton (2002) for discussion of its significance in this context.

to the E-E link, definitely *not* have value o_i with respect to state $|\psi(t)\rangle$ iff $\langle\psi(t)|\,\widehat{\Pi}_i\,|\psi(t)\rangle = 0$. So Hegerfeldt's theorem can be rephrased as

Hegerfeldt's theorem (indefiniteness form): Given a system evolving unitarily over some interval of time under a Hamiltonian whose spectrum is bounded below, a given property is either (a) definitely *not* possessed at every time in that interval, or (b) *not* definitely not possessed at almost every time in that interval.

Put another way: suppose there is some property that, at some time in the indefinitely distant future, the system might have some probability to be found to possess. Then, according to the E-E link, it is *immediately*—that is, within an arbitrarily short window of time—indefinite whether the system has that property.

Put yet another way: anything that might at some future point be indefinite will be indefinite immediately. This seems to render the E-E link fairly useless as a description of ontology. We might have imagined that systems begin having some definite value of a given quantity, then gradually evolve so as to be indefinite across several values of that quantity, and in due course become completely indefinite with respect to that quantity (perhaps until some wavefunction collapse restores definiteness). But dynamically, that can't happen: indefiniteness is immediate if it is going to happen at all.

(I should, in fairness, acknowledge one context in which we seem to be able to get some content out of the E-E link even given Hegerfeldt's theorem. In the specific case of angular momentum (including both orbital angular momentum of some bound system, and the intrinsic spin of a particle), we could imagine the angular momentum *precessing*, so that the state at time t is an eigenstate of angular momentum with respect to some angle $\Omega(t)$. The system would then at all times have a definite angular momentum even though it would only be definite for an instant with respect to angular momentum in a given direction. But this relies on special features of the angular-momentum case, in particular does not generalise to position and momentum, and looks highly likely to be unstable once angular momentum couples to other degrees of freedom.)

To push the consequences of Hegerfeldt's theorem further (and also provide a rigorous justification of the claims of the previous section), suppose $\langle\psi(t)|\,\widehat{\Pi}\,|\psi(t)\rangle = 0$ for all t, and consider

$$\mathcal{S} = \text{Span}\{|\psi(t)\rangle\}. \tag{17.30}$$

$\mathcal{S}$ is time invariant, and so must be spanned by (possibly improper) eigenstates of the Hamiltonian. And of course any element of $\mathcal{S}$ is an eigenstate of $\widehat{\Pi}$ with eigenvalue 0. So we can conclude that:

Complete indefiniteness corollary: Unless some operator has some (possibly improper) eigenstates in common with the Hamiltonian, its associated quantity is completely indefinite at almost every time.

(Readers uncomfortable with my casual use of improper eigenstates can just rephrase the requirement as "$\widehat{O}$ has an eigensubspace invariant under the Hamiltonian".)

As an application of this result, suppose that the Hamiltonian is non-degenerate (that is: has no two eigenstates with the same eigenvalue). Then a necessary and sufficient condition for a quantity not to be almost always completely indefinite is that it is a function of the Hamiltonian.

As another, consider some collection of scalar particles interacting via some potential:

$$\widehat{H} = \sum_i \frac{\widehat{P}_i^2}{2m_i} + V(\widehat{X}_1, \ldots \widehat{X}_n), \tag{17.31}$$

for some smooth function V. In every case I know, and in particular in the case of free particles, the eigenfunctions of this Hamiltonian have only isolated zeroes. (This is easily provable in the case where V is a polynomial or other holomorphic function, so that the eigenfunctions themselves are holomorphic; it also follows in one dimension from the uniqueness theorem for solutions of ordinary differential equations.) But in that case, no eigenfunction has compact support, so no projector onto localised states is time-invariant. It follows that every particle has a completely indefinite position almost always.

The underlying problem here is a radical mismatch between the E-E link and the way quantum mechanics *actually* handles the idea of a system's becoming more spread out (speaking loosely) with respect to a given quantity. QM handles the latter through probabilities: the likelihood of a particle localised at x being found very far from x is initially negligibly small and only gradually increases—and, depending on the dynamics, may never increase beyond negligible levels. But the E-E link is all-or-nothing: as soon as the system has any probability, even $10^{-10^{20}}$, of being found in some region, it is completely indefinite whether it is in that region.

I conclude that statements about a system's properties that rely on the E-E link convey essentially no information about a system between measurements (and, at the instant of measurement, everything empirically salient is coded in the Born rule in any case). As such, the E-E link cannot plausibly play a role in orthodox QM.

17.8 Quantum Mechanics in Practice: The Lab View

Neither the collapse rule nor the E-E link can be part of orthodox, i.e. actually-used-in-practice, QM. It isn't that they are ultimately unsatisfactory on philosophical grounds, but rather that they are not even *prima facie* satisfactory, and lead to nonsense (all systems maximally indefinite all the time) or violation of empirical predictions (all measurements non-disturbing; continuous measurement impossible to define without Zeno freezing).

So what *do* physicists do, if they don't do "orthodox" QM as it is usually understood? In the rest of the paper I shall consider two paradigms to describe orthodox QM. The first—the "lab view"—is not uncommonly found in more careful foundational discussions of QM in the physics literature (especially in quantum information) but is not ultimately satisfactory to account for physical practice; the second—the "decoherent view"—does, I think, provide an adequate fit to physical practice.

In the Lab View (my presentation here is modelled on Peres 1993), any application of QM should be understood as applying to some experimental setup, and that setup in turn is broken into three processes:

1. State preparation;
2. Dynamics;
3. State measurement.

The first and last of these are *primitive*: the question of how the system is prepared in a given state, and how it is measured, are external to the experiment and so not modelled in the physics. Only the second is regarded as a modelled physical process.

In quantum mechanics, in particular:

1. The system is prepared in a state represented by some (pure or mixed) Hilbert-space state;
2. It evolves under the Schrödinger equation for some fixed period of time;
3. The outcome of the measurement is given by the Born Rule.

The Lab View itself does not force a unique interpretation of the underlying physics, but it is often presented in parallel with a particular interpretation, each of which is sometimes claimed as 'orthodoxy'. Particularly prominent examples include:

Straight operationalism: There is nothing more to quantum mechanics than a calculus that connects preparation processes (conceived of macroscopically and phenomenologically) to measurement processes (likewise conceived of); physics neither needs, nor can accommodate, any microscopic story linking the two.
Straight operationalism is perhaps the closest realisation in mainstream physics of the old logical-positivist conception of the philosophy of science; it seems to have been more or less Heisenberg's preferred approach, and has been advocated more recently by Peres (Peres 1993, pp. 373–429; Fuchs and Peres 2000). The 'quantum Bayesianism' or "QBism" of Fuchs et al. (Fuchs 2002; Fuchs et al. 2014; Fuchs and Schack 2015) has much in common with straight operationalism, although it holds out for some objective physical description at a deeper level (see Timpson (2010, pp. 188–235) for a critique).

Complementarity: It *is* possible to describe a physical system at the microscopic level, but the appropriate description depends on the experimental context in question. Is an electron a wave or a particle? If you're carrying out a two-slit experiment, it's a wave; if you change experimental context to check which slit it went down, it's a particle.
Niels Bohr is the most famous proponent of complementarity, though he tended to describe it in qualitative terms and engaged little with modern (Schrödinger-Heisenberg-Dirac) quantum mechanics. Saunders (2005) provides a rational reconstruction of complementarity in modern terminology; the approaches of

Omnes (1988, 1992, 1994) and Griffiths (1984, 1993, 1996) are very much in the spirit of complementarity.

But most relevant for our purposes is:

Measurement-induced collapse: The system can be described in microscopic terms, and in a way independent of the measurement process: the physical quantities of the system are represented by the state, via the E-E link. But at the final moment of measurement at the end of the experiment, the Projection Postulate is applied, jumping the system into an eigenstate of the quantity being measured.

Here we might seem to find a rehabilitation of orthodoxy. But consider: (i) as we have seen, the E-E link in practice tells us nothing about the physical state of the system between preparation and measurement, for it is almost certain that the system is maximally indefinite with respect to any quantity of interest pretty much throughout its evolution; (ii) collapse, occurring as it does at the very end of the physical process described by the Lab View, can do no actual work in physical predictions beyond what we already get from the Born rule.

(As a terminological aside, although contemporary physics often uses "Copenhagen interpretation" to refer to measurement-induced collapse, the historical views developed under that name are closer to complementarity and to straight instrumentalism. See Cushing (1994) and Saunders (2005) for further discussion.)

But in any case, the Lab View is itself insufficient to do justice to actual applications of QM, once they transcend the prepare-evolve-measure framework we have considered so far.

17.9 Limitations of the Lab View

We have already considered situations that go beyond what the Lab View, strictly speaking, can handle: those when the measurement process is not the end of our interaction with the system, where measurements are repeated or continuous. Furthermore, and even outside these cases, the Lab View's stipulation that measurement is primitive is itself in conflict with physical practice: measurement devices are physical systems, made from atoms and designed and built on the assumption that their behaviour is governed by physical laws.

Which laws? Back in the glory days of the Copenhagen interpretation, perhaps it was possible to suppose that the workings of lab equipment should be analysed classically, but in these days of quantum optics, superconducting supercolliders and gravity-wave-sensitive laser interferometers, we cannot avoid making extensive reference to quantum theory itself to model the workings of our apparatus. And now a regress beckons: if we can understand quantum theory only with respect to some experimental context, what is the context in which we understand the application of quantum theory to the measurement itself?

The method used is in each case the same (and we have already seen it play out in our discussion of the projection postulate):

1. Insofar as the physics of the measurement process are relevant, we expand the analysis to include the apparatus itself as part of the quantum system. (In quantum information this move has come to be known as 'the Church of the Larger Hilbert Space'.)
2. We avoid infinite regress by treating the Born-Rule-derived probability distribution over *macroscopic* degrees of freedom not as a probability of *getting certain values on measurement*, but as a probability of *certain values already being possessed.*

The need for an objective, non-quantum, macroscopically applicable language to describe the physics of measurement was already recognised by Bohr (and is acknowledged in more sophisticated operationalist accounts of quantum theory; cf. Peres (1993, 423–427)). But it is really a special case of a more general requirement, for modern applications of quantum mechanics go beyond cases where measurement is repeated or continuous and embrace cases where we cannot really avoid interpreting the QM probabilities as entirely separate from a formal 'measurement' process.

This is particularly clear in cosmology, where it has long been suggested that Lab View quantum mechanics is unsuitable simply because cosmology concerns the whole Universe, and so there is no 'outside measurement context'; indeed, it was for exactly these reasons that Hugh Everett developed his approach to quantum theory in the first place (Everett 1957).

However, this slightly misidentifies the problem. Cosmology is concerned with the Universe on its largest scales, but not with every last feature of the Universe: realistic theories in cosmology concern particular degrees of freedom of the universe (the distribution of galaxies, for instance) and we can perfectly well treat these degrees of freedom as being measured via their interaction with other degrees of freedom outside the scope of those theories (Fuchs and Peres 2000).

But there is a problem nonetheless. Namely: the processes studied in cosmology cannot, even in the loosest sense, be forced into the Lab View. They are (treated as) objective, ongoing historical processes, tested indirectly via their input into other processes; they are neither prepared in some state at the beginning, nor measured at the end, and indeed in many cases they are ongoing.

Nor is the issue specific to cosmology. The luminosity of the Sun, for instance, is determined in part via quantum mechanics: in particular, via the quantum tunneling processes that control the rate of nuclear fusion in the Sun's core as a function of its mass and composition. We can model this fairly accurately and, on the basis of that model, can deduce how the Sun's luminosity has increased over time. Astrophysicists pass that information to climate scientists, geologists, and paleontologists, who feed it into their respective models of prehistoric climates, geological processes, and ecosystems. All good science—but only in the most Procrustean sense can we realistically regard a successful fit to data in a paleoclimate model as being a measurement of the nuclear fusion processes in the Sun a billion years ago.[9]

Issues of this kind abound whenever we apply quantum theory outside stylised lab contexts. (Is the increased incidence of cancer due to Cold War nuclear-weapons tests a quantum measurement of the decay processes in the fallout products of those tests? Again, only in the most Procrustean sense.) In each case we seem to have extracted objective facts about the unobserved world from quantum theory, not merely to be dealing with a mysterious microworld that gets its meaning only when observed. But they are particularly vivid in cosmology, which is a purely observational science, and a science chiefly concerned not with repeating events in the present but with the historical evolution of the observed Universe as a whole.

As perhaps the most dramatic example available—and probably the most important application of quantum theory in contemporary cosmology—consider the origin of structure in the Universe. Most of that story is classical: we posit a very small amount of randomly-distributed inhomogeneity in the very-early Universe, and then plug that into our cosmological models to determine both the inhomogeneity in the cosmic microwave background and the present-day distribution of galaxies. The latter, in particular, requires very extensive computer modelling that takes into account astrophysical phenomena on a great many scales; it cannot except in the most indirect sense be regarded as a 'measurement' of primordial inhomogeneity. Quantum theory comes in as a proposed source of the inhomogeneity: the posited scalar field (the 'inflaton field') responsible for cosmic inflation is assumed to be in a simple quantum state in the pre-inflationary Universe (most commonly the ground state) and quantum fluctuations in that ground state, time-evolved through the inflationary era, are identified with classical inhomogeneities. Quantum-mechanical predictions thus play a role in our modelling of the Universe's history, but not a role that the Lab View seems remotely equipped to handle.

17.10 Quantum Mechanics in Practice: The Decoherent View

Once again: the point is not that the orthodoxy of the Lab View is conceptually inadequate, and so we must seek an unorthodox alternative; it is that physicists manifestly *are* doing quantum mechanics in regimes beyond the reach of the Lab View, so they must *already* have a method for applying it that goes beyond the Lab View.

In fact, the method is fairly obvious. The probability distribution over certain degrees of freedom—solar energy density, radiation rate, modes of the inflaton field—is simply treated as objective, as a probability distribution over actually-existing facts, and not merely as something that is realised when an experiment

[9]This is an instance of Quine's classic objection to logical positivism (Quine 1951)—the empirical predictions of particular applications of quantum mechanics cannot be isolated from the influence of myriad other parts of our scientific world-view.

is performed. So we can say, for instance, not merely that a given mode of the primordial inflaton field *would have had* probability such-and-such of having a given amplitude if we were to measure it (whatever that means operationally), but that it *actually did have* probability such-and-such of that amplitude.

Now, it's tempting to imagine extending this objective take on quantum probabilities to *all* such probabilities: to interpret a quantum system as having some objectively-possessed value of every observable, and the quantum state as simply an economical way of coding a probability distribution over those observables. But of course, this cannot straightforwardly be done. A collection of formal results—the Kochen-Specker theorem (Kochen and Specker 1967; Bell 1966; Redhead 1987, pp. 119–152; Mermin 1993), Gleason's theorem (Gleason 1957; Redhead 1987, pp. 27–9; Peres 1993, pp. 190–195; Caves et al. 2004), the Bub-Clifton theorem (Bub and Clifton 1996; Bub et al. 2000), the PBR theorem and its relatives (Pusey et al. 2011; Maroney 2012; Leifer 2014)—establish that reading quantum mechanics along these lines as bearing the same relation to some underlying objective theory as classical statistical mechanics bears to classical mechanics is pretty much[10] impossible.

In fact, the central problem can be appreciated without getting into the details of these results. To take an objective view of some physical quantity is to suppose that the quantity has a definite value at each instant of time, so that we can consider the various possible *histories* of that quantity (that is: the various ways it can evolve over time) and assign probabilities to each. But the phenomena of interference means that this does not generically work in quantum mechanics. The quantum formalism for (say) the two slit experiment assigns a well-defined probability $P_1(x)$ to the history where the particle goes through Slit One and then hits some point x on the screen, and a similarly-well-defined probability $P_2(x)$ to it hitting point x via Slit Two, but of course the probability of it hitting point x at all (irrespectively of which slit it goes through) is not in general $P_1(x) + P_2(x)$. So the 'probabilities' assigned to these two histories do not obey the probability calculus. And things that don't obey the probability calculus are not probabilities at all.

At a fundamental level, the problem is that quantum mechanics is a dynamical theory about amplitudes, not about probabilities. The *amplitudes* of the two histories in the two-slit experiment sum perfectly happily to give the amplitude of the particle reaching the slit, but amplitudes are not probabilities, and in giving rise to probabilities they can cancel out or reinforce.

However, in most physical applications of quantum theory we are *not* working 'at a fundamental level', which is to say that we are not attempting the usually-impossible task of deducing (far less interpreting) the evolution of the full quantum state over time. Rather, we are interested in finding higher-level, emergent dynamics, whereby we can write down dynamical equations for, and make predictions about,

[10]A more precise statement would be "Impossible unless that underlying objective theory has a number of extremely pathological-seeming features." It is not universally accepted that this rules out such theories, though; see, e.g., Spekkens (2007) and Leifer (2014) for further discussion.

certain degrees of freedom of a system without having to keep track of all the remaining degrees of freedom. In the examples of the previous section, for instance, we have considered:

- The robust relations between macrostates of measuring devices and states of the system being measured, abstracting over the microscopic details of the measuring devices
- The bulk thermal properties of the core of the Sun, abstracting over the vast number of microstates compatible with those bulk thermal properties
- The low-wavelength modes of the inflaton field which are responsible for primordial inhomogeneities, abstracting over the high-wavelength degrees of freedom and the various other fields present.

In each case, we can derive from the quantum-mechanical dynamics an autonomous system of dynamical equations for these degrees of freedom. In each case, we can also derive from the Born Rule a time-dependent probability distribution over the values of those degrees of freedom. And in each case, that probability distribution defines a probability over histories that obeys the probability calculus. In each case, then, we are justified—at least formally, if perhaps not philosophically—in studying the autonomous dynamical system in question as telling us how these degrees of freedom are actually evolving, quite independently of our measurement processes.

To put the position intentionally crudely: orthodox QM, I am suggesting, consists of shifting between two different ways of understanding the quantum state according to context: interpreting quantum mechanics realistically in contexts where interference matters, and probabilistically in contexts where it does not. *Obviously* this is conceptually unsatisfactory (at least on any remotely realist construal of QM)—it is more a description of a *practice* than it is a stable *interpretation*. But why should that be surprising? Philosophers have spent decades complaining that physicists' approach to QM is philosophically unsatisfactory, after all. In a way, philosophers' version of 'orthodoxy' does physicists too much credit in providing a self-consistent realist account of QM that just lacks a satisfactory account of exactly when collapse happens, even as it does them too little credit in failing to recognise the unsuitability of the orthodox version of orthodoxy to physical practice.

And in fact, physics has made considerable progress in clarifying just when we can, and cannot, get away with a probabilistic interpretation of the quantum state, and in particular in helping us understand why we can reliably get away with it in macroscopic contexts. The *decoherence theory* developed by, *inter alia*, Joos and Zeh (Joos and Zeh 1985; Zeh 1993), Zurek (1991, 1998), Gell-Mann and Hartle (1993), Omnes (1988, 1992), and Griffiths (1984, 1993) in the 1980s and 1990s is concerned precisely with when the quantum state can be treated as probabilistic, understood either (in the environment-induced decoherence framework of Joos, Zeh and Zurek) because interference is suppressed with respect to some basis, or directly (in the consistent-histories framework of Griffiths, Omnes, and Gell-Mann and Hartle) by finding a consistent rule to assign probabilities to histories.[11] Hence my name for this position: the "decoherent view".

And where is collapse in all this? Well, the condition of decoherence can be reinterpreted as a condition for when we can impose an explicit collapse rule without empirically contradicting quantum theory. But that 'condition' is precisely the condition in which we can get away with treating the quantum state probabilistically, and from that perspective, "collapse" is just probabilistic conditionalisation. Of course, we can continue to think of the quantum state non-probabilistically, and use decoherence as a condition for when a *physical* collapse process can be introduced, but now we are well outside the assessment of orthodoxy, and well along the path towards a proposed *solution* of the measurement problem.

17.11 Two Applications of the Decoherent View

To illustrate the efficacy of the decoherent view in doing justice to physical practice, I consider two examples, from radically different sectors of physics: Stern-Gerlach type experiments, and the emergence of structure in the early Universe from primordial fluctuations of the inflaton field.

The Stern-Gerlach-type experiments I have in mind proceed as follows:

1. A beam of silver atoms emerges from a furnace.
2. That beam is split by a magnetic field and the beam corresponding to spin down in the z direction is discarded.
3. The beam is subjected to a series of interference experiments.
4. The spin of the atoms in the beam in (say) the x direction is measured by once again splitting the beam and measuring what fraction of atoms are in each beam.

Initially, the spin degrees of freedom of a silver atom is in a mixed state

$$\rho_1 = \frac{1}{2}(|+_z\rangle\langle +_z| + |-_z\rangle\langle -_z|). \tag{17.32}$$

(The justification of this state comes from quantum statistical mechanics and lies outside the scope of this paper.) After the magnetic field is applied, the particles spin and position degrees of freedom become entangled, having state:

$$\rho_2 = \frac{1}{2}(|+_z\rangle\langle +_z| \otimes |\varphi_+(t)\rangle\langle\varphi_+(t)| + |+_z\rangle\langle +_z| \otimes |\varphi_-(t)\rangle\langle\varphi_-(t)|) \tag{17.33}$$

[11] Appreciating that this is the task being performed by decoherence in contemporary physical practice also goes some way to explaining why the physics community has regarded decoherence as a major step towards understanding the interpretation of QM, something not generally shared by philosophers (Barrett (1999, p. 230) is typical: "That decoherence destroys simple interference effects does not solve the measurement problem since it does not explain the determinateness of our measurement records ... In order to observe a single determinate record there must somewhere be a single determinate record.)."

where $|\varphi_+(t)\rangle$ and $|\varphi_-(t)\rangle$ are wavepacket states of negligible overlap. The mixed state cannot be used for interference experiments, so we can get away with treating it probabilistically. We now discard the $-$ part of the beam, and continue to operate only on the $+$ part; conditional on the silver atom still being in the apparatus, its spin state must be $+_z$, and so we update the state by conditionalising, to the pure state

$$|\psi_3\rangle = |+_z\rangle \otimes |\varphi_+(t)\rangle \tag{17.34}$$

(A more realistic treatment might allow for slight overlap of wavepackets, so that there is still some admixture of $|-_z\rangle$.)

Now we do a series of interference experiments with the system. At this point, treating it probabilistically will get us into trouble, so we avoid doing so: we continue to evolve the state unitarily and abjure probabilistic conditionalising.

Finally, we split the beam again, so it has form

$$|\psi_4\rangle = \alpha_+ \, |+_z\rangle \otimes |\chi_+(t)\rangle + \alpha_- \, |-_z\rangle \otimes |\chi_-(t)\rangle \tag{17.35}$$

(with the values of $\alpha_\pm$ depending on the details of the interference processes, and with $|\chi_+(t)\rangle$ and $|\chi_-(t)\rangle$ again having negligible overlap.) We once again treat this probabilistically (since we are going to do no further interference experiments, and indeed are about to entangle the system with a macroscopically large measurement device) and interpret $|\alpha_\pm|^2$ as the probability that the particle's spin is in fact $\pm$.

As for primordial structure formation, it works as follows (here I follow Weinberg (2008, pp. 470–474), and excerpt a more detailed discussion in Wallace 2016). A quantum field theory (the inflaton field) is coupled to spacetime geometry in a perturbative fashion, and allowed to evolve in time. In the very early Universe the system is inherently quantum-mechanical and the mod-squared amplitudes of the various modes of the field cannot consistently be interpreted probabilistically. But as the universe expands, the various mechanisms of decoherence come into play and—still very early in the Universe's history—we reach the point at which a probabilistic interpretation is consistent. At that point, we interpret those mod-squared amplitudes as probabilities of the actual modes of the inflaton field having various values; this determines a probability distribution over various possible inhomogeneities in the density of the early Universe, and that distribution is fed into cosmological simulations of structure formation. There is no measurement here, and no natural point for a collapse—only a quantum state which, in due course, we can get away with treating probabilistically.

The examples probably strike the reader as uncomfortably opportunistic, even ad hoc. Indeed, they should so strike the reader. The ad hoc, opportunistic approach that physics takes to the interpretation of the quantum state, and the lack, in physical practice, of a clear and unequivocal understanding of the state—*this* is the quantum measurement problem, once the distractions of collapse and the E-E link are removed.

17.12 Conclusion: The Measurement Problem from the Perspective of Contemporary Physics

Quantum mechanics, as actually practiced in mainstream physics, makes no use of the eigenstate-eigenvector link, nor of the collapse postulate. Its dynamics are unitary; the unitarily evolving quantum state is interpreted inchoately, as describing physical goings on in regimes where interference is important and as describing probabilities in regimes where it can be neglected. On pain of failure to account for interference, we cannot (it seems) consistently treat the state as probabilistic; on pain of failure to account for the probability rule, and more generally of failing to make contact with observation, we cannot (it seems) consistently treat the state as representational. The "measurement problem" from this perspective, is the task of taking this inchoate practice and showing how it can be justified given, as starting point, a well-defined physical theory—where what counts as a "well-defined physical theory" will depend on one's general stance on scientific realism and the philosophy of science. Perhaps we can do so by showing how an ultimately physical superposition nonetheless appears emergently probabilistic (Everett's strategy); perhaps we can do so by showing that interference can after all be made sense of on probabilistic grounds (the quantum Bayesian strategy, and the ψ-epistemic one); perhaps we can do so by adding additional representational structure (the Bohmian strategy) or by changing the dynamics to introduce a stochastic element (the dynamical-collapse strategy) or by adopting a conception of scientific theories that diverges from standard realism (the complementarity, quantum-logic, and instrumentalist strategies; perhaps the quantum-Bayesian strategy too).

From this perspective, the distinction between 'pure interpretations' that leave the formalism of QM alone, and modificatory strategies that modify or supplement it, is clear. The (real) Copenhagen interpretation, quantum Bayesianism, and the Everett interpretation (whatever their strengths or weaknesses otherwise) fall into the former category, as would a (hypothetical) ψ-epistemic interpretation: their dynamics is unitary throughout, their formalism unsupplemented by hidden variables. Dynamical-collapse and hidden-variable theories are in the latter category, being committed to adding additional variables and/or to modifying the Schrödinger equation.

From this perspective, too, the "orthodox interpretation"—that is, the theory obtained by adding the E-E link and the projection postulate to unitary quantum mechanics, and deriving the Born rule from them—is just one more modificatory strategy, and a strikingly implausible and unattractive one to boot. Perhaps some better attempt to solve the measurement problem will incorporate one or both, perhaps in modified and improved form—but it is time to retire the theory that is based on them as a starting point for discussions of the measurement problem.

Acknowledgements This paper has benefitted greatly from conversations with Simon Saunders and Chris Timpson, and from feedback when it was presented at the 2016 Michigan Foundations of Modern Physics workshop, especially from Jeff Barrett, Gordon Belot, and Antony Leggett.

References

Albert, D. Z. (1992). *Quantum mechanics and experience*. Cambridge, MA: Harvard University Press.

Albert, D. Z., & Loewer, B. (1996). Tails of Schrödinger's Cat. In R. Clifton (Ed.), *Perspectives on quantum reality* (pp. 81–92). Dordrecht: Kluwer Academic Publishers.

Barrett, J. A. (1999). *The quantum mechanics of minds and worlds*. Oxford: Oxford University Press.

Bell, J. S. (1966). On the problem of hidden variables in quantum mechanics. *Reviews of Modern Physics, 38*, 447–452. Reprinted in Bell (1987), pp. 1–13.

Bell, J. S. (1987). *Speakable and unspeakable in quantum mechanics*. Cambridge: Cambridge University Press.

Belot, G., Earman, J., & Ruetsche, L. (1999). The Hawking information loss paradox: the anatomy of a controversy. *British Journal for the Philosophy of Science, 50*, 189–229.

Bokulich, A. (2014). Metaphysical indeterminacy, properties, and quantum theory. *Res Philosophica, 91*, 449–475.

Bub, J. (1997). *Interpreting the quantum world*. Cambridge: Cambridge University Press.

Bub, J., & Clifton, R. (1996). A uniqueness theorem for "no collapse" interpretations of quantum mechanics. *Studies in the History and Philosophy of Modern Physics, 27*, 181–219.

Bub, J., Clifton, R., & Goldstein, S. (2000). Revised proof of the uniqueness theorem for 'no collapse' interpretations of quantum mechanics. *Studies in the History and Philosophy of Modern Physics, 31*, 95.

Busch, P., Lahti, P. J., & Mittelstaedt, P. (1996). *The quantum theory of measurement* (2nd revised ed.). Berlin: Springer.

Caves, C., Fuchs, C., Manne, K., & Renes, J. (2004). Gleason-type derivations of the quantum probability rule for generalized measurements. *Foundations of Physics, 34*, 193.

Cushing, J. T. (1994). *Quantum mechanics: Historical contingency and the Copenhagen hegemony*. Chicago: University of Chicago Press.

Darby, G. (2010). Quantum mechanics and metaphysical indeterminacy. *Australasian Journal of Philosophy, 88*, 227–245.

DeWitt, B., & Graham, N. (Eds.) (1973). *The many-worlds interpretation of quantum mechanics*. Princeton: Princeton University Press.

Dirac, Paul (1930): *The principles of quantum mechanics*. Oxford: Oxford University Press.

Einstein, A., Podolsky, B., & Rosen, N. (1935). Can quantum-mechanical description of reality be considered complete? *Physical Review, 47*, 777–780.

Elitzur, A. C., & Vaidman, L. (1993). Quantum mechanical interaction-free measurements. *Foundations of Physics, 23*, 987–997.

Everett, H. I. (1957). Relative state formulation of quantum mechanics. *Review of Modern Physics, 29*, 454–462. Reprinted in DeWitt and Graham (1973).

Fuchs, C. (2002). Quantum mechanics as quantum information (and only a little more). Available online at http://arXiv.org/abs/quant-ph/0205039.

Fuchs, C., & Peres, A. (2000). Quantum theory needs no "interpretation". *Physics Today, 53*(3), 70–71.

Fuchs, C. A., Mermin, N. D., & Schack, R. (2014). An introduction to QBism with an application to the locality of quantum mechanics. *American Journal of Physics, 82*, 749–754.

Fuchs, C. A., & Schack, R. (2015). QBism and the Greeks: Why a quantum state does not represent an element of physical reality. *Physica Scripta, 90*, 015104.

Gell-Mann, M., & Hartle, J. B. (1993). Classical equations for quantum systems. *Physical Review D, 47*, 3345–3382.

Gleason, A. (1957). Measures on the closed subspaces of a Hilbert space. *Journal of Mathematics and Mechanics, 6*, 885–893.

Griffiths, R. (1993). Consistent interpretation of quantum mechanics using quantum trajectories. *Physical Review Letters, 70*, 2201–2204.

Griffiths, R. B. (1984). Consistent histories and the interpretation of quantum mechanics. *Journal of Statistical Physics, 36*, 219–272.
Griffiths, R. B. (1996). Consistent histories and quantum reasoning. *Physical Review A, 54*, 2759–2773.
Halvorson, H., & Clifton, R. (2002). No place for particles in relativistic quantum theories? *Philosophy of Science, 69*, 1–28.
Hawking, S. W. (1976). Black holes and thermodynamics. *Physical Review D, 13*, 191–197.
Hegerfeldt, G. A. (1998a). Causality, particle localization and positivity of the energy. In A. Böhm (Ed.), *Irreversibility and causality* (pp. 238–245). New York: Springer.
Hegerfeldt, G. A. (1998b). Instantaneous spreading and Einstein causality. *Annalen der Physik, 7*, 716–725.
Home, D., & Whitaker, M. A. B. (1997). A conceptual analysis of quantum Zeno: Paradox, measurement and experiment. *Annals of Physics, 258*, 237–285.
Joos, E., & Zeh, H. (1985). The emergence of classical particles through interaction with the environment. *Zeitschrift fur Physik, B59*, 223–243.
Kochen, S., & Specker, E. (1967). The problem of hidden variables in quantum mechanics. *Journal of Mathematics and Mechanics, 17*, 59–87.
Leifer, M. (2014). Is the quantum state real? An extended review of ψ-ontology theorems. *Quanta, 3*, 67–155.
Maroney, O. (2012). How statistical are quantum states? Available online at http://arxiv.org/abs/1207.6906.
Mermin, N. D. (1993). Hidden variables and the two theorems of John Bell. *Reviews of Modern Physics, 65*, 803–815.
Mermin, N. D. (2004). Could Feynman have said this? *Physics Today, 57*, 10.
Misra, B., & Sudarshan, E. C. G. (1977). The Zeno's paradox in quantum theory. *Journal of Mathematical Physics, 18*, 756.
Omnes, R. (1988). Logical reformulation of quantum mechanics. I. Foundations. *Journal of Statistical Physics, 53*, 893–932.
Omnes, R. (1992). Consistent interpretations of quantum mechanics. *Reviews of Modern Physics, 64*, 339–382.
Omnes, R. (1994). *The interpretation of quantum mechanics*. Princeton: Princeton University Press.
Page, D. (1994). Black hole information. In R. Mann & R. McLenaghan (Eds.), *Proceedings of the 5th Canadian conference on general relativity and relativistic astrophysics* (pp. 1–41). Singapore: World Scientific.
Penrose, R. (1989). *The Emperor's new mind: Concerning computers, brains and the laws of physics*. Oxford: Oxford University Press.
Peres, A. (1993). *Quantum theory: Concepts and methods*. Dordrecht: Kluwer Academic Publishers.
Pusey, M. F., Barrett, J., & Rudolph, T. (2011). On the reality of the quantum state. *Nature Physics, 8*, 476. arXiv:1111.3328v2.
Quine, W. V. O. (1951). Two dogmas of empiricism. *Philosophical Review, 60*, 20–43.
Redhead, M. (1987). *Incompleteness, nonlocality and realism: A prolegomenon to the philosophy of quantum mechanics*. Oxford: Oxford University Press.
Rudin, W. (1991). *Functional analysis* (2nd ed.). New York: McGraw-Hill.
Ruetsche, L. (2011). *Interpreting quantum theories*. Oxford: Oxford University Press.
Saunders, S. (2005). Complementarity and scientific rationality. *Foundations of Physics, 35*, 347–372.
Skow, B. (2010). Deep metaphysical indeterminacy. *Philosophical Quarterly, 58*, 851–858.
Spekkens, R. W. (2007). In defense of the epistemic view of quantum states: A toy theory. *Physical Review A, 75*, 032110.
Timpson, C. (2010). *Quantum information theory and the foundations of quantum mechanics*. Oxford: Oxford University Press.

von Neumann, J. (1955). *Mathematical foundations of quantum mechanics*. Princeton: Princeton University Press.
Wallace, D. (2012). *The emergent multiverse: Quantum theory according to the Everett interpretation*. Oxford University Press.
Wallace, D. (2013). Inferential vs. dynamical conceptions of physics. Available online at http://arxiv.org/abs/1306.4907.
Wallace, D. (2016, forthcoming). Interpreting the quantum mechanics of cosmology. In A. Ijjas & B. Loewer (Eds.), *Introduction to the philosophy of cosmology*. Oxford University Press.
Weinberg, S. (2008). *Cosmology*. Oxford: Oxford University Press.
Wilson, J. (2016). Quantum metaphysical indeterminacy. Talk to the Jowett Society, Oxford, 26 Feb 2016.
Wolff, J. (2015). Spin as a determinable. *Topoi, 34*, 379–386.
Zeh, H. D. (1993). There are no quantum jumps, nor are there particles! *Physics Letters, A172*, 189.
Zurek, W. H. (1991). Decoherence and the transition from quantum to classical. *Physics Today, 43*, 36–44. Revised version available online at http://arxiv.org/abs/quant-ph/0306072.
Zurek, W. H. (1998). Decoherence, einselection, and the quantum origins of the classical: The rough guide. *Philosophical Transactions of the Royal Society of London, A356*, 1793–1820. Available online at http://arxiv.org./abs/quant-ph/98050.